Dr. John Chung's

Advanced Placement

Calculus

AB/BC

Fifth Edition

Copyright © 2017 by Dr. John Chung

Dear Students,

This book is designed to help students prepare for the AP Calculus Examinations. Over the past two decades of teaching, I have written and compiled hundreds of sample questions of varying levels of difficulty. This book contains concise notes on each topic covered by the AP Exams, and is intended to be used alongside your textbook and class notes to clarify areas of weakness. I have also provided you with eight full-length practice tests. There are easy-to-follow worked-out solutions for every example in this book.

I have also written books to prepare students for SAT I Math, SAT II Math Level 1, and SAT II Math Level 2. My books are written with the goal of giving students a solid understanding of basic mathematical concepts. By mastering these concepts, you will be able to do well on any standardized test. At the request of my students all over the world, I am also working on a web video series to supplement my books.

If you have any suggestions, find a typo, or need clarification, please send me an email at drjcmath@gmail.com. I am committed to the highest standards of excellence and am happy to assist you in any way that I can.

Thank you to all of my readers. I wish you luck on the exam!

Sincerely,

Dr. John Chung

Contents

CH 1 Limits and Their Properties — Page

- A The Definition of a Limit — 7
- B Limit of a Polynomial Function — 8
- C Limit of Rational Function 1 — 8
- D Limit of Rational Function 2 — 9
- E Limit of Rational Function 3 — 11
- F Limit of a Function Involving a Radical — 12
- G Limit of a Function as x Approaches ∞ — 13
- H Limit of $f(x)$ on a Graph — 14
- I Limit of Trigonometric Functions — 15

CH 2 Continuity and Discontinuity

- A The Definition of Continuity — 17
- B Types of Discontinuities — 17
- C Removable and Non-Removable Discontinuity — 18
- D Intermediate Value Theorem — 22

CH 3 Differentiation

- A The Definition of the Average Slope — 23
- B The Definition of the Slope of a Tangent Line — 23
- C Notation of Derivatives — 23
- D Differentiability and Continuity — 25
- E Rules for Differentiation — 28
- F Derivatives of Trigonometric Functions — 28
- G Derivatives of Inverse Trigonometric Functions — 30
- H Derivatives of Exponential and Logarithmic Functions — 32
- I Explicit and Implicit Differentiation — 33

CH 4 Applications of Differentiation

- A Equation of a tangent Line and Normal Line — 35
- B The Mean Value Theorem for Derivatives — 36
- C Rolle's Theorem — 37
- D Motion (Position, Velocity, and Acceleration) — 39
- E Related Rates — 44

CH 5 Curve Sketching

A	Extrema of a Function	48
B	Types of Relative Extrema	48
C	Critical Points	49
D	Increasing and Decreasing Functions	54
E	The First Derivative Test	57
F	Concavity	59
G	The Second Derivative Test	59
H	Points of Inflection	60

CH 6 More Applications of Differentiation

A	How to Sketch the Graph of a Function	63
B	Optimization	64
C	Local Linear Approximation	68
D	Error Propagation	70
E	Newton's Method	72
F	L'Hôpital's Rule	74
G	Evaluating the Derivative of an Inverse Function	77

CH 7 Integration

A	Antiderivatives	79
B	U-Substitution in Order to Use the Power Rule	81
C	Definite Integral (The Fundamental Theorem of Calculus)	83
D	Approximating Integrals (Finding the Area Under a Curve)	85
E	Theorem of Calculus	93
F	Exponential and Logarithmic Functions	96
G	The Integrals of Trigonometric Functions	98
H	Integration by Parts	100
I	Tabular Integration by Parts	102
J	Partial Fractions	104
K	Improper Integrals	106
L	Length of a Curve	107

CH 8 **Finding the Area Between Two Curves**

    A   Finding the Area Between Two Curves (Vertical) ———————— 109
    B   Finding the Area Between Two Curves (Horizontal) ——————— 112

CH 9 **Volume of a Solid**

    A   The Volume of a Solid of Revolution (Disk) ————————— 114
    B   Finding the Volume of a Solid of Revolution Using the Washer Method   116
    C   Finding the Volume of a Solid of Revolution Using the Cylindrical Shell Method   118
    D   Volumes of Solids with Known Cross Sections ———————— 122

CH 10 **Differential Equations**

    A   Definitions and Basic Concepts ———————————————— 124
    B   Separable Differential Equation ———————————————— 125
    C   Exponential Growth and Decay ———————————————— 129
    D   Radio Active Decay ———————————————————————— 130
    E   Newton's Law of Cooling ———————————————————— 131
    F   Logistic Growth ———————————————————————————— 132
    C   Euler's Method for Approximating Differential Equations ———— 138
    H   Slope Field ———————————————————————————————— 140

CH 11 **Infinite Series**

    A   Telescoping Series ———————————————————————— 152
    B   Sequence Convergence ———————————————————————— 146
    C   Harmonic Series ———————————————————————————— 151
                                                                                                                            154

          Convergence Tests   154
    A   Convergence and Divergence Test ———————————————— 155
    B   Integral Test ———————————————————————————————— 157
    C   *P*-Series Test ———————————————————————————————— 158
    D   Comparison Test ———————————————————————————— 159
    E   The Limit Comparison Test ———————————————————— 161
    F   Alternating Series Test ———————————————————————— 163
    G   Absolute Convergence Test ———————————————————— 164
    H   Ratio Test ———————————————————————————————— 166
    I   Ratio Test for Absolute Convergence ———————————————— 167
    J   Root Test ———————————————————————————————— 167
    K   Expressing the Riemann Sum as a Definite Integral —————— 177

CH 12 Maclaurin and Taylor Series

    A Power Series — 180
    B Radius of Interval of Convergence — 181
    C Derivative and Integral of Power Series — 184
    D Taylor Series — 186
    E Some Important Maclaurin Series — 189
    F Approximation by Taylor Polynomial — 191
    G Function represented by Taylor Series — 194
    H Lagrange Upper Error Bound — 195

CH 13 Polar Coordinates

    A Polar Coordinates — 198
    B Slope of Tangent — 201
    C Calculus of Parametric Equation — 202
    D Arc Length for Polar Curves — 205
    E Area in Polar Coordinates — 207

PRACTICE TESTS

AP Calculus AB Test 1 — 211
AP Calculus AB Test 2 — 247
AP Calculus AB Test 3 — 285
AP Calculus AB Test 4 — 325

AP Calculus BC Test 1 — 363
AP Calculus BC Test 2 — 399
AP Calculus BC Test 3 — 443
AP Calculus BC Test 4 — 487

Chapter 1. Limit
A. The Definition of a Limit

Two-sided limits
If $f(x)$ becomes arbitrarily close to L as x approaches c from either side, then the limit of $f(x)$ is L.
$$\lim_{x \to c} f(x) = L$$

One-sided limits: If $f(x)$ approaches L as x approaches c from the right, then
$$\lim_{x \to c^+} f(x) = L$$
If $f(x)$ approaches L as x approaches c from the left, then
$$\lim_{x \to c^-} f(x) = L$$

The relationship between one-sided limits and two sided limits
If $\lim_{x \to c^+} f(x) \neq \lim_{x \to c^-} f(x)$, then $\lim_{x \to c} f(x)$ does not exist. (DNE)
If $\lim_{x \to c^+} f(x) = \lim_{x \to c^-} f(x)$, then $\lim_{x \to c} f(x) = L$.

Infinite Limits: If $\lim_{x \to c^+} f(x) = +\infty$ and $\lim_{x \to c^-} f(x) = +\infty$, then $\lim_{x \to c} f(x) = +\infty$. Similarly,
If $\lim_{x \to c^+} f(x) = -\infty$ and $\lim_{x \to c^-} f(x) = -\infty$, then $\lim_{x \to c} f(x) = -\infty$.

▶ **Example**

Find the limit.

1. $\lim_{x \to 0} \dfrac{|x|}{x}$

2. $\lim_{x \to 3} |x-3| + 2$

Find the limit at $x = a$.

3.

4.

5.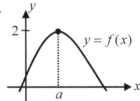

Answer
1. $\lim_{x \to 0^+} \dfrac{|x|}{x} = \lim_{x \to 0^+} \dfrac{x}{x} = 1$ and $\lim_{x \to 0^-} \dfrac{|x|}{x} = \lim_{x \to 0^+} \dfrac{-x}{x} = -1$ → Limit does not exist.
2. Since $\lim_{x \to 3^+} |x-3| + 2 = \lim_{x \to 3^+} x - 3 + 2 = 2$ and $\lim_{x \to 3^-} |x-3| + 2 = \lim_{x \to 3^-} -x + 3 + 2 = 2$, $\lim_{x \to 3} |x-3| + 2 = 2$.
3. $\lim_{x \to a} f(x) = 2$ 4. $\lim_{x \to a} f(x) = 2$ 5. $\lim_{x \to a} f(x) = 2$

Chapter 1. Limit

B. Limit of Polynomial Function

THEOREM. If $f(x)$ is a polynomial function and c is a real number, then
$$\lim_{x \to c} f(x) = f(c)$$

▶ **Example**

Find the limit.

1. $\lim\limits_{x \to -2} \left(2x^2 - 3x + 4\right) =$

2. $\lim\limits_{x \to 3} \left(x^2 - 5\right) =$

3. $\lim\limits_{x \to 1} (x^3 - 2x^2) =$

4. $\lim\limits_{x \to -1} 5x^3 - 4x - 6 =$

Answer
1. $f(-2) = 18$ 2. $f(3) = 4$ 3. $f(1) = -1$ 4. $f(-1) = -7$

C. Limit of Rational Function (1)

THEOREM. If $f(x)$ is a rational function given by $f(x) = \dfrac{p(x)}{q(x)}$ and c is a real number such that $q(c) \neq 0$, then
$$\lim_{x \to c} f(x) = \frac{p(c)}{q(c)}$$

▶ **Example**

Find the limit.

1. $\lim\limits_{x \to 2} \dfrac{x^2 + x + 3}{x + 1} =$

2. $\lim\limits_{x \to -1} \dfrac{x^2 + 1}{x - 1} =$

3. $\lim\limits_{x \to 0} \dfrac{x^3 - 3x + 5}{x^2 + 1} =$

3. $\lim\limits_{x \to 1} \dfrac{x^2 - 1}{x + 1} =$

Answer
1. $f(2) = \dfrac{p(2)}{q(2)} = \dfrac{9}{3} = 3$ 2. $f(-1) = \dfrac{p(-1)}{q(-1)} = \dfrac{2}{-2} = -1$ 3. $f(0) = \dfrac{p(0)}{q(0)} = \dfrac{5}{1} = 5$ 4. $f(1) = \dfrac{p(1)}{q(1)} = \dfrac{0}{2} = 0$

Chapter 1. Limit

D. Limit of Rational Functions (2)

THEOREM. If $f(x)$ is a rational function given by $f(x) = \dfrac{p(x)}{q(x)}$ and c is a real number such that $p(x) \neq 0$ and $q(c) = 0$, then check their one-sided limit. The answer will be one of the followings.

1) If $\begin{cases} \lim\limits_{x \to c^+} f(x) = \infty \text{ and } \lim\limits_{x \to c^-} f(x) = -\infty \\ \text{or} \\ \lim\limits_{x \to c^+} f(x) = -\infty \text{ and } \lim\limits_{x \to c^-} f(x) = \infty \end{cases}$, then limit does not exist (DNE).

2) If $\lim\limits_{x \to c^+} f(x) = +\infty$ and $\lim\limits_{x \to c^-} f(x) = +\infty$, then $\lim\limits_{x \to c} f(x) = +\infty$

3) If $\lim\limits_{x \to c^+} f(x) = -\infty$ and $\lim\limits_{x \to c^-} f(x) = -\infty$, then $\lim\limits_{x \to c} f(x) = -\infty$

▶ **Example**

Find the limit.

1. $\lim\limits_{x \to 0} \dfrac{1}{x} =$

2. $\lim\limits_{x \to 1} \dfrac{1}{x-1} =$

3. $\lim\limits_{x \to 0} \dfrac{1}{x^2} =$

4. $\lim\limits_{x \to 0} \dfrac{-1}{x^2} =$

5. $\lim\limits_{x \to 0^+} \log x =$

6. $\lim\limits_{x \to 1} \dfrac{x}{x-1} =$

Answer

1. $\lim\limits_{x \to 0^+} \dfrac{1}{x} = +\infty$ or $\lim\limits_{x \to 0^-} \dfrac{1}{x} = -\infty$

   Since the limit from the right and the limit from the left are different, the limit does not exist.

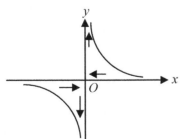

2. $\lim\limits_{x \to 1^+} \dfrac{1}{x-1} = +\infty$ and $\lim\limits_{x \to 1^-} \dfrac{1}{x-1} = -\infty$

   Since the limit from the right and the limit from the left are different, the limit does not exist.

Chapter 1. Limit

3. $\lim\limits_{x \to 0^+} \dfrac{1}{x^2} = +\infty$ and $\lim\limits_{x \to 0^-} \dfrac{1}{x^2} = +\infty$

   Since the limit from the right and the limit from the right are the same, the limit is $+\infty$.

4. $\lim\limits_{x \to 0^+} \dfrac{-1}{x^2} = -\infty$ and $\lim\limits_{x \to 0^-} \dfrac{-1}{x^2} = -\infty$

   Since the limit from the right and the limit from the right are the same, the limit is $-\infty$.

5. When x approaches to 0 from the right, y goes to $-\infty$.

   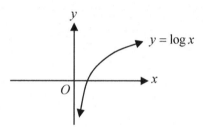

6. Since $\lim\limits_{x \to 1^+} \dfrac{x}{x-1} = +\infty$ and $\lim\limits_{x \to 1^-} \dfrac{x}{x-1} = -\infty$, limit does not exist.

Chapter 1. Limit

E. Limit of Rational Function (3)

THEOREM. If $f(x)$ is a rational function given by $f(x) = \dfrac{p(x)}{q(x)}$ and c is a real number such that $q(c) = 0$ and $P(c) = 0$, then

$$\lim_{x \to c} \frac{p(x)}{q(x)} = \lim_{x \to c} k(x), \text{ where } \frac{p(x)}{q(x)} = k(x).$$

▶ **Example**

Find the limit.

1. $\lim\limits_{x \to 1} \dfrac{x^2 - 3x + 2}{x - 1} =$

2. $\lim\limits_{x \to 1} \dfrac{x^2 + x - 2}{x^2 - 1} =$

3. $\lim\limits_{x \to 2} \dfrac{x^3 - 8}{x - 2} =$

4. $\lim\limits_{x \to -1} \dfrac{2x^2 - x - 3}{x + 1} =$

Answer

1. $\lim\limits_{x \to 1} \dfrac{x^2 - 3x + 2}{x - 1} = \lim\limits_{x \to 1} \dfrac{(x-2)(x-1)}{(x-1)} = \lim\limits_{x \to 1}(x - 2) = -1$

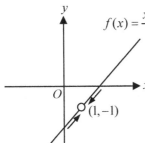

$f(x) = \dfrac{x^2 - 3x + 2}{x - 1} = x - 2 = k(x)$ and $x \neq 1$

Since $f(x)$ becomes arbitrarily close to -1 as x approaches 1 from **either side**, the limit is -1.

2. $\lim\limits_{x \to 1} \dfrac{x^2 + x - 2}{x^2 - 1} = \lim\limits_{x \to 1} \dfrac{(x+2)\cancel{(x-1)}}{(x+1)\cancel{(x-1)}} = \lim\limits_{x \to 1} \dfrac{x+2}{x+1} = \dfrac{3}{2}$

3. $\lim\limits_{x \to 2} \dfrac{x^3 - 8}{x - 2} = \lim\limits_{x \to 2} \dfrac{\cancel{(x-2)}(x^2 + 2x + 4)}{\cancel{(x-2)}} = \lim\limits_{x \to 2}(x^2 + 2x + 4) = 12$

4. $\lim\limits_{x \to -1} \dfrac{2x^2 - x - 3}{x + 1} = \lim\limits_{x \to -1} \dfrac{(2x-3)\cancel{(x+1)}}{\cancel{(x+1)}} = \lim\limits_{x \to -1}(2x - 3) = -5$

Dr. John Chung's AP Calculus AB/BC

Chapter 1. Limit

F. Limit of Functions Involving a Radical

THEOREM.

$$\lim_{x \to c} \sqrt[n]{x} = \sqrt[n]{c}$$

If n is a positive even integer, then $c > 0$. If n is a positive odd integer, then c is all real number.

1. $\lim\limits_{x \to 3} \sqrt{x^2 - 5} = \lim\limits_{x \to 3} \sqrt{3^2 - 5} = 2$

2. $\lim\limits_{x \to -1} \sqrt[3]{x^2 - 9} = -2$

THEOREM.

$$f(x) = \frac{p(x)}{q(x)}$$

If $p(c) = 0$ and $q(c) = 0$ $\left(f(c) = \dfrac{0}{0}\right)$, then multiply its conjugate.

3. $\lim\limits_{x \to 9} \dfrac{x-9}{\sqrt{x}-3} = \lim\limits_{x \to 9} \dfrac{(x-9)(\sqrt{x}+3)}{(\sqrt{x}-3)(\sqrt{x}+3)} = \lim\limits_{x \to 9} \dfrac{(x-9)(\sqrt{x}+3)}{(x-9)} = \lim\limits_{x \to 9}(\sqrt{x}+3) = \sqrt{9} + 3 = 6$

4. $\lim\limits_{x \to 0} \dfrac{\sqrt{x+1}-1}{x} = \lim\limits_{x \to 0} \dfrac{(\sqrt{x+1}-1)(\sqrt{x+1}+1)}{x(\sqrt{x+1}+1)} = \lim\limits_{x \to 0} \dfrac{x}{x(\sqrt{x+1}+1)} = \lim\limits_{x \to 0} \dfrac{1}{\sqrt{x+1}+1} = \dfrac{1}{\sqrt{0+1}+1} = \dfrac{1}{2}$

▶ **Example**

1. $\lim\limits_{x \to 3} \dfrac{\sqrt{x+1}}{x-4} =$

2. $\lim\limits_{x \to 1} \dfrac{\sqrt{x+8}-3}{x-1} =$

3. $\lim\limits_{x \to 0} \sqrt{x^2 + 3x + 2} =$

4. $\lim\limits_{x \to 1} \dfrac{x-1}{\sqrt{x+3}-2} =$

Answer

1. $\lim\limits_{x \to 3} \dfrac{\sqrt{x+1}}{x-4} = \lim\limits_{x \to 3} \dfrac{\sqrt{3+1}}{3-4} = -2 \quad \left(\because f(4) \neq \dfrac{0}{0}\right)$

2. $\lim\limits_{x \to 1} \dfrac{\sqrt{x+8}-3}{x-1} = \lim\limits_{x \to 1} \dfrac{(\sqrt{x+8}-3)(\sqrt{x+8}+3)}{(x-1)(\sqrt{x+8}+3)} = \lim\limits_{x \to 1} \dfrac{x+8-9}{(x-1)(\sqrt{x+8}+3)} = \lim\limits_{x \to 1} \dfrac{x-1}{(x-1)(\sqrt{x+8}+3)}$

   $= \lim\limits_{x \to 1} \dfrac{1}{(\sqrt{x+8}+3)} = \dfrac{1}{\sqrt{9}+3} = \dfrac{1}{6} \quad \left(\because f(1) = \dfrac{0}{0}\right)$

3. $\lim\limits_{x \to 0} \sqrt{x^2 + 3x + 2} = \lim\limits_{x \to 0} \sqrt{0^2 + 3 \cdot 0 + 2} = \sqrt{2} \quad \left(\because f(0) \neq \dfrac{0}{0}\right)$

Chapter 1. Limit

4. $\lim\limits_{x \to 1} \dfrac{x-1}{\sqrt{x+3}-2} = \lim\limits_{x \to 1} \dfrac{(x-1)(\sqrt{x+3}+2)}{(\sqrt{x+3}-2)(\sqrt{x+3}+2)} = \lim\limits_{x \to 1} \dfrac{(x-1)(\sqrt{x+3}+2)}{x-1} = \lim\limits_{x \to 1} (\sqrt{x+3}+2) = 4$

$\left(\because f(x) = \dfrac{0}{0} \right)$

G. Limit of a Function as x approaches infinity

THEOREM. Let $\delta(p)$ be the degree of the numerator and $\delta(q)$ be the degree of the denominator of
$$f(x) = \dfrac{p(x)}{q(x)}.$$

1) If $\delta(p) < \delta(q)$, then $\lim\limits_{x \to \infty} \dfrac{p(x)}{q(x)} = 0$.

2) If $\delta(p) > \delta(q)$, then $\lim\limits_{x \to \infty} \dfrac{p(x)}{q(x)} = \infty$ or $-\infty$.

3) If $\delta(p) = \delta(q)$, then $\lim\limits_{x \to \infty} \dfrac{p(x)}{q(x)} =$ Constant (Ratio of the leading cofficients of $p(x)$ and $q(x)$)

▶ **Example**

Find the limit.

1. $\lim\limits_{x \to \infty} \dfrac{3x+1}{2x^3 - 3x - 1} =$

2. $\lim\limits_{x \to \infty} \dfrac{3x^3 - 5x + 1}{x^2 + 3} =$

3. $\lim\limits_{x \to \infty} \dfrac{4x^3 - 3x - 1}{2x^3 + 3} =$

4. $\lim\limits_{x \to \infty} \dfrac{2x}{\sqrt{x^2 + 1} - 2} =$

Answer

1. $\lim\limits_{x \to \infty} \dfrac{3x+1}{2x^3 - 3x - 1} = 0 \quad (\because \delta(p) < \delta(q))$

2. $\lim\limits_{x \to \infty} \dfrac{3x^3 - 5x + 1}{x^2 + 3} = \infty$

3. $\lim\limits_{x \to \infty} \dfrac{4x^3 - 3x - 1}{2x^3 + 3} = \dfrac{4}{2} = 2 \quad (\because \delta(p) = \delta(q))$

4. $\lim\limits_{x \to \infty} \dfrac{2x}{\sqrt{x^2 + 1} - 2} = \lim\limits_{x \to \infty} \dfrac{\frac{2x}{x}}{\sqrt{\dfrac{x^2+1}{x^2}} - \dfrac{2}{x}} = \lim\limits_{x \to \infty} \dfrac{2}{\sqrt{1 + \dfrac{1}{x^2}} - \dfrac{2}{x}} = 2$

Chapter 1. Limit

H. Limit of $f(x)$ on a Graph

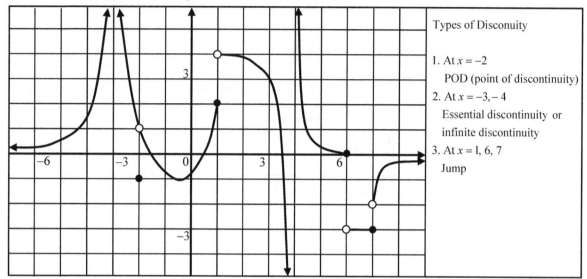

Types of Disconuity

1. At $x = -2$
   POD (point of discontinuity)
2. At $x = -3, -4$
   Essential discontinuity or infinite discontinuity
3. At $x = 1, 6, 7$
   Jump

Example

Given the graph of $f(x)$ above, find the limit.

1. $\lim_{x \to -\infty} f(x) =$
2. $\lim_{x \to \infty} f(x) =$
3. $\lim_{x \to -3} f(x) =$
4. $\lim_{x \to -2^-} f(x) =$
5. $\lim_{x \to -2^+} f(x) =$
6. $\lim_{x \to -2} f(x) =$
7. $\lim_{x \to 1^-} f(x) =$
8. $\lim_{x \to 1^+} f(x) =$
9. $\lim_{x \to 1} f(x) =$
10. $\lim_{x \to 4^-} f(x) =$
11. $\lim_{x \to 4^+} f(x) =$
12. $\lim_{x \to 4} f(x) =$
13. $\lim_{x \to 6^-} f(x) =$
14. $\lim_{x \to 6^+} f(x) =$
15. $\lim_{x \to 6} f(x) =$
16. $\lim_{x \to 7^-} f(x) =$
17. $\lim_{x \to 7^+} f(x) =$
18. $\lim_{x \to 7} f(x) =$
19. $f(-2) =$
20. $f(1) =$
21. Find the x-values at which $f(x)$ is not continuous and describe type of discontinuity.

Answer

1. $\lim_{x \to -\infty} f(x) = 0$
2. $\lim_{x \to \infty} f(x) = 0$
3. $\lim_{x \to -3} f(x) = \infty$
4. $\lim_{x \to -2^-} f(x) = 1$
5. $\lim_{x \to -2^+} f(x) = 1$
6. $\lim_{x \to -2} f(x) = 1$
7. $\lim_{x \to 1^-} f(x) = 2$
8. $\lim_{x \to 1^+} f(x) = 4$
9. $\lim_{x \to 1} f(x) =$ Limit does not exist (from 7 and 8).
10. $\lim_{x \to 4^-} f(x) = -\infty$
11. $\lim_{x \to 4^+} f(x) = \infty$
12. $\lim_{x \to 4} f(x) =$ Limit does not exist (from 10 and 11).
13. $\lim_{x \to 6^-} f(x) = 0$
14. $\lim_{x \to 6^+} f(x) = -3$
15. $\lim_{x \to 6} f(x) =$ Limit does not exist (from 13 and 14).
16. $\lim_{x \to 7^-} f(x) = -3$
17. $\lim_{x \to 7^+} f(x) = -2$
18. $\lim_{x \to 7} f(x) =$ Limit does not exist (from 16 and 17).
19. $f(-2) = -1$
20. $f(1) = 2$

Find the x-values at which $f(x)$ is not continuous and describe type of discontinuity.
1) $x = -3, 4$ (Essential discontinuity or infinite discontinuity occurs at vertical asymptotes)
2) $x = -2$ (POD: Point of discontinuity occurs when a curve has a hole)
3) $x = 1, 6, 7$ (A jump discontinuity occurs when the curve breaks at the point and starts somewhere else)

Chapter 1. Limit

I. Limit of Trigonometric Function

1. $\lim\limits_{x \to 0} \dfrac{\sin x}{x} = 1$
2. $\lim\limits_{x \to 0} \dfrac{1 - \cos x}{x} = 0$
3. $\lim\limits_{x \to 0} \dfrac{\tan x}{x} = 1$
4. $\lim\limits_{x \to 0} \dfrac{\cos x}{x} = \text{DNE}$
5. $\lim\limits_{x \to 0} \dfrac{x}{\cos x} = 0$
6. $\lim\limits_{x \to \infty} \dfrac{\sin x}{x} = 0$
7. $\lim\limits_{x \to \infty} \dfrac{x}{\sin x} = \text{DNE}$
8. $\lim\limits_{x \to \infty} \dfrac{\cos x}{x} = 0$
9. $\lim\limits_{x \to \infty} \dfrac{x}{\cos x} = \text{DNE}$
10. $\lim\limits_{x \to \infty} \dfrac{\tan x}{x} = \text{DNE}$
11. $\lim\limits_{x \to \infty} \dfrac{x}{\tan x} = \text{DNE}$

DNE = Limit does not exist.

▶ **Practice**

Find the limit.

1. $\lim\limits_{x \to \frac{\pi}{2}} \dfrac{1 - \sin x}{\cos^2 x} =$

2. $\lim\limits_{x \to 0} \dfrac{1 - \cos x}{x} =$

3. $\lim\limits_{x \to 0} x \cos \dfrac{1}{x} =$

4. $\lim\limits_{x \to 0} \dfrac{\sin 3x}{\tan 6x} =$

5. $\lim\limits_{x \to \frac{\pi}{2}} \dfrac{1 - \sin x}{\cos x} =$

6. $\lim\limits_{x \to 0} \dfrac{1 - \cos 2x}{x^2} =$

Answer

1. $\lim\limits_{x \to \frac{\pi}{2}} \dfrac{1 - \sin x}{\cos^2 x} = \lim\limits_{x \to \frac{\pi}{2}} \dfrac{(1 - \sin x)(1 + \sin x)}{\cos^2 x (1 + \sin x)} = \lim\limits_{x \to \frac{\pi}{2}} \dfrac{1 - \sin^2 x}{\cos^2 x (1 + \sin x)} = \lim\limits_{x \to \frac{\pi}{2}} \dfrac{\cos^2 x}{\cos^2 x (1 + \sin x)}$

$= \lim\limits_{x \to \frac{\pi}{2}} \dfrac{1}{1 + \sin x} = \dfrac{1}{2}$

Or,

$\lim\limits_{x \to \frac{\pi}{2}} \dfrac{1 - \sin x}{\cos^2 x} = \lim\limits_{x \to \frac{\pi}{2}} \dfrac{1 - \sin x}{(1 - \sin^2 x)} = \lim\limits_{x \to \frac{\pi}{2}} \dfrac{\cancel{1 - \sin x}}{(1 + \sin x)\cancel{(1 - \sin x)}} = \lim\limits_{x \to \frac{\pi}{2}} \dfrac{1}{(1 + \sin x)} = \dfrac{1}{2}$

2. $\lim\limits_{x \to 0} \dfrac{1 - \cos x}{x} = \lim\limits_{x \to 0} \dfrac{1 - \cos x}{x} \cdot \dfrac{1 + \cos x}{1 + \cos x} = \lim\limits_{x \to 0} \dfrac{1 - \cos^2 x}{x(1 + \cos x)} = \lim\limits_{x \to 0} \dfrac{\sin^2 x}{x(1 + \cos x)} = \lim\limits_{x \to 0} \dfrac{\sin x \sin x}{x(1 + \cos x)} =$

$\lim\limits_{x \to 0} \dfrac{\sin x}{1 + \cos x} = \dfrac{0}{2} = 0$

3. $\lim\limits_{x \to 0} x \cos \dfrac{1}{x} = \lim\limits_{x \to 0} \dfrac{\cos \dfrac{1}{x}}{\dfrac{1}{x}}$ Let $y = \dfrac{1}{x}$, then $\lim\limits_{y \to \infty} \dfrac{\cos y}{y} = 0$ ($\because |\cos y| \le 1$)

Chapter 1. Limit

4. $\lim\limits_{x\to 0}\dfrac{\sin 3x}{\tan 6x}=\lim\limits_{x\to 0}\dfrac{\dfrac{\sin 3x}{3x}\cdot 3x}{\dfrac{\tan 6x}{6x}\cdot 6x}=\dfrac{3}{6}=\dfrac{1}{2}$

5. $\lim\limits_{x\to\frac{\pi}{2}}\dfrac{1-\sin x}{\cos x}=\lim\limits_{x\to\frac{\pi}{2}}\dfrac{(1-\sin x)(1+\sin x)}{\cos x(1+\sin x)}=\lim\limits_{x\to\frac{\pi}{2}}\dfrac{1-\sin^2 x}{\cos x(1+\sin x)}=\lim\limits_{x\to\frac{\pi}{2}}\dfrac{\cos^2 x}{\cos x(1+\sin x)}$

$=\lim\limits_{x\to\frac{\pi}{2}}\dfrac{\cos x}{1+\sin x}=\dfrac{0}{1+1}=0$

6. $\lim\limits_{x\to 0}\dfrac{1-\cos 2x}{x^2}=\lim\limits_{x\to 0}\dfrac{(1-\cos 2x)(1+\cos 2x)}{x^2(1+\cos 2x)}=\lim\limits_{x\to 0}\dfrac{1-\cos^2(2x)}{x^2(1+\cos 2x)}=\lim\limits_{x\to 0}\dfrac{\sin^2(2x)}{x^2(1+\cos 2x)}$

$=\lim\limits_{x\to 0}\dfrac{\dfrac{\sin^2(2x)}{(2x)(2x)}\cdot 4x^2}{x^2(1+\cos 2x))}=\lim\limits_{x\to 0}\dfrac{4x^2}{x^2(1+\cos 2x)}=\lim\limits_{x\to 0}\dfrac{4}{(1+\cos 2x)}=\dfrac{4}{1+1}=2$

Or

Since $\cos 2x=1-2\sin^2 x$, $\lim\limits_{x\to 0}\dfrac{1-\cos 2x}{x^2}=\lim\limits_{x\to 0}\dfrac{1-(1-2\sin^2 x)}{x^2}=\lim\limits_{x\to 0}\dfrac{2\sin^2 x}{x^2}=2$.

Chapter 2. Continuity and Discontinuity

A. The Definition of Continuity

THEOREM. A function $f(x)$ is continuous at c, if
$$\lim_{x \to c} f(x) = f(c).$$

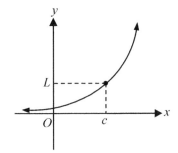

1. $\lim_{x \to c} f(x) = L$: The limit of $f(x)$ exists at $x = c$.
2. $f(c) = L$: The function is defined at $x = c$.
3. $\lim_{x \to c} f(x) = f(c)$

∴ The function is continuous at $x = c$.

B. Types of Discontinuities

THEOREM. A function $f(x)$ is discontinuous at c as follows.

1. Pod (Point of Discontinuity)

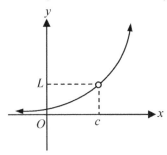

1. $\lim_{x \to c} f(x) = L$: The limit of $f(x)$ exists at c.
2. $f(c)$ is undefined : The function is not defined at $x = c$.

∴ The function is not continuous at $x = c$.

2. Jump

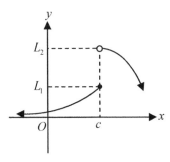

1. $\lim_{x \to c} f(x) = \text{DNE}$: The limit of $f(x)$ does not exists at c.
2. $f(c) = L_1$: The function is defined at $x = c$.

∴ The function is not continuous at $x = c$.

Chapter 2. Continuity and Discontinuity

3. Essential Discontinuity (Infinite Discontinuity)

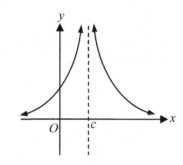

1. $\lim_{x \to c} f(x) = \infty$
2. $f(c)$ is not defined at $x = c$.

∴ The function is not continuous at $x = c$.

C. Removable and Non-Removable Discontinuity

THEOREM. *Removable Discontinuity*

A function has removable discontinuity at $x = c$, if f can be made continuous *by filling in a point*.

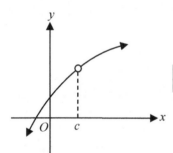

Pod is a removable discontinuity at $x = c$.

A function f has a removable discontinuity at $x = c$, if $\lim_{x \to c} f(x)$ exists but f is not continuous.

Non-Removable Discontinuity
A function has a non-removable discontinuity at $x = c$, if there is no way it can be continuous by filling in a point.

Jump

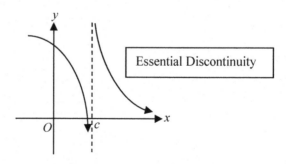

Essential Discontinuity

Chapter 2. Continuity and Discontinuity

▶ **Example**

Find the continuity of the following function.

1. $f(x) = \dfrac{1}{x}$
2. $f(x) = \dfrac{2}{x-3}$
3. $f(x) = \dfrac{x^2-1}{x+1}$
4. $g(x) = \dfrac{x}{x^2+1}$
5. $f(x) = \dfrac{x-2}{x^2-4}$
6. $h(x) = x + \sin x$
7. $r(x) = \dfrac{|x-2|}{x-2}$
8. $k(x) = [x]$
9. $g(x) = \dfrac{x+2}{x^2-3x-10}$

Answer

1. Essential discontinuity at $x = 0$
2. Essential discontinuity at $x = 3$
3. $f(x) = \dfrac{x^2-1}{x+1} = \dfrac{(x+1)(x-1)}{(x+1)} = x-1$, POD at $x = -1$ (Removable discontinuity)
4. Continuous on its entire domain
5. $f(x) = \dfrac{x-2}{x^2-4} = \dfrac{(x-2)}{(x+2)(x-2)} = \dfrac{1}{x+2}$

   POD at $x = 2$ (Removable discontinuity) and essential discontinuity at $x = -2$ (Non-removable discontinuity)

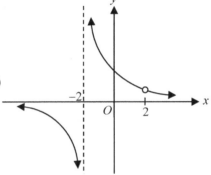

6. Continuous on its entire domain

7. $r(x) = \dfrac{|x-2|}{x-2} = \begin{cases} 1, & x > 2 \\ -1, & x < 2 \end{cases}$ Jump at $x = 2$ (Non-removable discontinuity)

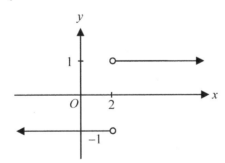

8. $k(x) = [x]$ Greatest integer function : Jump at $x =$ an integer

9. $g(x) = \dfrac{x+2}{x^2-3x-10} = \dfrac{(x+2)}{(x-5)(x+2)} = \dfrac{1}{x-5}$

   POD at $x = -2$ and essential discontinuity at $x = 5$

Dr. John Chung's AP Calculus AB/BC

Chapter 2. Continuity and Discontinuity

▶ **Example**

Find the continuity of the following function.

10. $f(x) = \begin{cases} x, & x \leq 1 \\ x^2, & x > 1 \end{cases}$

11. $g(x) = \begin{cases} -2x+6, & x < 1 \\ x^2+1, & x \geq 1 \end{cases}$

12. $f(x) = \begin{cases} 2x^2-4, & x < 3 \\ 14, & x = 3 \\ 5x-1, & x > 3 \end{cases}$

Answer

10. $f(x) = \begin{cases} x, & x \leq 1 \\ x^2, & x > 1 \end{cases}$

$f(1) = 1$, $\lim\limits_{x \to 1^+} x^2 = 1$, and $\lim\limits_{x \to 1^+} f(x) = f(1)$ ⇒ Continuous at $x = 1$

Therefore, $f(x)$ is continuous on its entire domain.

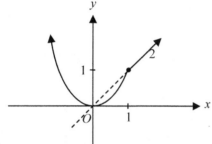

11. $g(x) = \begin{cases} -2x+6, & x < 1 \\ x^2+1, & x \geq 1 \end{cases}$

$\lim\limits_{x \to 1^+}(-2x+6) = 4$, $g(1) = 2$, and $\lim\limits_{x \to 1^+} f(x) \neq f(1)$ ⇒ Jump at $x = 1$

12. $f(x) = \begin{cases} 2x^2-4, & x < 3 \\ 14, & x = 3 \\ 5x-1, & x > 3 \end{cases}$

$\lim\limits_{x \to 3^-}(2x^2-4) = 14$, $f(3) = 14$, and $\lim\limits_{\delta x \to 3^+}(5x-1) = 14$

Since $\lim\limits_{x \to 3^-} f(x) = \lim\limits_{x \to 3^+} f(x) = f(3)$, $f(x)$ is continuous on its entire domain.

▶ **Example**

Find the constant k, a, and b so that the function is continuous on the entire real line.

13. $f(x) = \begin{cases} x^3+3, & x \leq 1 \\ x^2-k, & x > 1 \end{cases}$

14. $f(x) = \begin{cases} 3, & x \leq -1 \\ ax+b, & -1 < x < 3 \\ -5, & x \geq 3 \end{cases}$

Chapter 2. Continuity and Discontinuity

15. $g(x) = \begin{cases} \dfrac{5\sin x}{x}, & x < 0 \\ k - 3x, & x \geq 0 \end{cases}$

16. $g(x) = \begin{cases} \dfrac{x^2 - k^2}{x - k}, & x \neq k \\ 10, & x = k \end{cases}$

Answer

13. $f(x) = \begin{cases} x^3 + 3, & x \leq 1 \\ x^2 - k, & x > 1 \end{cases}$ $\quad \lim\limits_{x \to 1^+}(x^2 - k) = 1 - k$ and $f(1) = 1^3 + 3 = 4$

    $\lim\limits_{x \to 1^+} f(x) = f(1) \Rightarrow 1 - k = 4 \quad k = -3$

14. $\lim\limits_{x \to -1^+}(ax + b) = -a + b$ and $f(-1) = 3 \quad \therefore -a + b = 3$

    $\lim\limits_{x \to 3^-}(ax + b) = 3a + b$ and $f(3) = -5 \quad \therefore 3a + b = -5$

    Solving the equations, $a = -2$ and $b = 1$.

15. $g(x) = \begin{cases} \dfrac{5\sin x}{x}, & x < 0 \\ k - 3x, & x \geq 0 \end{cases}$ $\quad \lim\limits_{x \to 0^-} \dfrac{5\sin x}{x} = 5$ and $f(0) = k \quad \therefore k = 5$

16. $g(x) = \begin{cases} \dfrac{x^2 - k^2}{x - k}, & x \neq k \\ 10, & x = k \end{cases}$

    $\lim\limits_{x \to k} \dfrac{(x-k)(x+k)}{(x-k)} = \lim\limits_{x \to k} \dfrac{(x+k)}{1} = 2k$ and $g(k) = 10 \quad \rightarrow \quad 2k = 10 \Rightarrow k = 5$

▶ **Example**

Find removable discontinuity for the following functions.

17. $f(x) = \dfrac{x^2 - 9}{x + 3}$

18. $f(x) = \dfrac{x^2 - 7x + 12}{x^2 + x - 20}$

Answer

17. $f(x) = \dfrac{x^2 - 9}{x + 3} = \dfrac{\cancel{(x+3)}(x-3)}{\cancel{(x+3)}}$: Removable discontinuity at $x = -3$.

18. $f(x) = \dfrac{x^2 - 7x + 12}{x^2 + x - 20} = \dfrac{(x-3)\cancel{(x-4)}}{(x+5)\cancel{(x-4)}}$:

    Removable discontinuity at $x = 4$ and Non-removable infinite discontinuity at $x = 5$

Chapter 2. Continuity and Discontinuity

D. Intermediate Value Theorem

THEOREM. Intermediate Value Theorem (IVT):

If f is *continuous* on a closed interval $[a, b]$ and k is any number between $f(a)$ and $f(b)$, then there exists at least one number of c on (a, b) such that $f(c) = k$.

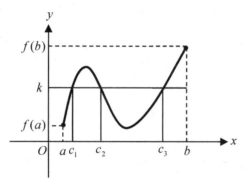

▶ **Example**

1. Use Intermediate Value Theorem to show that $f(x) = x^3 - 5x - 1$ has a zero on the interval $(1, 3)$.

2. Use Intermediate Value Theorem to verify that $f(x) = \dfrac{x^2 + x}{x - 1}$ has the number of c such that $f(c) = 7$ on the interval $(2, 5)$.

Answer
1. a) f is continuous on the closed interval $[1, 3]$, and b) $f(1) = -5$ and $f(3) = 11$ → It follows that 0 is between $f(1)$ and $f(3)$. Therefore, there is at least one number c on $(1, 3)$ such that $f(c) = 0$ by the Intermediate Value Theorem.
2. a) f is continuous on the closed interval $[2, 5]$. b) $f(2) = 6$, $f(5) = 7.5$, and $f(2) < 7 < f(5)$. There exists at least one c on $(2, 5)$ such that $f(c) = 7$ by the Intermediate Value Theorem.

Chapter 3. Differentiation

A. THE DEFINITION OF THE AVERAGE RATE OF CHANGE

THEOREM. The average rate of change of y (slope m) with respect to x over the interval $[x_1, x_2]$ is given by

$$r = \frac{y_2 - y_1}{x_2 - x_1} = \frac{f(x_2) - f(x_1)}{x_2 - x_1} = \frac{f(x_1 + h) - f(x_1)}{h} = \text{The slope of the secant line } \ell$$

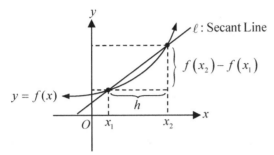

B. THE DEFINITION OF THE SLOPE OF A TANGENT LINE

THEOREM. The instantaneous rate of change of y with respect to x at x_1 given by

$$r_{inst} = \lim_{x_2 \to x_1} \frac{y_2 - y_1}{x_2 - x_1} = \lim_{x_2 \to x_1} \frac{f(x_2) - f(x_1)}{x_2 - x_1} = \lim_{h \to 0} \frac{f(x_1 + h) - f(x_1)}{h} = \text{The slope of the tangent line } \ell$$

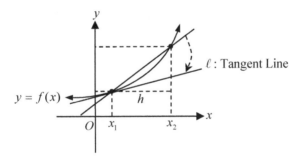

Now replace x_1 (constant) with x (variable).

$$f'(x) = \lim_{h \to 0} \frac{f(x + h) - f(x)}{h} = \text{Derivative of a Function}$$

The Derivative is the Slope of the Tangent Line.
Differentiation is the Processes of Finding the Derivative of a Function.

C. NOTATION FOR DERIVATIVE

THEOREM. The Slope of a tangent line at $x = x_1$ given by

$$f'(x), \quad \frac{dy}{dx}, \quad y', \quad \frac{d(f(x))}{dx}$$

Chapter 3. Differentiation

▶ Example

Use the definition of the derivative to find the derivative of a function.

1. $f(x) = 5$
2. $f(x) = 3x - 1$
3. $f(x) = -3$
4. $f(x) = x^2 - 2x + 3$
5. $f(x) = x^3 - 3x$
6. $f(x) = \dfrac{1}{x-1}$

Answer

1. $f'(x) = \lim\limits_{h \to 0} \dfrac{f(x+h) - f(x)}{h} = \lim\limits_{h \to 0} \dfrac{5 - 5}{h} = 0$

2. $f'(x) = \lim\limits_{h \to 0} \dfrac{f(x+h) - f(x)}{h} = \lim\limits_{h \to 0} \dfrac{[3(x+h) - 1] - [3x - 1]}{h} = \lim\limits_{h \to 0} \dfrac{3\cancel{h}}{\cancel{h}} = 3$

3. $f'(x) = \lim\limits_{h \to 0} \dfrac{(-3) - (-3)}{h} = 0$

4. $f'(x) = \lim\limits_{h \to 0} \dfrac{\left[(x+h)^2 - 2(x+h) + 3\right] - \left[x^2 - 2x + 3\right]}{h} = \lim\limits_{h \to 0} \dfrac{(x^2 + 2hx + h^2 - 2x - 2h + 3) - x^2 + 2x - 3}{h}$

   $= \lim\limits_{h \to 0} \dfrac{2hx + h^2 - 2h}{h} = \lim\limits_{h \to 0} 2x + h - 2 = 2x - 2$

5. $f'(x) = \lim\limits_{h \to 0} \dfrac{\left[(x+h)^3 - 3(x+h)\right] - \left[x^3 - 3x\right]}{h} = \lim\limits_{h \to 0} \dfrac{x^3 + 3x^2h + 3xh^2 + h^3 - 3x - 3h - (x^3 - 3x)}{h}$

   $= \lim\limits_{h \to 0} \dfrac{3x^2h + 3xh^2 + h^3 - 3h}{h} = \lim\limits_{h \to 0} (3x^2 + 3xh + h^2 - 3) = 3x^2 - 3$

6. $f'(x) = \lim\limits_{h \to 0} \dfrac{\dfrac{1}{(x+h)-1} - \dfrac{1}{x-1}}{h} = \lim\limits_{h \to 0} \dfrac{\dfrac{(x-1) - (x+h-1)}{(x+h-1)(x-1)}}{h} = \lim\limits_{h \to 0} \dfrac{-h}{h(x+h-1)(x-1)}$

   $= \lim\limits_{h \to 0} \dfrac{-1}{(x+h-1)(x-1)} = \dfrac{-1}{(x-1)^2}$

Chapter 3. Differentiation

D. DIFFERENTIABILITY AND CONTINUITY

DEFINITION. A function said to be differentiable at $x = x_0$ if the limit
$$f'(x_0) = \lim_{h \to 0} \frac{f(x_0 + h) - f(x_0)}{h}$$
exists. And similarly, a function is differentiable at $x = x_0$, if the derivatives from the right and from the left exist and are equal.

THEOREM. If f is *differentiable* at $x = x_0$, then f *is continuous* at $x = x_0$.

THEOREM. If f is *continuous* at $x = x_0$, then f is *not necessarily differentiable* at $x = x_0$.

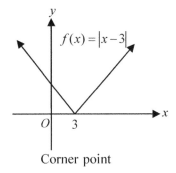

Corner point

$\lim\limits_{x \to 3^+} \dfrac{f(x) - f(3)}{x - 3} = \lim\limits_{x \to 3^+} \dfrac{(x-3) - 0}{x-3} = 1$ and

$\lim\limits_{x \to 3^-} \dfrac{f(x) - f(3)}{x - 3} = \lim\limits_{x \to 3^-} \dfrac{-(x-3) - 0}{x-3} = -1$

$\therefore f(x)$ is not differentiable at $x = 3$, because the one-sided derivatives are not equal.

$\lim\limits_{x \to 3^+} \dfrac{f(x) - f(3)}{x - 3} \neq \lim\limits_{x \to 3^-} \dfrac{f(x) - f(3)}{x - 3}$

The graph of f is continuous, but f is not differentiable at $x = 3$ and f does not have a tangent line at the point.

---

1. A Graph with a Cusp or a Sharp Corner (Sharp Turn) is not differentiable at the point.
2. A Graph with a Vertical Tangent Line is not differentiable at the vertical line.

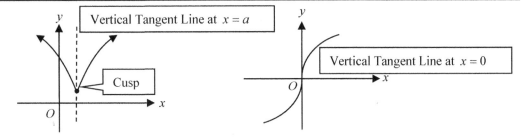

Note: The graph of a function f has a vertical tangent at the point (x_0, y_0) if and if only $f'(x_0) = +\infty$ or $f'(x_0) = -\infty$. The graph of the function becomes vertical and then virtually doubles back on itself. In general, we say that the graph of f has a vertical cusp at (x_0, y_0).

Note: The graph of $f(x) = |x - 3|$ has no vertical tangent and no vertical cusp at $x = 3$. The phenomenon this function shows at $x = 3$ is called a corner (or sharp corner). $f'(3)$ does not exist. In fact, you have left and right derivatives with $f'_-(3) = -1$ and $f'_+(3) = 1$.

Chapter 3. Differentiation

▶ **Example**

1. The Graph of a function is defined on the interval $[-6, 12]$. Find every x-value at which the function is not differentiable on $(-6, 12)$

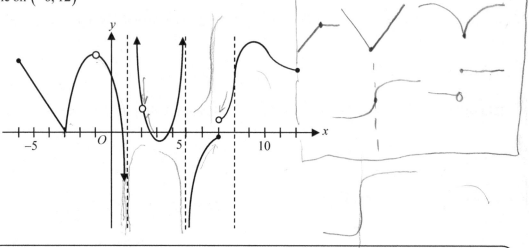

Answer
1. $x = -3$ (Sharp corner), -1 (Pod), 1 (Vertical asymptote), 2 (Pod)
   5 (Vertical asymptote), 7 (Jump), 8 (Vertical tangent line)

▶ **Example**

Determine continuity and differentiability of the given functions at $x = 0$.

2. $f(x) = x|x|$ 3. $f(x) = x^2 - 3|x| + 2$

---

2. A) Continuity

$f(x) = x|x| \begin{cases} x^2, & x \geq 0 \\ -x^2, & x < 0 \end{cases}$ $f(0) = 0^2 = 0$ and $\lim_{x \to 0^-} f(x) = \lim_{x \to 0^-} x|x| = \lim_{x \to 0^-} -x^2 = 0$

Since $\lim_{x \to 0^-} f(x) = f(0)$, f is continuous at $x = 0$.

B) Differentiability

$f'(x) = \begin{cases} 2x, & x \geq 0 \rightarrow f'(0) = 0 \\ -2x, & x < 0 \rightarrow f'(0) = 0 \end{cases}$ The derivatives are equal. Therefore, f is differentiable at $x = 0$.

3. A) Continuity

$f(x) = \begin{cases} x^2 - 3x + 2, & x \geq 0 \rightarrow f(0) = 2 \\ x^2 + 3x + 2, & x < 0 \rightarrow \lim_{x \to 0^-}(x^2 + 3x + 2) = 2 \end{cases}$ f is continuous at $x = 0$.

B) Differentiability

$f'(x) = \begin{cases} 2x - 3, & x \geq 0 \rightarrow f'(0) = -3 \\ 2x + 3, & x < 0 \rightarrow f'(0) = 3 \end{cases}$ Since the one sided derivatives are not equal, f is not differentiable at $x = 0$.

Chapter 3. Differentiation

▶ Example

Determine whether the function is differentiable at $x = 1$.

4. $f(x) = |x-1|$

5. $f(x) = \sqrt{1-x^2}$

6. $f(x) = \begin{cases} x+1, & x \leq 1 \\ x^2+1, & x > 1 \end{cases}$

7. $f(x) = \begin{cases} (x-1)^2, & x \leq 1 \\ (x-1)^3, & x > 1 \end{cases}$

Answer

4. Not differentiable: $f(x) = |x-1|$ has a sharp corner at $x = 1$.

5. Since $f(x) = \sqrt{1-x^2}$ is defined on the interval $[-1, 1]$, f is not differentiable at $x = 1$.

6. $f(x) = \begin{cases} x+1, & x \leq 1 \\ x^2+1, & x > 1 \end{cases}$

$\lim_{x \to 1^+}(x^2+1) = 2$ and $f(1) = 2$ → Continuous

$f'(x) = \begin{cases} \lim_{x \to 1^+} \dfrac{f(x)-f(1)}{x-1} = \lim_{x \to 1^+} \dfrac{x^2+1-(1+1)}{x-1} = \lim_{x \to 1^+} \dfrac{x^2-1}{x-1} = \lim_{x \to 1^+}(x+1) = 2 \\ \lim_{x \to 1^-} \dfrac{f(x)-f(1)}{x-1} = \lim_{x \to 1^+} \dfrac{x+1-(1+1)}{x-1} = \lim_{x \to 1^+} \dfrac{x-1}{x-1} = \lim_{x \to 1^+} 1 = 1 \end{cases}$

Since one sided derivatives are not equal, f is not differentiable at $x = 1$.

∴ f is continuous, but not differentiable.

7. $f(x) = \begin{cases} (x-1)^2, & x \leq 1 \\ (x-1)^3, & x > 1 \end{cases}$

$\lim_{x \to 1^+}(x-1)^3 = 0$ and $f(1) = 0$ (This implies $\lim_{x \to 1^-} f(x) = 0$)

Since $\lim_{x \to 1} f(x) = f(1)$, f is continuous.

$f'(x) = \begin{cases} \lim_{x \to 1^+} \dfrac{(x-1)^3 - (1-1)^3}{x-1} = \lim_{x \to 1^+} \dfrac{(x-1)^3 - (1-1)^3}{x-1} = \lim_{x \to 1^+}(x-1)^2 = 0 \\ \lim_{x \to 1^-} \dfrac{(x-1)^2 - (1-1)^2}{x-1} = \lim_{x \to 1^-} \dfrac{(x-1)^2 - (1-1)^2}{x-1} = \lim_{x \to 1^-}(x-1) = 0 \end{cases}$

∴ f is continuous at $x = 1$ and differentiable at the point.

Chapter 3. Differentiation

E. RULES FOR DIFFERENTIATION

1. **The Constant Rule:** The derivative of a constant function is 0.

   $$\frac{d}{dx}[k] = 0, \text{ where } k \text{ is a real}$$

2. **The Power Rule:**

   $$\frac{d}{dx}[x^n] = nx^{n-1}$$

3. **The Power Rule with a constant: (This is also called "The Constant Multiple Rule")**

   $$\frac{d}{dx}[kf(x)] = k\frac{d}{dx}[f(x)] = kf'(x)$$

4. **The Sum and Difference Rules:**

   $$\frac{d}{dx}[f(x) + g(x)] = \frac{d}{dx}[f(x)] + \frac{d}{dx}[g(x)] = f'(x) + g'(x)$$

   $$\frac{d}{dx}[f(x) - g(x)] = \frac{d}{dx}[f(x)] - \frac{d}{dx}[g(x)] = f'(x) - g'(x)$$

5. **The Product Rule:**

   $$\frac{d}{dx}[f(x)g(x)] = f(x)\frac{d}{dx}[g(x)] + g(x)\frac{d}{dx}[f(x)] = f(x)g'(x) + g(x)f'(x)$$

6. **The Quotient Rule:**

   $$\frac{d}{dx}\left[\frac{f(x)}{g(x)}\right] = \frac{g(x)f'(x) - f(x)g'(x)}{[g(x)]^2}$$

7. **Chain Rule:**

   $$\frac{dy}{dx} = \frac{dy}{du} \cdot \frac{du}{dx}$$

F. Derivatives of Trigonometric Functions

1. $\frac{d}{dx}[\sin x] = \cos x$
2. $\frac{d}{dx}[\cos x] = -\sin x$
3. $\frac{d}{dx}[\tan x] = \sec^2 x$
4. $\frac{d}{dx}[\sec x] = \sec x \tan x$
5. $\frac{d}{dx}[\csc x] = -\csc x \cot x$
6. $\frac{d}{dx}[\cot x] = -\csc^2 x$

▶ **Example**

Find the derivative of the function.

1. $y = 5$
2. $f(x) = -3$
3. $f(x) = 2x + 5$
4. $g(x) = 5x^2 - 4x + 1$
5. $f(x) = t^3 - 2t^2 + 6$
6. $y = x^{10}$
7. $g(x) = \frac{1}{x^3} + \frac{1}{2}\sin x$
8. $f(x) = (3x^2 + 5x + 2)^3$
9. $y = (x^2 + 1)(x^3 - 1)^2$
10. $f(x) = \pi x + 5$

Chapter 3. Differentiation

Answer
1. $f'(x) = 0$ 2. $f'(x) = 0$ 3. $f'(x) = 2$ 4. $g'(x) = 10x - 4$
5. $f(x) = 3t^2 - 4t$ 6. $\dfrac{dy}{dx} = 10x^9$ 7. $g(x) = x^{-3} + \dfrac{1}{2}\sin x \Rightarrow g'(x) = -3x^{-4} + \dfrac{1}{2}\cos x = -\dfrac{3}{x^4} + \dfrac{1}{2}\cos x$
8. $f'(x) = 3(3x^2 + 5x + 2)^2(6x + 5)$
9. $y' = (x^3 - 1)^2 \left[(x^2 + 1)\right]' + (x^2 + 1)\left[(x^3 - 1)^2\right]' = (x^3 - 1)^2(2x) + (x^2 + 1)2(x^3 - 1)3x^2$
   $= (x^3 - 1)^2(2x) + 6x^2(x^2 + 1)(x^3 - 1)$
10. $y' = \pi$

▶ **Example**

Find the derivative.

11. $g(x) = \sec^2(x^3)$
12. $y = \tan x \cdot \sec x$
13. $g(x) = \csc^3(2x + 1)$
14. $y = \sqrt{\cot 2x}$

Answer
11. $g'(x) = 2\sec(x^3) \cdot \left[\sec(x^3)\tan(x^3)\right] \cdot (3x^2) = 6x^2 \cdot \sec^2(x^3) \cdot \tan(x^3)$
12. $y' = \sec x(\tan x)' + \tan x(\sec x)' = \sec x \sec^2 x + \tan x \sec x \tan x = \sec^3 x + \sec x \tan^2 x$
13. $g'(x) = 3\csc^2(2x + 1)\left[-\csc(2x + 1)\cot(2x + 1)\right](2) = -6\csc^3(2x + 1)\cot(2x + 1)$
14. $y = (\cot 2x)^{\frac{1}{2}} \rightarrow y' = \dfrac{1}{2}(\cot 2x)^{-\frac{1}{2}}(-\csc^2 2x)(2) = -\dfrac{\csc^2 2x}{\sqrt{\cot 2x}}$

▶ **Example**

15. Find a and b such that
$$f(x) = \begin{cases} x^3 - 2x^2 + ax, & x \leq 1 \\ bx^2 - 6x + 4, & x > 1 \end{cases}$$ is differentiable everywhere.

16. Find a and b such that
$$f(x) = \begin{cases} ax^3 + 2, & x \leq 2 \\ x^2 + b, & x > 2 \end{cases}$$ is differentiable at $x = 2$.

Chapter 3. Differentiation

Answer

15. $f(x) = \begin{cases} x^3 - 2x^2 + ax, & x \le 1 \\ bx^2 - 6x + 4, & x > 1 \end{cases}$

   a) $f(x)$ should be continuous at $x = 1$.

   $\lim\limits_{x \to 1^+} bx^2 - 6x + 4 = b - 6 + 4 = b - 2$ and $f(1) = 1 - 2 + a = a - 1$

   $\therefore b - 2 = a - 1 \Rightarrow a = b - 1$ -----(1)

   b) One sided derivatives should be equal.

   For $x \le 1$, $f'(x) = 3x^2 - 4x + a \Rightarrow f'(1) = 3 - 4 + a = a - 1$

   For $x > 1$, $f'(x) = 2bx - 6 \Rightarrow f'(1) = 2b - 6$

   $\therefore 2b - 6 = a - 1 \Rightarrow a = 2b - 5$ -----(2)

   From (1) and (2)

   $b - 1 = 2b - 5 \Rightarrow b = 4$ and $a = 3$

16. $f(x) = \begin{cases} ax^3 + 2, & x \le 2 \\ x^2 + b, & x > 2 \end{cases}$

   a) $f(x)$ should be continuous at $x = 2$.

   $\lim\limits_{x \to 2^+} x^2 + b = 4 + b$ and $f(2) = 8a + 2 \Rightarrow 4 + b = 8a + 2 \Rightarrow b = 8a - 2$

   b) One sided derivatives should be equal.

   $f'(x) = 3ax^2 \Rightarrow f'(2) = 12a$ and $f'(x) = 2x \Rightarrow f'(2) = 2(2) = 4$

   $12a = 4 \Rightarrow a = \dfrac{1}{3}$ and $b = 8a - 2 = 8\left(\dfrac{1}{3}\right) - 2 = \dfrac{2}{3}$

G. DERIVATIVES OF INVERSE TRIGONOMETRIC FUNCTIONS

1. $\dfrac{d}{dx}\left[\sin^{-1} x\right] = \dfrac{1}{\sqrt{1-x^2}}$ 2. $\dfrac{d}{dx}\left[\cos^{-1} x\right] = \dfrac{-1}{\sqrt{1-x^2}}$ 3. $\dfrac{d}{dx}\left[\tan^{-1} x\right] = \dfrac{1}{1+x^2}$

4. $\dfrac{d}{dx}\left[\sec^{-1} x\right] = \dfrac{1}{|x|\sqrt{x^2-1}}$ 5. $\dfrac{d}{dx}\left[\csc^{-1} x\right] = \dfrac{-1}{|x|\sqrt{x^2-1}}$ 6. $\dfrac{d}{dx}\left[\cot^{-1} x\right] = \dfrac{-1}{1+x^2}$

If u is a differentiable function of x, then

1. $\dfrac{d}{dx}\left[\sin^{-1} u\right] = \dfrac{u'}{\sqrt{1-u^2}}$ 2. $\dfrac{d}{dx}\left[\cos^{-1} u\right] = \dfrac{-u'}{\sqrt{1-u^2}}$ 3. $\dfrac{d}{dx}\left[\tan^{-1} u\right] = \dfrac{u'}{1+u^2}$

4. $\dfrac{d}{dx}\left[\sec^{-1} u\right] = \dfrac{u'}{|u|\sqrt{u^2-1}}$ 5. $\dfrac{d}{dx}\left[\csc^{-1} u\right] = \dfrac{-u'}{|u|\sqrt{u^2-1}}$ 6. $\dfrac{d}{dx}\left[\cot^{-1} u\right] = \dfrac{-u'}{1+u^2}$

Chapter 3. Differentiation

▶ **Example**

Find the derivative of the function.

1. $g(x) = \arcsin x + \sqrt{1-x^2}$
2. $f(x) = 2\arcsin(x-1)$
3. $f(x) = 3\arccos\dfrac{x}{2}$
4. $f(x) = \arctan\dfrac{x}{2}$

Solution

1. $g'(x) = \dfrac{1}{\sqrt{1-x^2}} - \dfrac{x}{\sqrt{1-x^2}} = \dfrac{1-x}{\sqrt{1-x^2}}$

2. $f'(x) = \dfrac{2}{\sqrt{1-(x-1)^2}} = \dfrac{2}{\sqrt{2x-x^2}}$

3. $f'(x) = \dfrac{-3\left(\dfrac{1}{2}\right)}{\sqrt{1-\left(\dfrac{x}{2}\right)^2}} = \dfrac{-3\left(\dfrac{1}{2}\right)}{\dfrac{1}{2}\sqrt{4-x^2}} = \dfrac{-3}{\sqrt{4-x^2}}$

4. $f'(x) = \dfrac{\dfrac{1}{2}}{1+\left(\dfrac{x}{2}\right)^2} = \dfrac{2}{4+x^2}$

Chapter 3. Differentiation

H. DERIVATIVES OF EXPONENTIAL AND LOGARITHMIC FUNCTIONS

1. $\lim_{x \to 0}(1+x)^{\frac{1}{x}} = e$
2. $\frac{d}{dx}[\log_a x] = \frac{1}{x \ln a}, (a > 0, a \neq 1)$
3. $\frac{d}{dx}[\ln x] = \frac{1}{x}$
4. $\frac{d}{dx}[\ln|x|] = \frac{1}{x}$
5. $\frac{d}{dx}[\log_a |x|] = \frac{1}{x \ln a}, (a > 0, a \neq 1)$
6. $\frac{d}{dx}[e^x] = e^x$
7. $\frac{d}{dx}[a^x] = a^x \ln a$

▶ **Example**

Find the derivative of the function.

1. $y = e^{-3x}$
2. $y = 10^x$
3. $y = 5^{-3x}$
4. $y = \ln|\sin x|$
5. $y = \log_3(x^2 + 4)$
6. $y = \ln|\ln x|$
7. $y = \ln\left|\frac{x-2}{x+2}\right|$
8. $y = x^x \ (x > 0)$
9. $y = \ln(\cos^2 x)$
10. $y = e^{\sin x}$
11. $y = \frac{e^x - 1}{e^x + 1}$
12. $y = \frac{\ln x}{e^x}$

Answer

1. $y' = e^{-3x}(-3) = -3e^{-3x}$
2. $y' = 10^x \ln 10$
3. $y' = 5^{-3x}(-3) = -3 \cdot (\ln 5) 5^{-3x}$
4. $y' = \frac{\cos x}{\sin x}$
5. $y' = \left[\frac{\ln(x^2+4)}{\ln 3}\right]' = \frac{1}{\ln 3}\left(\frac{2x}{x^2+4}\right) = \frac{2x}{(x^2+4)\ln 3}$
6. $y' = \frac{(\ln x)'}{\ln x} = \frac{1}{x \ln x}$
7. $y = \ln|x-2| - \ln|x+2| \Rightarrow y' = \frac{1}{x-2} - \frac{1}{x+2} = \frac{4}{x^2-4}$
8. $y = x^x \Rightarrow \ln y = x \ln x \Rightarrow (\ln y)' = (x \ln x)' \Rightarrow \frac{y'}{y} = x(\ln x)' + (\ln x)(x)' = 1 + \ln x$
   $y' = y(1 + \ln x) = x^x(1 + \ln x)$
9. $y' = \frac{(\cos^2 x)'}{\cos^2 x} = \frac{-2 \cos x \sin x}{\cos^2 x} = \frac{-2 \sin x}{\cos x} = -2 \tan x$
10. $y' = e^{\sin x}(\cos x)$
11. $y' = \frac{(e^x+1)(e^x-1)' - (e^x-1)(e^x+1)'}{(e^x+1)^2} = \frac{(e^x+1)e^x - (e^x-1)e^x}{(e^x+1)^2} = \frac{2e^x}{(e^x+1)^2}$
12. $y' = \frac{e^x(\ln x)' - (\ln x)(e^x)'}{(e^x)^2} = \frac{\frac{e^x}{x} - (\ln x)e^x}{e^{2x}} = \frac{e^x(1 - x \ln x)}{xe^{2x}} = \frac{1 - x \ln x}{xe^x}$

Chapter 3. Differentiation

I. EXPLICIT AND IMPLICIT DIFFERENTIATION

Explicit Functions: Function y is written explicitly as function of x (Explicit Form).

$$1)\ y = 3x^2 + 5x + 2\cos x \quad \Rightarrow \quad \frac{d}{dx}\left[3x^2 + 5x + 2\cos x\right] = 6x + 5 - 2\sin x$$

$$2)\ y = \frac{1}{x} \quad \Rightarrow \quad \frac{d}{dx}\left[\frac{1}{x}\right] = \frac{-1}{x^2}$$

Implicit Differentiation: Function y is differentiable function of x and the equation is written implicitly and involve both x and y.

$$1)\ 2x + 3y = y^2 \quad \Rightarrow \quad \frac{d}{dx}[2x + 3y] = \frac{d}{dx}[y^2] \quad \Rightarrow \quad 2 + 3\frac{dy}{dx} = 2y\frac{dy}{dx}$$

$$\Rightarrow \quad (2y - 3)\frac{dy}{dx} = 2 \quad \Rightarrow \quad \frac{dy}{dx} = \frac{2}{2y - 3}$$

You can find $\frac{dy}{dx}$ as follows. 1) Differentiate both sides of the equation with respect to x.

2) Express $\frac{dy}{dx}$ in terms of x and y.

▶ **Example**

Find $\frac{dy}{dx}$ by implicit differentiation.

Note: $\frac{dy}{dx} = y'$

1. $x^2 + y^2 = 4$
2. $x^3 - xy + 2y^2 = 1$
3. $\sin x + \cos^2 y = 3$
4. $xy = 5$
5. $x - y = \cot y$
6. $x = \sec y$

Answer

1. $y' = \dfrac{-x}{y}$ 2. $y' = \dfrac{y - 3x^2}{4y - x}$ 3. $y' = \dfrac{\cos x}{2\sin y \cos y}$ 4. $\dfrac{dy}{dx} = -\dfrac{y}{x}$

5. $x - y = \cot y \Rightarrow 1 - \dfrac{dy}{dx} = -\csc^2 y \cdot \dfrac{dy}{dx} \Rightarrow 1 = \dfrac{dy}{dx} - \csc^2 y \cdot \dfrac{dy}{dx} \Rightarrow 1 = \dfrac{dy}{dx}(1 - \csc^2 y)$

$\Rightarrow \dfrac{dy}{dx} = \dfrac{1}{1 - \csc^2 y} = \dfrac{1}{-\cot^2 y} = -\tan^2 x$

Note: $1 + \cot^2 x = \csc^2 x \Rightarrow 1 - \csc^2 x = -\cot^2 x$

6. $\dfrac{dy}{dx} = \dfrac{1}{\sec y \tan y}$

Chapter 3. Differentiation

▶ **Example**

Find the derivative at the given point.

1. $x^3 - y^3 = 0$ at (2, 2)
2. $x \sin y = 1$ at $\left(2, \dfrac{\pi}{3}\right)$
3. $x^3 + y^3 - 2xy = 0$ at (1, 1)
4. $x^2 y + 4y = 8$ at (2, 1)

Answer

1. $x^3 - y^3 = 0 \;\Rightarrow\; 3x^2 - 3y^2 \dfrac{dy}{dx} = 0 \;\Rightarrow\; \dfrac{dy}{dx} = \dfrac{x^2}{y^2} \;\Rightarrow\; \left.\dfrac{dy}{dx}\right|_{(2,2)} = \dfrac{4}{4} = 1$

2. $x \sin y = 1 \;\Rightarrow\; (\sin y)(x)' + x(\sin y)' = (1)' \;\Rightarrow\; \sin y + x \cos y \cdot \dfrac{dy}{dx} = 0 \;\Rightarrow\; \dfrac{dy}{dx} = -\dfrac{\sin y}{x \cos y}$

    $\Rightarrow\; \left.\dfrac{dy}{dx}\right|_{\left(2,\frac{\pi}{3}\right)} = -\dfrac{\sin \frac{\pi}{3}}{2\cos \frac{\pi}{3}} = -\dfrac{\frac{\sqrt{3}}{2}}{2\left(\frac{1}{2}\right)} = -\dfrac{\sqrt{3}}{2}$

3. $x^3 + y^3 - 2xy = 0 \;\Rightarrow\; 3x^2 + 3y^2 y' - 2(y + xy') = 0 \;\Rightarrow\; y' = \dfrac{2y - 3x^2}{3y^2 - 2x}$

    $\Rightarrow\; \left.y'\right|_{(1,1)} = \dfrac{2(1) - 3(1)^2}{3(1)^2 - 2(1)} = \dfrac{-1}{1} = -1$

4. $x^2 y + 4y = 8 \;\Rightarrow\; 2xy + x^2 y' + 4y' = 0 \;\Rightarrow\; y' = \dfrac{-2xy}{x^2 + 4} \;\Rightarrow\; \left.y'\right|_{(2,1)} = \dfrac{-2(2)(1)}{2^2 + 4} = -\dfrac{1}{2}$

Chapter 4. Applications of Differentiation

A. FINDING THE EQUATION OF THE TANGENT LINE AND THE NORMAL LINE

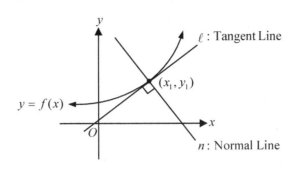

The equation of the tangent line through point (x_1, y_1) is
$$y - y_1 = m(x - x_1), \text{ where } m = f'(x_1).$$

The equation of the normal line through point (x_1, y_1) is
$$y - y_1 = -\frac{1}{m}(x - x_1), \text{ where } m = f'(x_1).$$

The *normal line* is defined as the line that is perpendicular to the tangent line at the point of tangency.

▶ **Example**

Find the equation of the tangent line and the normal line at the indicated point.

1. $f(x) = x + \dfrac{2}{x}$ at $(-1, -3)$
2. $f(x) = 2\ln x$ at $(e, 2)$
3. Find the equation of the tangent line to the graph of $y = e^x$ and passing through the origin.
4. If the graph of $f(x) = \sqrt{x} + k$ is tangent to the graph of $y = x$, what is the value of k?

---

1. Tangent line: $y + 3 = -(x+1)$, Normal line: $y + 3 = (x+1)$
2. $f'(x) = \dfrac{2}{x} \rightarrow f'(e) = \dfrac{2}{e}$, Tangent line: $y - 2 = \dfrac{2}{e}(x - e)$, Normal line: $y - 2 = -\dfrac{e}{2}(x - e)$
3. The equation of the tangent line at (c, e^c) is $y - e^c = f'(c)(x - c)$ and $f'(c) = e^c$.
   $y - e^c = e^c(x - c) \Rightarrow y = e^c x - ce^c + e^c$
   Since the tangent line passes through the origin, $0 = -ce^c + e^c \Rightarrow e^c(1 - c) = 0 \Rightarrow c = 1$.
   Therefore, $y = e^x$.
4. $f'(x) = \dfrac{1}{2\sqrt{x}}$
   Since the slope of $y = x$ is 1 at (a, a) (point of tangency), $f'(a) = \dfrac{1}{2\sqrt{a}} = 1 \Rightarrow \sqrt{a} = \dfrac{1}{2} \Rightarrow a = \dfrac{1}{4}$.

   The tangent passes through point (a, a). $a = \sqrt{a} + k \Rightarrow \dfrac{1}{4} = \sqrt{\dfrac{1}{4}} + k \Rightarrow \dfrac{1}{4} = \dfrac{1}{2} + k \Rightarrow k = -\dfrac{1}{4}$

Chapter 4. Applications of Differentiation

B. THE MEAN-VALUE THEOREM FOR DERIVATIVES

THEOREM. The Mean-Value Theorem (MVT)

If f is *continuous* on the closed interval $[a, b]$ and *differentiable* on the open interval (a, b), then there exists at least one number c between a and b such that

$$\frac{f(b)-f(a)}{b-a} = f'(c)$$

The slope of the secant line is equal to the slope of the tangent line.

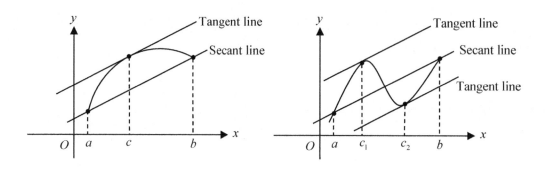

▶ **Example**

1. Given $f(x) = x^2$ on the interval $[-1, 3]$, find all values of c in the open interval $(-1, 3)$ such that $f'(c) = \dfrac{f(3)-f(-1)}{3-(-1)}$.

2. Given $f(x) = \ln x$ on the interval $[1, 3]$, find all values of c in the open interval $(1, 3)$ satisfying the Mean-Value Theorem.

3. The function $f(x) = \dfrac{1}{4}x^3 + 1$ satisfies the Mean-Value Theorem over the interval $[0, 2]$. Find all values of c in the interval $(0, 2)$ at which the tangent line to the graph of f is parallel to the secant line joining the points $(0, f(0))$ and $(2, f(2))$.

Chapter 4. Applications of Differentiation

1. $\dfrac{f(3)-f(-1)}{3-(-1)} = \dfrac{9-1}{4} = \dfrac{8}{4} = 2$ and $f'(x) = 2x$. $f'(c) = 2c \rightarrow 2c = 2 \rightarrow c = 1$

   There exists one value of c in the interval $(-1, 3)$

2. $f'(x) = \dfrac{1}{x}$ and $f'(c) = \dfrac{1}{c} = \dfrac{\ln 3 - \ln 1}{3-1} \rightarrow \dfrac{1}{c} = \dfrac{\ln 3}{2} \rightarrow c = \dfrac{2}{\ln 3}$ in the interval $(1, 3)$

3. The function f is continuous and differentiable everywhere because it is a polynomial. In particular, f is continuous on [0, 2] and differentiable on (0, 2), so the hypotheses of the Mean-Value Theorem are satisfied with $a = 0$ and $b = 2$. But $f(0) = 1$, $f(2) = 3$, $f'(x) = \dfrac{3x^2}{4}$, and $f'(c) = \dfrac{3c^2}{4}$. Therefore,

   $\dfrac{3c^2}{4} = \dfrac{3-1}{2-0} \rightarrow 3c^2 = 4 \rightarrow c = \pm \dfrac{2}{\sqrt{3}} \approx \pm 1.25$. However, only $c = \dfrac{2}{\sqrt{3}} \approx 1.25$ lies in the interval $(0, 2)$.

C. ROLLE'S THEOREM

THEOREM. Rolle's Theorem (Special case of the Mean Value Theorem)

Let f be continuous on the closed interval $[a, b]$ and differentiable on the open interval (a, b).

If $f(b) = f(a)$, then there exists at least one number c between a and b such that

$$\dfrac{f(b)-f(a)}{b-a} = f'(c) = 0.$$

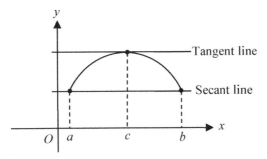

Rolle's Theorem will guarantee the existence of extreme value (relative maximum or relative minimum) in the interval.

Remember: The differentiability requirement in Rolle's Theorem is critical. If f fails to be differentiable at even one place in the interval (a,b), then the conclusion of the theorem may not hold. For example, the function $f(x) = |x| - 2$ graphed in figure has roots at $x = -2$ and $x = 2$, yet there is no horizontal tangent to the graph of f over the interval $(-2, 2)$.

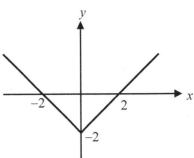

Dr. John Chung's AP Calculus AB/BC

Chapter 4. Applications of Differentiation

▶ **Example**

Find all values of c on the given interval such that $f'(c) = 0$.

1. $f(x) = x^3 - 3x^2 + 1$ on the interval $[-1, 3]$
2. $f(x) = (x-1)(x-2)$ on the interval $[1, 2]$
3. $f(x) = \dfrac{x^2 - 1}{x}$ on the interval $[-1, 1]$
4. $f(x) = \tan x$ on the interval $[0, 2\pi]$
5. $f(x) = \dfrac{x^2 - 2x - 3}{x + 2}$ on the interval $[-1, 3]$
6. $f(x) = 3 - |x - 3|$ on the interval $[0, 6]$
7. Find the two x-intercepts of $f(x) = x^2 - 5x + 4$ and confirm that $f'(c) = 0$ at some point c between those intercepts.

---

1. $f'(x) = 3x^2 - 6x = 0 \Rightarrow 3x(x-2) = 0 \Rightarrow x = 0$ or 2
   The polynomial f is continuous and differentiable everywhere. Therefore, in the interval $[-1, 3]$, $c = 0, 2$.
2. $f'(x) = 2x - 3 = 0 \Rightarrow x = 1.5$
   Therefore, in the interval, $c = 1.5$.
3. Since $f(x)$ is not continuous at $x = 0$, Rolle's Theorem cannot be applied to f on the interval.
4. Since $f(x)$ is not continuous at $\dfrac{\pi}{2}, \dfrac{3\pi}{2}$, Rolle's Theorem cannot be applied to f on the interval.
5. $f'(x) = \dfrac{(x+2)(2x-2) - (x^2 - 2x - 3)(1)}{(x+2)^2} = 0 \Rightarrow \dfrac{x^2 + 4x - 1}{(x+2)^2} = 0 \Rightarrow x^2 + 4x - 1 = 0$
   $x = \dfrac{-4 \pm \sqrt{20}}{2} = -2 \pm \sqrt{5}$
   Therefore, in the interval, $c = -2 + \sqrt{5}$.
6. $f(0) = 0$ and $f(6) = 0 \Rightarrow f(0) = f(3)$ but $f(x)$ is not differentiable at $x = 3$. Rolle's Theorem cannot be applied to f.
7. $x^2 - 5x + 4 = (x-1)(x-4)$ so the x-intercepts are $x = 1$ and $x = 4$. Since the polynomial f is continues and differentiable everywhere, the hypotheses of Rolle's Theorem are satisfied on the interval $[1, 4]$. Thus we are guaranteed the existence of at least one point c in the interval $(1, 4)$ such that $f'(c) = 0$.

Chapter 4. Applications of Differentiation

D. MOTION (POSITION, VELOCITY, ACCLERATION)

DEFINITION.
1. A velocity function can be obtained by taking the derivative of a position function.
2. An acceleration function can be obtained by taking the derivative of a velocity function.

$s(t)$: Position function
$v(t) = s'(t)$: Velocity function
$a(t) = v'(t) = s''(t)$: Acceleration function
$|v(t)| = |s'(t)|$: Speed function

THEOREM. SPEEDING UP AND SPEEDIND DOWN

1. If the velocity is positive, the particle is moving to the right.
2. If the velocity is negative, the particle is moving to the left.
3. If the velocity and the acceleration have the same sign, the particle's speed is increasing.
4. If the velocity and the acceleration have opposite signs, the particle's speed is decreasing.
5. If the velocity is zero and the acceleration is not zero, the particle is changing direction.

▶ **Example**

1. The position of a particle moving along the x-axis at any time $t > 0$ is given by the function
$x(t) = t^3 - 9t^2 + 24t + 5$, $t > 0$, where t is time in seconds.

a) When is the particle at rest?
b) When does the particle change direction?
c) When does the particle speed up?
d) When does the particle slow down?

a) $v(t) = 0 \Rightarrow \dfrac{dx}{dt} = 3t^2 - 18t + 24 = 3(t-2)(t-4) = 0 \ \ t = 2, 4$

b) At $t = 2$, the particle changes direction from right to left.
   At $t = 4$, the particle changes direction from left to right.

c) $v(t) = 3(t-2)(t-4) = 0$ and $a(t) = v'(t) = 6t - 18 = 6(t-3) = 0$

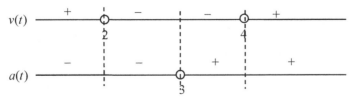

$v(t)$ and $a(t)$ have the same sign: The particle speeds up at $(2, 3)$ and $(4, \infty)$.

d) $v(t)$ and $a(t)$ have the opposite sign: The particle slows down at $(0, 2)$ and $(3, 4)$.

Chapter 4. Applications of Differentiation

Observe behavior of the particle about the position versus time curve.

Graph	At $t = t_0$	Description
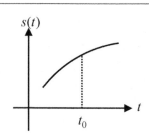	At $t = t_0$ • Curve has positive slope. • Curve is concave down. • $s(t_0) > 0$ • $s'(t_0) = v(t_0) > 0$ • $s''(t_0) = a(t_0) < 0$	• Particle is on the positive side of the origin. • Particle is moving in the positive direction. • Velocity is decreasing. • Particle is slowind down. $v(t_0) > 0$ and $a(t_0) < 0$
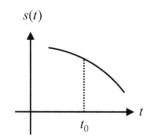	At $t = t_0$ • Curve has negative slope. • Curve is concave down. • $s(t_0) > 0$ • $s'(t_0) = v(t_0) < 0$ • $s''(t_0) = a(t_0) < 0$	• Particle is on the positive side of the origin. • Particle is moving in the negative direction. • Velocity is decreasing. • Particle is speeding up $v(t_0) < 0$ and $a(t_0) < 0$
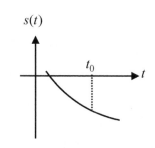	At $t = t_0$ • Curve has negative slope. • Curve is concave up. • $s(t_0) < 0$ • $s'(t_0) = v(t_0) < 0$ • $s''(t_0) = a(t_0) > 0$	• Particle is on the negative side of the origin. • Particle is moving in the negative direction. • Velocity is increasing. • Particle is slowing down. $v(t_0) < 0$ and $a(t_0) > 0$
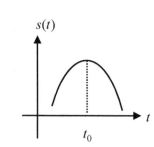	At $t = t_0$ • Curve has zero slope. • Curve is concave down. • $s(t_0) > 0$ • $s'(t_0) = v(t_0) = 0$ • $s''(t_0) = a(t_0) < 0$	• Particle is on the positive side of the origin. • Particle is momentarily stopped.

Chapter 4. Applications of Differentiation

▶ Example

2. A particle moves along the x-axis so that its position function at time t is given by
$$s(t) = -t - e^{1-t}.$$
a) Find the velocity function.
b) Find the acceleration function.
c) Is the speed of the particle increasing at time $t = 4$?
d) Find all values of t at which the particle changes direction.
e) Find the total distance traveled by the particle over the time interval $0 \le t \le 4$.

a) $v(t) = s'(t) = -1 - e^{1-t}(-1) = -1 + e^{1-t}$ b) $a(t) = s''(t) = v'(t) = e^{1-t}(-1) = -e^{1-t}$

c) $v(4) = -1 + e^{-3} = 1 - \dfrac{1}{e^3} < 0$ and $a(4) = -e^{1-4} = -e^{-3} < 0$

By rule #4, the velocity and the acceleration at $t = 4$ are both negative, therefore the speed is increasing.

d) $v(t) = 0 \Rightarrow -1 + e^{1-t} = 0 \Rightarrow e^{1-t} = 1 \Rightarrow t = 1$
The particle changes direction at $t = 1$.

e) Since $v(t) > 0$ for $t < 1$ and $v(t) < 0$ for $t > 1$,
Total distance $= [s(1) - s(0)] + [s(1) - s(4)] = [-2 + e] + [-2 - (-4 - e^{-3})] = -2 + e + 2 + e^{-3} = e + e^{-2}$

▶ Example

3. If the position of a particle at time t is given by $x(t) = t^3 - 4t^2 + 12t$, find the velocity and the acceleration of the particle at time $t = 5$. Is the speed of the particle increasing at time $t = 5$?

$v(t) = x'(t) = 3t^2 - 8t + 12 \Rightarrow v(5) = 75 - 40 + 12 = 47$
$a(t) = v'(t) = 6t - 8 \Rightarrow a(5) = 30 - 8 = 22$
Since $v(5) > 0$ and $a(5) > 0$, the speed is increasing at time $t = 5$.

Chapter 4. Applications of Differentiation

▶ **Example**

4. A particle moves along the *x*-axis so that the position of the particle at time t is given by

$$x(t) = t^3 - 6t^2 + 9t + 3.$$

a) For $0 \leq t \leq 6$, find all times t during the particle is moving to the right.

b) Find the acceleration at time $t = 4$. Is the speed of the particle increasing at $t = 4$?

c) Find all times t in the open interval $0 < t < 6$ when particle changes direction.

d) Find the total distance traveled by the particle from $t = 0$ until time $t = 6$.

e) During the interval $0 \leq t \leq 4$, what is the greatest distance between the particle from the origin?

---

a) The velocity is positive.
   $v(t) = x'(t) = 3t^2 - 12t + 9 > 0 \implies 3(t^2 - 4t + 3) > 0 \implies 3(t-1)(t-3) > 0$
   $v(t) > 0$ for $0 < t < 1$ and $3 < t < 6$
   Therefore, the particle is moving to the right for $0 < t < 1$ and $3 < t < 6$

b) $v(4) = 3(4^2) - 12(4) + 9 = 9$
   $a(t) = v'(t) = 6t - 12 \implies a(4) = 12$
   Since $v(4) > 0$ and $a(4) > 0$, the particle is speeding up. (Rule # 4)

c) $v(t) = 0 \implies t = 1$ and $t = 3$.

d) Since $v(t) > 0$ for $0 < t < 1$ and $3 < t < 6$, and $v(t) < 0$ for $1 < t < 3$.
   Total traveled distance $= [x(1) - x(0)] + [x(1) - x(3)] + [x(6) - x(3)] = [7-3] + [7-3] + [57-3] = 62$

e) $x(1) = 7$, $x(3) = 3$, and $x(4) = 7$. The greatest distance from the origin is 7.

Chapter 4. Applications of Differentiation

▶ Example

5. The position function of a particle moving on a horizontal x-axis is shown in the accompanying figure. Points $P, Q,$ and R are the points of inflection.

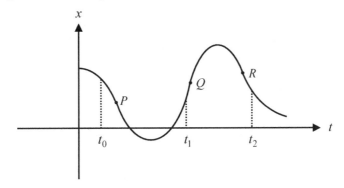

a) Is the particle moving left or right at time t_0?
b) Is the acceleration positive or negative at time t_0?
c) Is the particle speeding up or slowing down at time t_0?
d) Is the particle moving left or right at time t_1?
e) Is the acceleration positive or negative at time t_1?
f) Is the particle speeding up or slowing down at time t_1?
g) Is the particle speeding up or slowing down at time t_2?

---

a) At time t_0, $v(t_0) < 0$ → moving left   b) At time t_0, concave down → $a(t_0) < 0$
c) At time t_0, velocity and acceleration have the same sign. → speeding up
d) At time t_1, $v(t_0) > 0$ → moving right   e) At time t_1, concave up → $a(t_0) > 0$
f) At time t_1, velocity and acceleration have the same sign. → speeding up
g) At time t_2, $v(t_2) < 0$ and $a(t_2) > 0$ → velocity and acceleration have opposite signs.
   → slowing down

Chapter 4. Applications of Differentiation

E. RELATED RATES

The way to solve related rates problems is as follows:

1. Identify the known variables, including rates of change and the rate of change that is to be found. Construct an equation relating the quantities whose rates of change are known to the quantity whose rate of change is to be found.
2. Implicitly differentiate both sides of the equation with respect to time.
3. Substitute the known rates of change and the known quantities into the equation.
4. Solve for the required rate of change.

 Example

Suppose that liquid is to be cleared of sediment by allowing it to drain through a conical filter that is 12 cm high and has a radius of 6 cm at the top. If the liquid is forced out of the cone at a constant rate of $3 \text{cm}^3 / \text{min}$, how fast is the depth of the liquid decreasing at the instant when the liquid in the cone is 4 cm deep?

Answer
Step 1. Define the primary equation between variables.
V = volume of liquid at time t h = depth of the liquid at time t

r = radius of the liquid at time t $\dfrac{dV}{dt} = -3$

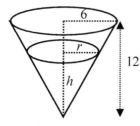

Primary equation *between volume and depth is*
$$V = \frac{\pi r^2 h}{3}$$
We need replace h in terms of r from the similar figures.
$$\frac{6}{12} = \frac{r}{h} \quad \rightarrow \quad r = \frac{h}{2}$$
Now the primary equation between two variables is
$$V = \frac{\pi \left(\dfrac{h}{2}\right)^2 \cdot h}{3} = \frac{\pi h^3}{12}$$

Differentiating both sides with respect to t we obtain
$$\frac{dV}{dt} = \frac{\pi h^2}{4}$$
At $r = 4$, $h = 2r = 8$.
$$\frac{dV}{dt} = \frac{\pi h^2}{4} \cdot \frac{dh}{dt} \quad \rightarrow \quad -2 = \frac{\pi (8)^2}{4} \cdot \frac{dh}{dt} \quad \rightarrow \quad \frac{dh}{dt} = -\frac{1}{8\pi} \text{cm/min}$$

Chapter 4. Applications of Differentiation

▶ **Example**

If x and y are both differentiable functions of t and are related by the equation $y = x^2 + 5$ and $\frac{dx}{dt} = 3$ when $x = 2$. Find $\frac{dy}{dt}$ when $x = 2$.

$$\frac{dy}{dt} = 2x\frac{dx}{dt}, \quad \frac{dx}{dt} = 2 \text{ and at } x = 2, \quad \frac{dy}{dt} = 3(2)(2) = 12$$

▶ **Example**

A water tank has the shape of a cylinder with radius 5 meters, as shown in the figure below. Let h be the depth of the water in the tank, measured in meters, where h is a function of time, t, measured in seconds. The volume V of the water tank is changing at the rate of -15π cubic meters per second. Find $\frac{dh}{dt}$.

Since $V = \pi r^2 h$, then $\frac{dV}{dt} = \pi r^2 \frac{dh}{dt} \Rightarrow -15\pi = \pi(5)^2 \frac{dh}{dt} \Rightarrow \frac{dh}{dt} = \frac{-15\pi}{25\pi} = \frac{-3}{5}$ meters/second.

▶ **Example**

If the radius r of a sphere is increasing at a rate of 10 inches per minute, what is the rate of change of the volume when $r = 5$ inches?

$$V = \frac{4}{3}\pi r^3 \text{ and } \frac{dr}{dt} = 10 \Rightarrow \frac{dV}{dt} = 4\pi r^2 \frac{dr}{dt} \Rightarrow \frac{dV}{dt} = 4\pi(5)^2(10) \text{ at } r = 5$$

Therefore, $\frac{dV}{dt} = 1000\pi$ cubic inches/minute

Chapter 4. Applications of Differentiation

▶ **Example**

A 25 feet- long ladder is leaning against the wall of a house and sliding away from the wall at a rate of 3 feet per second. How fast is the top of the ladder moving down the wall when the base of the ladder is 15 feet?

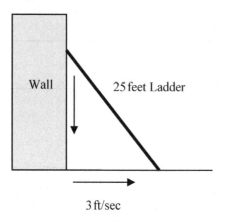

Let x = distance from the foot of the ladder to the base of tha wall and y = distance from the top of the ladder to the ground.

$\frac{dx}{dt} = 3$ feet/second and $x^2 + y^2 = 25$, $2x\frac{dx}{dt} + 2y\frac{dy}{dt} = 0$

Solve for $\frac{dy}{dt}$ at $x = 15$. When $x = 15$, $y = \sqrt{25^2 - 15^2} = 20$.

Substitute into the equation.

$2(15)(3) + 2(20)\frac{dy}{dt} = 0 \Rightarrow \frac{dy}{dt} = -\frac{9}{4}$ feet/second

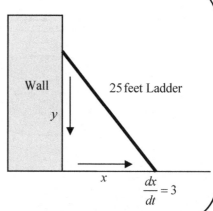

▶ **Example**

Oil is leaking from a pipeline on the surface of a water tank and forms an oil slick whose volume increases at a constant rate of 1200π cubic inches per minute. The oil slick takes the form of a right circular cylinder with both its radius and height changing with time. The radius is increasing at the rate of 5 inches per minute when the radius of the oil slick is 10 inches and the height is 8 inches. At this instant, what is the rate of change of the height of the oil slick with respect to time?

Chapter 4. Applications of Differentiation

> The volume of the oil slick: $V = \pi r^2 h$, where $\dfrac{dV}{dt} = 1200\pi \text{ in}^3/\text{min}$ and $\dfrac{dr}{dt} = 5 \text{ in/min}$.
>
> Take the derivative of the equation.
>
> $\dfrac{dV}{dt} = \pi\left(2rh\dfrac{dr}{dt} + r^2\dfrac{dh}{dt}\right)$
>
> Substitute all numbers into the derivative equation at $r = 10$ and $h = 8$.
>
> $1200\pi = \pi\left(2(10)(8)(5) + (10)^2\dfrac{dh}{dt}\right) \;\Rightarrow\; 1200 = 800 + 100\dfrac{dh}{dt}$
>
> $\dfrac{dh}{dt} = \dfrac{400}{100} = 4$ inches/minute

▶ **Example**

A space shuttle is rising vertically at a rate of 2 miles per second. A television camera of an observer on the ground is filming the lift-off of the space shuttle and the camera is 5 miles from the launch point. If θ is the angle of elevation, find the rate of change in the angle of elevation of the camera at 15 seconds after launch.

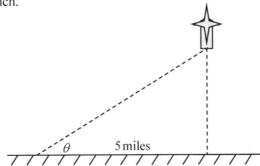

> The equation is $\tan\theta = \dfrac{y}{5}$.
>
> Take the derivative. $\sec^2\theta \dfrac{d\theta}{dt} = \dfrac{1}{5}\dfrac{dy}{dt}$ and $\dfrac{dy}{dt} = 2$ miles per second
>
> From the figure, $\sec^2\theta = 1 + \tan^2\theta$ and $\tan\theta = \dfrac{y}{5}$.
>
> $\sec^2\theta = 1 + \dfrac{y^2}{25}$ and $y = 2\times 15 = 30$ miles.
>
> $\sec^2\theta = 1 + \dfrac{900}{25} = 37$
>
> Substitute all numbers into the derivative equation.
>
> $37\dfrac{d\theta}{dt} = \dfrac{1}{5}(2) \;\Rightarrow\; \dfrac{d\theta}{dt} = \dfrac{2}{185}$ radians/second

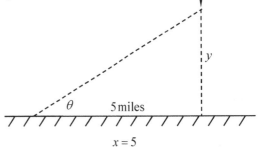

Chapter 5. Curve Sketching (The Derivative Test)

A. EXTREMUM OF A FUNCTION

DEFINITION. The extremum of a Function are the values of the maximum (or absolute maximum) and minimum (or absolute minimum) of a function on the interval $[a,b]$.

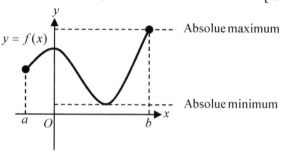

Relative Extremum:
$\begin{cases} \text{relative maximum (local maximum)} = \text{This is a maximum value relative to the points that are close} \\ \qquad \text{to it on the graph.} \\ \text{relative minimum (local minimum)} = \text{This is a minimum value relative to the points that are close} \\ \qquad \text{to it on the graph.} \end{cases}$

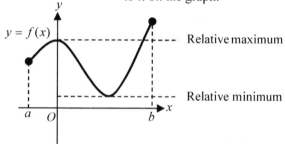

B. TYPE OF RELATIVE MAXIMUM AND RELATIVE MINIMUM

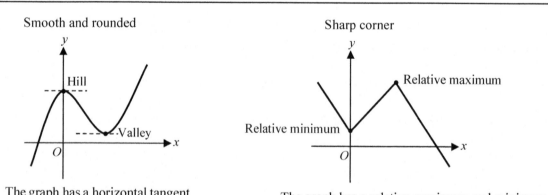

The graph has a horizontal tangent line at these two points.
The graph has a relative maximum and minimum at these points and are differentiable.

The graph has a relative maximum and minimum at these points, but they are not differentiable.

48

Chapter 5. Curve Sketching (The Derivative Test)

C. CRITICAL POINTS

THEOREM. CRITICAL POINT. A point on the interior of the domain of a function f at which $f'(c) = 0$ or *undefined* (does not exist) *is a critical point of* f.

Critical Point (c) $\begin{cases} \text{Sharp Corner} \text{ --------------- } f'(c_1) \text{ is undefined} \\ \text{Smooth and Rounded ----- } f'(c_4) = 0 \, (\text{Relative maximum}) \\ \text{Smooth and Rounded ----- } f'(c_5) = 0 \, (\text{Relative extrema is undefined}) \end{cases}$

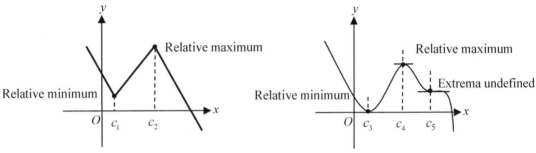

Sharp corner — $f'(c_1)$ and $f'(c_2)$ are undefined.

Smooth and Rounded — $f'(c_3)=0, f'(c_4)=0,$ and $f'(c_5)=0$

Points $c_1, c_2, c_3, c_4,$ and c_5 are all critical points.

Note: The derivative of $f(x) = \dfrac{1}{x^2 - 1}$ is $f'(x) = \dfrac{-2x}{(x-1)^2 (x+1)^2}$.

At $x = \pm 1$, f' is undefined, but $x = \pm 1$ *are not critical points because* $x = \pm 1$ *are not on the interior of the domain.*

THEOREM. STATIONARY POINT

If $f'(c) = 0$ at a critical point at $x = c$, then it is also said to *be a stationary point*.

In the figure above, critical points $c_3, c_4,$ and c_5 are also said to be *stationary points*.

Chapter 5. Curve Sketching (The Derivative Test)

▶ **Example**

Find all critical points and the extrema of $f(x) = x^4 - 2x^3$ on the interval $[-1, 3]$.

$f'(x) = 4x^3 - 6x^2 = 0 \implies f'(x) = 2x^2(2x - 3) = 0$

$f'(x) = 0$ at $x = 0$ and $x = \dfrac{3}{2}$. Therefore, $x = 0$ and $x = \dfrac{3}{2}$ are critical points.

Critical points: $\begin{cases} \text{at } x = \dfrac{3}{2},\ f\left(\dfrac{3}{2}\right) = \left(\dfrac{3}{2}\right)^4 - 2\left(\dfrac{3}{2}\right)^3 = -\dfrac{27}{16} \\ \text{at } x = 0,\ \text{no relative extrema} \end{cases}$

At the end points: $f(-1) = 3$ and $f(3) = 27$

Absolute Minimum $= -\dfrac{27}{16}$ at $x = \dfrac{3}{2}$, and absolute Maximum $= 27$ at $x = 3$

▶ **Example**

Find critical points of the function on the graph.

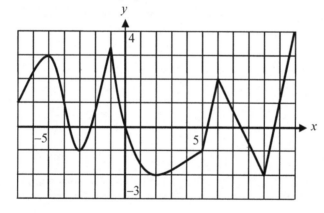

$f'(c) = 0 \rightarrow x = -5$ (relative maximum), $x = -3$ (relative minimum), $x = 2$ (relative minimum)

$f'(c) = $ undefined $\rightarrow x = -1$ (relative maximum), $x = 5$ (no relatitive extremum)

$x = 6$ (relative maximum), $x = 9$ (relative minimum)

Note: At $x = 5$, f is not differentiable and does not yield a relative minimum or a relative maximum.

Chapter 5. Curve Sketching (The Derivative Test)

Remember Types of Critical Points

1.

Hill
Critical point
Stationary point
Relative maximum
$f'(c) = 0$

2.

Valley
Critical point
Stationary point
Relative minimum
$f'(c) = 0$

3.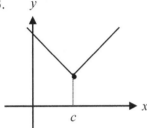

Sharp corner
Critical point
Not a stationary point
Relative minimum
$f'(c)$ = undefined

4.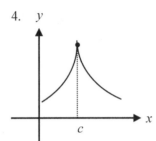

Cusp
Critical point
Not a stationary point
Relative maximum
$f'(c)$ = undefined

5.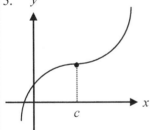

Critical point
Stationary point
Inflection point
Not relative extremum
$f'(c) = 0$

6.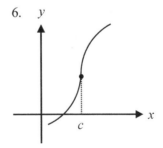

Critical point
Not stationary point
Inflection point
Not relative extremum
$f'(c)$ = undefined

Chapter 5. Curve Sketching (The Derivative Test)

▶ **Example**

Find the absolute extrema of $3x^{\frac{2}{3}} - 2x$ on the closed interval $[-1, 2]$.

Differentiate the given function to find critical points, where $f'(x) = 0$ or $f'(x) = $ undefined.

1) $f'(x) = 0$

$$f'(x) = 2x^{-\frac{1}{3}} - 2 = 0 \rightarrow x^{-\frac{1}{3}} = 1 \Rightarrow x = 1$$

2) $f'(x) = $ undefined

$$f'(x) = \frac{2}{x^{\frac{1}{3}}} - 2 \text{ is undefined at } x = 0.$$

Left endpoint $x = -1$	↘	Critical point $x = 0$	↗	Critical point $x = 1$	↘	Right endpoint $x = 2$
$f(-1) = 5$		$f(0) = 0$ Relative minimum		$f(1) = 1$ Relative maximum		$f(2) \approx 0.7622$

$f(-1) = 3(-1)^{\frac{2}{3}} - 2(-1) = 5$ $f(1) = 3(1)^{\frac{2}{3}} - 2(1) = 1$, $f(2) = 3(2)^{\frac{2}{3}} - 2(2) \approx 0.7622$

Absolute maximum $f(-1) = 5$ and absolute minimum $f(0) = 0$

▶ **Example**

Find the critical points and the relative extrema of the following function.

$$f(x) = (x-3)^{\frac{2}{3}} + 5$$

Critical point at $x = 3$, because $f'(x) = \dfrac{2}{3(x-3)^{\frac{1}{3}}}$ and undefined at $x = 3$.

But f has a relative minimum 5 at $x = 3$.

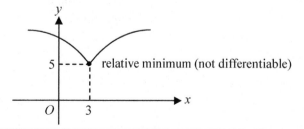

relative minimum (not differentiable)

Chapter 5. Curve Sketching (The Derivative Test)

▶ **Example**

Find the critical points of the equation $f(x) = \dfrac{x^2 - 9}{x^2 + 10x + 25}$.

$$f'(x) = \dfrac{(x^2 + 10x + 25)(2x) - (x^2 - 9)(2x + 10)}{(x^2 + 10x + 25)^2} = \dfrac{(x+5)^2(2x) - 2(x^2 - 9)(x+5)}{(x^2 + 10x + 25)^2}$$

$$= \dfrac{2x(x+5) - 2(x^2 - 9)}{(x+5)^3} = \dfrac{2(5x + 9)}{(x+5)^3} = 0$$

Critical point at $x = -\dfrac{9}{5}$

Note: At $x = -5$, f' is undefined. You may think that $x = -5$ is a critical point. But it is not a critical point, because $x = -5$ is not in the domain of f.

Asymptotes do not have critical point.

▶ **Example**

Find the critical point of $f(x) = 3x^{5/3} - 15^{2/3}$.

$$f'(x) = 5x^{2/3} - 10x^{-1/3} = 5x^{-1/3}(x - 2) = 0 \qquad f'(x) = \dfrac{5(x-2)}{\sqrt[3]{x}} = 0$$

Critical point: $x = 2$, $x = 0$ $\left(\text{At } x = 0,\ f'(0) = \text{undifined and } x = 0 \text{ in the domain}\right)$

Chapter 5. Curve Sketching (The Derivative Test)

D. INCREASING AND DECREASING FUNCTIONS

DEFINITION

If $f(x_1) < f(x_2)$ in the interval $x_1 < x_2$, then f is increasing.
If $f(x_1) > f(x_2)$ in the interval $x_1 < x_2$, then f is decreasing.

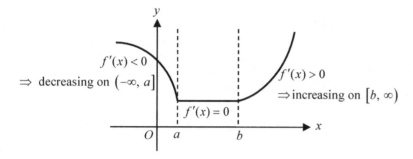

The function f is *continuous* on $[a, b]$ and *differentiable* on (a, b).
  If $f'(x) > 0$ in (a, b), then f is increasing on $[a, b]$.
  If $f'(x) < 0$ in (a, b), then f is decreasing on $[a, b]$.
  If $f'(x) = 0$ in (a, b), then f is constant on $[a, b]$.

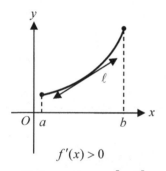

$f'(x) > 0$

f is increasing on $[a, b]$
The slope of tangent line ℓ is positive.

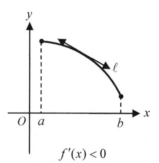

$f'(x) < 0$

f is decreasing on $[a, b]$
The slope of tangent line ℓ is negative.

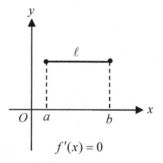

$f'(x) = 0$

f is constant on $[a, b]$
The slope of tangent line ℓ is zero.

▶ **Example**

Find the intervals on which $f(x) = x^3 - 3x^2$ is increasing or decreasing.

Chapter 5. Curve Sketching (The Derivative Test)

Note that f is continuous on the entire real line. To find the critical numbers of f, set $f'(x)$ equal to 0.
$f'(x) = 3x^2 - 6x = 0 \rightarrow 3x(x-2) = 0$ Critical numbers: $x = 0, 2$

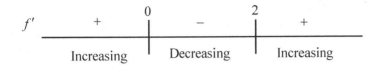

Or, we can use the graph of f'

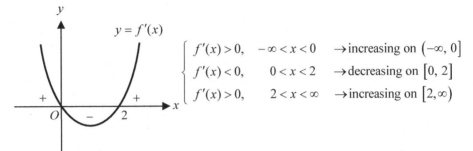

$\begin{cases} f'(x) > 0, & -\infty < x < 0 \quad \rightarrow \text{increasing on } (-\infty, 0] \\ f'(x) < 0, & 0 < x < 2 \quad \rightarrow \text{decreasing on } [0, 2] \\ f'(x) > 0, & 2 < x < \infty \quad \rightarrow \text{increasing on } [2, \infty) \end{cases}$

▶ **Example**

Find the intervals on which the function is increasing or decreasing.

$f(x) = x^4 - 2x^2$

$f(x) = x^4 - 2x^2 \Rightarrow f'(x) = 4x^3 - 4x = 4x(x^2 - 1) = 4x(x+1)(x-1) = 0$

[sign chart of $y = f'(x)$: negative on $(-\infty, -1)$, positive on $(-1, 0)$, negative on $(0, 1)$, positive on $(1, \infty)$]

From the graph,
$\begin{cases} f'(x) < 0 & \text{on } (-\infty, -1) \Rightarrow \text{decreasing on } (-\infty, -1] \\ f'(x) > 0 & \text{on } (-1, 0) \Rightarrow \text{increasing on } [-1, 0] \\ f'(x) < 0 & \text{on } (0, 1) \Rightarrow \text{decreasing on } [0, 1] \\ f'(x) > 0 & \text{on } (1, +\infty) \Rightarrow \text{increasing on } [1, \infty) \end{cases}$

Chapter 5. Curve Sketching (The Derivative Test)

▶ **Example**

Find the intervals on which the function is increasing or decreasing.

$$f(x) = \frac{x^2}{x+1}$$

$f(x) = \dfrac{x^2}{x+1} \quad \Rightarrow \quad f'(x) = \dfrac{(x+1)(2x) - x^2(1)}{(x+1)^2} = \dfrac{x^2 + 2x}{(x+1)^2} = \dfrac{x(x+2)}{(x+1)^2}$

Critical points at $x = 0, -2$
Asymptote at $x = -1$
Note: $x = -1$ is not the critical point, but we use this x-value to determine the test interval

Increasing on $(-\infty, -2]$ and $[0, \infty)$,
Decreasing on $[-2, -1)$ and $(-1, 0]$

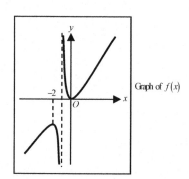

Graph of $f(x)$

Chapter 5. Curve Sketching (The Derivative Test)

E. THE FIRST DERIVATIVE TEST

THEOREM.
1. If $f'(x)$ changes from negative to positive at $x = k$, then $f(k)$ is a relative minimum of f.
2. If $f'(x)$ changes from positive to negative at $x = k$, then $f(k)$ is a relative maximum of f.

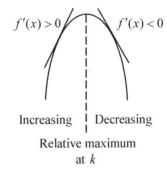

▶ **Example**

Find the relative maximum or relative minimum of $f(x) = \left(x^2 - 1\right)^{\frac{2}{3}}$.

$$f(x) = \left(x^2 - 1\right)^{\frac{2}{3}} \;\to\; f'(x) = \frac{2}{3}\left(x^2 - 1\right)^{-\frac{1}{3}}(2x) = \frac{4x}{3\sqrt[3]{x^2 - 1}} = \frac{4x}{3\sqrt[3]{(x+1)(x-1)}}$$

Find the critical points and the x-values where $f'(x)$ is undefined.

$f'(x) = 0 \;\to\; x = 0$: Critical point

$f'(x)$ is undefined at $x = 1, -1$: Critical point (Cusp), because $f(x)$ is continuous at $x = \pm 1$.

Relative maximum at $x = 0$ and relative minimum at $x = \pm 1$

Relative maximum: $f(0) = 1$

Relative minimum: $f(-1) = 0$, $f(1) = 0$

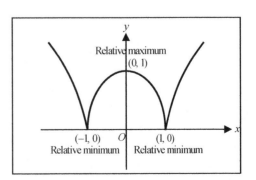

Chapter 5. Curve Sketching (The Derivative Test)

▶ Example

Find the relative extrema of $\dfrac{x^4+1}{x^2}$.

$f(x) = \dfrac{x^4+1}{x^2} \rightarrow f'(x) = \dfrac{x^2(4x^3)-(x^4+1)(2x)}{x^4} = \dfrac{2x^4-2}{x^3} = \dfrac{2(x^2+1)(x+1)(x-1)}{x^3}$

Critical points: $f'(x) = 0 \rightarrow x = -1, 1$

$f'(x)$ is undefined at $x = 0$ and $f(0)$ is undefined. Therefore, $x = 0$ is a vertical asymptote.
Note: $x = 0$ is not critical point.

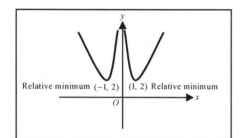

Relative minimum: $f(-1) = 2$, $f(1) = 2$

Chapter 5. Curve Sketching (The Derivative Test)

F. CONCAVITY

THEOREM 1.
1. If $f(x)$ is differentiable on an interval (a, b) and $f'(x)$ *is increasing on the interval*, the graph of f is concave up (upward).
2. If $f(x)$ is differentiable on an interval (a, b) and $f'(x)$ *is decreasing on the interval*, the graph of f is concave down (downward).

Concave Upward

$f'(x) < 0$ $f'(x) > 0$

$f'(x)$ is increasing.

Concave Downward

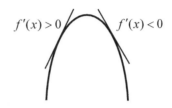

$f'(x) > 0$ $f'(x) < 0$

$f'(x)$ is decreasing.

THEOREM 2.
1. If $f''(x) > 0$ for all x in interval I, then the graph of f is concave up.
2. If $f''(x) < 0$ for all x in interval I, then the graph of f is concave down.
3. If $f''(x) = 0$ for all x in interval I, then the graph of f is linear (concavity is undefined).

G. SECOND DERIVATIVE TEST

THEOREM.
1. If $f'(x_0) = 0$ and $f''(x_0) > 0$, then f has a relative minimum at $x = x_0$.
2. If $f'(x_0) = 0$ and $f''(x_0) < 0$, then f has a relative maximum at $x = x_0$.
3. If $f'(x_0) = 0$ and $f''(x_0) = 0$, then the test is inconclusive at $x = x_0$.

▶ **Example**

Find the intervals on which the graph of the equation is concave up and the intervals on which the graph is concave down.

$$f(x) = x^4 - 4x^3 + 12$$

$f'(x) = 4x^3 - 12x^2$ and $f''(x) = 12x^2 - 24x = 12x(x - 2) = 0 \rightarrow x = 0, x = 2$

$f''(x)$	+		−		+
	concave up	$x = 0$	concave down	$x = 2$	concave up

Chapter 5. Curve Sketching (The Derivative Test)

H. POINTS OF INFLECTION

THEOREM.
Let f be a function whose second derivative exists on an open interval I.

If f is continuous at $x = c$, and $f''(c) = 0$ or $f''(c)$ is undefined, then $(c, f(c))$ is a point of inflection.

There are three different types of points of inflection:

1)

2)

3)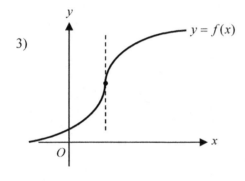

▶ **Example**

Find all points of inflection.
$$f(x) = x^3 - 12x$$

$f'(x) = 3x^2 - 12$ and $f''(x) = 6x$

$f''(x) \to$	−		+
	concave down	$x = 0$	concave up

$f(0) = 0 \;\to\;$ Inflection point is $(0, 0)$.

Chapter 5. Curve Sketching (The Derivative Test)

▶ **Example**

Find all points of inflection.

$$f(x) = (x-1)^{\frac{1}{3}}$$

$f'(x) = \frac{1}{3}(x-1)^{-2/3}$ and $f''(x) = -\frac{2}{9}(x-1)^{-5/3}$ → $f''(x) = -\frac{2}{9} \cdot \frac{1}{\sqrt[3]{(x-1)^5}}$

At $x=1$, $f''(x)$ is undefined. Note: $(x=1) \in$ Domain.
Test value: $f''(0) < 0$ and $f''(1) > 0$

```
                    |
  f''(x) →    −     |     +
                    |
          concave down   concave up
                  x = 1
```

At $x=1$, $f(1) = 0$. Change in concavity
Therefore, $(1, 0)$ is a point of inflection.

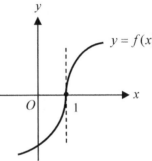

Chapter 5. Curve Sketching (The Derivative Test)

▶ Example

Sketch the graph of $f(x) = x^3 - 3x^2 - 24x + 32$.

Derivative Test

$f'(x) = 3x^2 - 6x - 24 \;\to\; 3(x-4)(x+2) = 0 \;\to\; x = 4$ and $x = -2$

$f''(x) = 6x - 6 \;\to\; 6(x-1) = 0 \;\to\; x = 1$

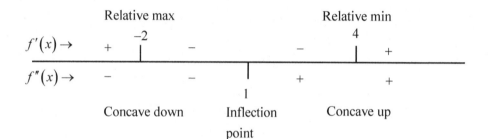

Relative max and min: $f(-2) = 60$ and $f(4) = -48$

Inflection point: $f(1) = 6$

y-intercept: $f(0) = 32$

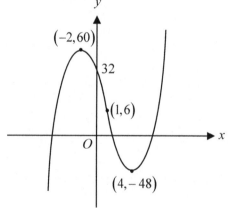

Graph of f

Chapter 6. More Applications of Differentiation

A. HOW TO SKETCH THE GRAPH OF A RATIONAL FUNCTION

When you are sketching the graph of a function, the following factors are useful.

1. Asymptotes
2. x- and y-intercepts
3. Critical points (These are found using the first derivative)
4. Points of inflection (These are found using the second derivative) and concavity

▶ **Example**

Sketch the graph of $f(x) = \dfrac{x^2 - 3x + 2}{x}$.

Asymptote: $x = 0$ (Vertical asymptote), $y = x - 3$ (Slant asymptote)

x-intercept: $(x-2)(x-1) = 0 \rightarrow x = 2, x = 1$ and no y-intercept

Critical point: $f'(x) = \dfrac{x^2 - 2}{x^2} \rightarrow \dfrac{x^2 - 2}{x^2} = 0 \rightarrow x = \pm\sqrt{2}$

$f(\sqrt{2}) = 2\sqrt{2} + 3$ and $f(-\sqrt{2}) = -2\sqrt{2} - 3$

Second derivative (Points of inflection):

$f''(x) = \dfrac{(x^2-2)'(2x) - (x^2-2)(2x)'}{x^4} = \dfrac{4}{x^3}$

$f''(x) = \dfrac{x^2(2x) - 2x(x^2-2)}{x^4} = \dfrac{4x}{x^4} = \dfrac{4}{x^3}$: At $x = 0$ $f''(0)$ is undefined but not inflection point.

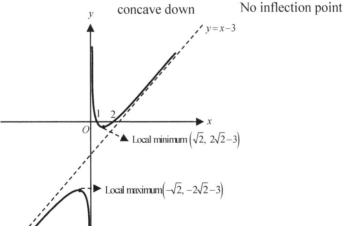

Chapter 6. More Applications of Differentiation

B. OPTIMAZATION

Optimization: Finding the largest value or the smallest value of a function subject to some kind of constraints (conditions)

Step 1) Define the primary equation for the quantity to be maximized.
Step 2) Identify the constraints.
Step 3) Reduce the primary equation to an equation with one variable.
Step 4) Use extrema to determine the desired maximum or minimum.

   Example

A building owner plans to fence in a rectangular field. The building is adjacent to one side of the field and so that side won't need fencing. If 1000 feet of fence is to be used, what are the dimensions of the field that will enclose the largest area of the field?

a) Define the primary equation to be maximized.
$$A = xy$$
b) Identify the constraints.
$$2x + y = 1000$$
c) Reduce the primary equation to one having one variable.
$$y = 1000 - 2x$$
Substitute into the primary equation.
$$A = x(1000 - 2x) = -2x^2 + 1000x, \text{ Domain: } x \in [0, 500]$$
d) Use extrema.
$$A'(x) = -4x + 1000 = 0 \Rightarrow 4x = 1000 \Rightarrow x = 250$$
Therefore, $A(x)$ has an absolute maximum area at $x = 250$ and $y = 500$ (feet).

Note:
Alternate method: Instead of using the first derivative test, the second derivative test can be used.
If $x = 250$ is the critical number and
If $A''(250) > 0$, then $A(250)$ is a relative minimum.
If $A''(250) < 0$, then $A(250)$ is a relative maximum.

Since $A''(x) = -4$, $A(x)$ is concave down. Therefore, $A(x)$ has a relative maximum (also it is absolute maximum, because $A''(x) < 0$ in entire interval I) at $x = 250$.

Chapter 6. More Applications of Differentiation

▶ **Example**

A window is constructed by adjoining a semicircle and a rectangle. If the total perimeter is 24 feet, what is the radius of the semicircle that will maximize the area of the window?

If $PQ = x$, then the perimeter of the window is $2x + 2r + \pi r$.

The primary equation to be maximized: $A = 2rx + \dfrac{\pi r^2}{2}$

Constraints: $2x + 2r + \pi r = 24$, $x = \dfrac{24 - 2r - \pi r}{2} = 12 - \dfrac{1}{2}(2 + \pi)r$, $r \in [0, 12]$

Reduce variable: $A = 2r\left(\dfrac{24 - 2r - \pi r}{2}\right) + \dfrac{\pi r^2}{2} = 24r - 2r^2 - \pi r^2$

Derivative: $A'(r) = 24 - 4r - 2\pi r = 24 - (4 + 2\pi)r = 0 \Rightarrow 12 = (\pi + 2)r \Rightarrow r = \dfrac{12}{\pi + 2}$

Therefore, the window has a maximum area at $r = \dfrac{12}{\pi + 2}$.

Note: Second Derivative Test
Since $A''(r) = -4 - 2\pi < 0$ for all real r, the window has a local maximum
It is also absolute maximum at $r = \dfrac{12}{\pi + 2}$.

▶ **Example**

A farmer has 160 feet of fencing enclosed with two congruent rectangular pastures. Find the dimensions that will maximize the enclosed area.

Primary equation: $A = 2xy$

Constraint: $4x + 3y = 160 \rightarrow 3y = 160 - 4x \rightarrow y = \dfrac{160 - 4x}{3}$,

Domain: $x \in [0, 40]$

Reduce variable and solve: $A = 2x\left(\dfrac{160 - 4x}{3}\right) = \dfrac{2}{3}(160x - 8x^2)$

$A' = \dfrac{2}{3}(160 - 16x) \rightarrow \dfrac{2}{3}(160 - 16x) = 0 \rightarrow x = 10$ and $y = 40$

$A'' = -\dfrac{32}{3} < 0$ for all x

Maximum at $x = 20$ and $y = 40$.

Dr. John Chung's AP Calculus AB/BC

Chapter 6. More Applications of Differentiation

▶ Example

A pipe line needs to be installed between point A and B. The point A is on the bank of a 5km-wide river. Point B is on the opposite side go the river 8 km downstream. If it costs $ 400 per km to install the pipe along the ground and $800 per km to install across the river. How far does point P have to be from point Q to minimize the cost of installing the pipe.

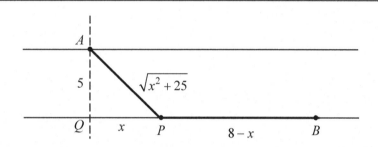

Primary equation: $C = (800)\sqrt{x^2 + 25} + (400)(8 - x)$, $x \in [0, 8]$

$\dfrac{dC}{dx} = 800\left(\dfrac{1}{2}\right)\dfrac{2x}{\sqrt{x^2+25}} - 400 \;\rightarrow\; \dfrac{dC}{dx} = 800\left(\dfrac{x}{\sqrt{x^2+25}} - \dfrac{1}{2}\right) = 0 \;\rightarrow\; \dfrac{x}{\sqrt{x^2+25}} = \dfrac{1}{2} \;\rightarrow\; 2x = \sqrt{x^2+25}$

$3x^2 = 25 \;\rightarrow\; x = \dfrac{5}{\sqrt{3}} = \dfrac{5\sqrt{3}}{3}$

Chapter 6. More Applications of Differentiation

▶ **Example**

A man is in a boat 3 miles from the closest point on the coast. He is going to reach a resting place located 5 miles down from the closest point. If he can row at 2 miles per hour and walk at 4 miles per hour, toward what point on the coast should he row in order to reach his resting place in the LEAST time?

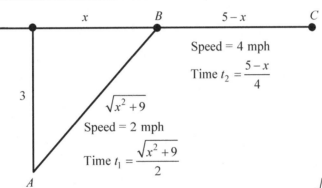

In the figure above, primary equation to be optimized is $T = t_1 + t_2 = \dfrac{\sqrt{x^2+9}}{2} + \dfrac{5-x}{4}$, $0 \le x \le 5$

Derivative: $T'(x) = \dfrac{(x^2+9)^{-\frac{1}{2}}(2x)}{4} - \dfrac{1}{4} = \dfrac{2x}{4\sqrt{x^2+9}} - \dfrac{1}{4} = \dfrac{2x - \sqrt{x^2+9}}{4\sqrt{x^2+9}}$

Critical number: $T'(x) = 0 \implies 2x = \sqrt{x^2+9} \implies 4x^2 = x^2+9 \implies 3x^2 = 9 \implies x = \sqrt{3}$

Therefore, $\sqrt{3}$ miles from the closest point.

Chapter 6. More Applications of Differentiation

C. LOCAL LINEAR APPROXIMATION

If a function is differentiable at $x = x_0$, the graph of the function can be approximated by a straight line, that is a tangent line at $(x_0, f(x_0))$.

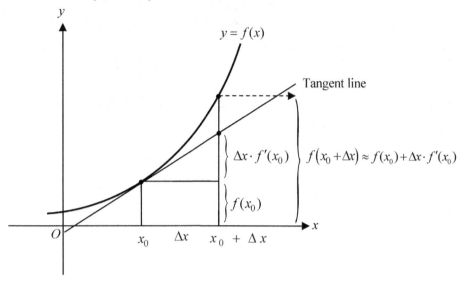

When Δx is small, $f(x_0 + \Delta x)$ can be approximated by a tangent line (linear function) as follows.
$$f(x_0 + \Delta x) \approx f(x_0) + \Delta x \cdot f'(x_0)$$

▶ **Example**

Find the linear approximation of $f(x) = 1 + \sin x$ at $x = 0.1$ using the equation of the tangent line at $x = 0$.

$f(x + \Delta x) = f(x) + \Delta x \cdot f'(x) \implies f(0 + 0.1) \approx f(0) + (0.1) f'(0)$
$f'(x) = \cos x \implies f'(0) = 1$ and $f(0) = 1 + \sin 0 = 1$
Therefore, $f(0.1) \approx f(0) + 0.1 f'(0) = 1 + 0.1 = 1.1$.
Note: Compare with the actual value $f(0.1) = 1 + \sin(0.1) = 0.0998334166...$

▶ **Example**

Estimate $\ln(1.02)$ using a linear approximation.

$(\ln x)' = \dfrac{1}{x}$, $f'(1) = 1$, $\ln(1) = 0$ and $\Delta x = 0.02$
$\ln(1.02) \approx \ln(1) + (0.02) f'(1) = 0 + (0.02)(1) = 0.02$
Note: Actual value $\ln(1.02) = 0.0198026273...$

Chapter 6. More Applications of Differentiation

▶ **Example**

Use differentials to approximate $\sqrt{9.2}$.

We can define $f(x) = \sqrt{x}$, $f(9) = \sqrt{9} = 3$, and $\Delta x = 0.2$.

$f'(x) = \dfrac{1}{2\sqrt{x}} \;\Rightarrow\; f'(9) = \dfrac{1}{2\sqrt{9}} = \dfrac{1}{6}$

The linear approximation: $f(9.2) \approx f(9) + 0.2 \cdot f'(9) \;\Rightarrow\; f(9.2) \approx 3 + 0.2\left(\dfrac{1}{6}\right) = 3.03$

Note: Actual value $\sqrt{9.2} = 3.033150178...$

▶ **Example**

Estimate $(4.2)^3$ using a linear approximation.

Define $f(x) = x^3$, $f(4) = 64$, and $\Delta x = 0.2$.

$f'(x) = 3x^2 \;\Rightarrow\; f'(4) = 3(4)^2 = 48$

The linear approximation: $f(4.2) \approx f(4) + (0.2)f'(4) \;\Rightarrow\; f(4.2) \approx 64 + 0.2(48) = 73.6$

▶ **Example**

If the radius of a circle increases from 5 to 5.2, what is the change in area using a linear approximation?

Since $A(r) = \pi r^2$, $A(5.2) \approx A(5) + 0.2 \times A'(5)$. $A'(r) = 2\pi r \;\Rightarrow\; A'(5) = 10\pi$

The change in area is $0.2 \times A'(5) = (0.2)(10\pi) = 2\pi$.

▶ **Example**

Estimate the $\arctan(0.9)$ using a linear approximation.

$f(x) = \arctan x$ and $(\arctan x)' = \dfrac{1}{1+x^2} \;\Rightarrow\; f'(1) = \dfrac{1}{1+1^2} = \dfrac{1}{2}$

$\arctan(1) = \dfrac{\pi}{4}$ and $\Delta x = -0.1$

$\arctan(0.9) \approx \arctan(1) + \dfrac{1}{2}(-0.1) = \dfrac{\pi}{4} - 0.05 \approx 0.735$

Note: Actual value of $\arctan(9.2) = 0.7328151018...$

Chapter 6. More Applications of Differentiation

D. ERROR PROPAGATION

From the differentials dx and dy (symbol dx and dy are called differentials),
We will get the differential form

$$dy = f'(x)dx \quad \text{or} \quad \frac{dy}{dx} = f'(x)$$

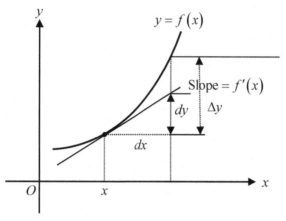

We define dx is much smaller than Δx. If Δx is very close to 0, we can see

$$dy \approx \Delta y \quad \text{or} \quad dy = f'(x)dx$$

Actually, small error invariably occurs in measured quantities. These errors produce error in the computed quantities. This is called *error propagation*.

THEOREM. If K is the true value of quantity and dK is the computed error, then the *relative error* in the measurement is $e = \dfrac{dK}{k}$.

▶ **EXAMPLE**

If the side of a equilateral triangle is measured with a ruler to be 20 cm with a measurement error of $\pm\dfrac{1}{40}$ cm, what will be the estimated error in the computed area of the triangle.

Let x be the exact length of a side and y the exact area so that $y = \dfrac{x^2\sqrt{3}}{4}$.

It follows that $dy = f'(x)dx = \dfrac{x\sqrt{3}}{2}dx$.

Substituting the measured value $x = 20 \to dy = \dfrac{20\sqrt{3}}{2}dx = 10\sqrt{3}\,dx$

The measurement error is $-\dfrac{1}{40} \leq dx \leq \dfrac{1}{40}$.

Now the propagated error is $10\sqrt{3}\left(-\dfrac{1}{40}\right) \leq 10\sqrt{3}\,dx \leq 10\sqrt{3}\left(\dfrac{1}{40}\right) \to -\dfrac{\sqrt{3}}{4} \leq dy \leq \dfrac{\sqrt{3}}{4}$.

Thus, the propagated error in the area is estimated to be within $\pm\dfrac{\sqrt{3}}{4}$.

Chapter 6. More Applications of Differentiation

 EXAMPLE

The side of a cube is measured with a percentage error within $\pm 0.02\%$. Estimate the percentage error in the calculated volume of the cube.

Volume of cube is $V = x^3$. So $dV = 3x^2 dx$
The relative error in V is approximately
$$\frac{dV}{V} = \frac{3x^2 dx}{x^3} = 3\frac{dx}{x}$$
The relative error in the measured value of x is ± 0.0002, which means that $-0.0002 \leq \frac{dx}{x} \leq 0.0002$

Thus, the estimated percentage error in the calculated value of V is
$$3(-0.0002) \leq 4\frac{dx}{x} \leq 3(0.0002) \quad \rightarrow \quad -0.0006 \leq \frac{dV}{V} \leq 0.0006$$
The percentage error in the calculated value of V is within $\pm 0.06\%$

Chapter 6. More Applications of Differentiation

E. NEWTON'S METHOD

1. When we encounter functions whose zeros are difficult to find using algebraic techniques, we use Newton's method.

2. Newton's method is useful for approximating real zeros only when the sequence converges to the limit (a zero of f). Newton's Method does not always yield a convergent sequence.

The equation of the tangent line is $y - y_1 = m(x - x_1)$ at point (x_1, y_1).
Let $y = 0$ and solve for x.

$$x - x_1 = \frac{0 - y_1}{m} \quad \rightarrow \quad x = x_1 - \frac{y_1}{m} \quad \rightarrow \quad x = x_1 - \frac{y_1}{f'(x_1)}$$

From this equation, we will get a new second estimate

$$x_2 = x_1 - \frac{y_1}{f'(x_1)} \quad \rightarrow \quad (x_2, y_2)$$

When you repeat this process, the third estimate is

$$x_3 = x_2 - \frac{y_2}{f'(x_2)} \quad \rightarrow \quad (x_3, y_3)$$

Generally,

$$x_{n+1} = x_n - \frac{y_n}{f'(x_n)} \quad \text{or}$$

$$x_{n+1} = x_n - \frac{f(x_n)}{f'(x_n)}, \text{ because } y_n = f(x_n) \quad \rightarrow \quad \text{Newton's Method}$$

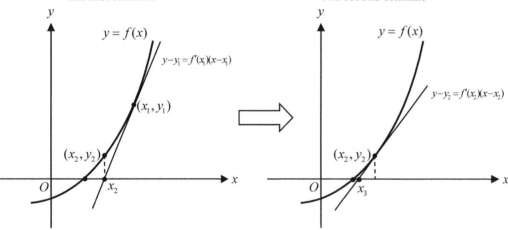

Continuing in this way we can see that a succession of values $x_1, x_2, x_3, \cdots$ approach the root of the equation.
The procedure for approximating the root of the equation is called Newton's Method.

Chapter 6. More Applications of Differentiation

Note
1) You can repeat the procedure until you get the desired accuracy.
2) Each successive application of the procedure is called iteration.
3) x_1 is called an initial value or seed value.
4) If $f'(x_n) = 0$, then Newton's Method fails.

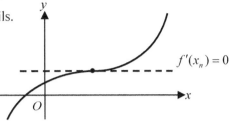

▶ **Example**

Use Newton's Method to approximate the real zeros of $f(x) = x^2 - 3$ using an initial value of $x_1 = 1$, then determine the accuracy of x_3.

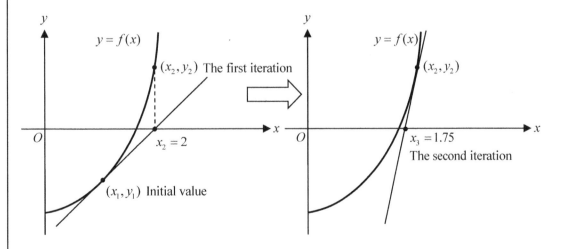

$f(x) = x^2 - 3$ and $f'(x) = 2x$. So, $f(x_n) = (x_n)^2 - 3$, and $f'(x_n) = 2x_n$

The Newton's Method: $x_{n+1} = x_n - \dfrac{f(x_n)}{f'(x_n)}$ becomes $x_{n+1} = x_n - \dfrac{x_n^2 - 3}{2x_n}$.

Initial value $x_1 = 1$

$x_2 = x_1 - \dfrac{x_1^2 - 3}{2x_1} = 1 - \dfrac{1-3}{2} = 2 \to x_2 = 2$ and $x_3 = x_2 - \dfrac{x_2^2 - 3}{2x_2} = 2 - \dfrac{4-3}{4} = \dfrac{7}{4} \to x_3 = \dfrac{7}{4}$

Note: The positive root of $x^2 - 3 = 0$ is $x = \sqrt{3} \approx 1.7321$. Therefore, $x_3 = 1.75$ is closer to the positive root than the initial vale $x_1 = 1$ and the first iteration $x_2 = 2$

Accuracy of $x_3 = |x_3 - \text{real zero}| = |1.75 - \sqrt{3}| \approx 0.0179492$

Chapter 6. More Applications of Differentiation

▶ **Example**

Explain why Newton's Method fails to calculate a zero of $f(x) = 2x^3 - 6x^2 + 6x - 1$ at $x_1 = 1$.

$f(x) = 2x^3 - 6x^2 + 6x - 1 \rightarrow f'(x) = 6x^2 - 12x + 6$

$x_2 = x_1 - \dfrac{f(x_1)}{f'(x_1)}$

We have $f'(1) = 0$.

Newton's Method fails to converge, because $f'(1) = 0$.

F. L'Hôpital's Rule

THEOREM. Let f and g be differentiable on an open interval (a,b) containing c. If the limit of $\dfrac{f(x)}{g(x)}$ as x approaches c is indeterminate, then

$$\lim_{x \to c} \frac{f(x)}{g(x)} = \lim_{x \to c} \frac{f'(x)}{g'(x)}$$

The limit of $\dfrac{f(x)}{g(x)}$ is indeterminate if $\dfrac{f(x)}{g(x)} = \dfrac{0}{0}, \dfrac{\pm\infty}{\pm\infty}, 0^0, 1^\infty, \infty^0$, and $0 \cdot \infty$.

Note Occasionally it is necessary to apply the rule more than once to remove an indeterminate form.

▶ **Example**

1. $\displaystyle\lim_{x \to 0} \dfrac{e^x - 1}{x} =$

2. $\displaystyle\lim_{x \to \infty} \dfrac{-\ln x}{x} =$

3. $\displaystyle\lim_{x \to \infty} \dfrac{2x^2}{e^{2x} - 1} =$

4. $\displaystyle\lim_{x \to \infty} e^{-x}\sqrt{2x} =$

---

1. Form $\dfrac{0}{0}$: $\displaystyle\lim_{x \to 0} \dfrac{e^x - 1}{x} = \lim_{x \to 0} \dfrac{e^x}{1} = 1$

2. Form $\dfrac{-\infty}{+\infty}$: $\displaystyle\lim_{x \to \infty} \dfrac{-\ln x}{x} = \lim_{x \to \infty} \dfrac{-\frac{1}{x}}{1} = \lim_{x \to \infty} \left(-\dfrac{1}{x}\right) = 0$

3. Form $\dfrac{\infty}{\infty}$: $\displaystyle\lim_{x \to \infty} \dfrac{2x^2}{e^{2x} - 1} = \lim_{x \to \infty} \dfrac{4x}{2e^{2x}} = \lim_{x \to \infty} \dfrac{4}{4e^{2x}} = 0$

4. Form $0 \cdot \infty$: $\displaystyle\lim_{x \to \infty} e^{-x}\sqrt{2x} = \lim_{x \to \infty} \dfrac{\sqrt{2x}}{e^x} = \lim_{x \to \infty} \dfrac{(2x)^{-\frac{1}{2}}}{e^x} = \lim_{x \to \infty} \dfrac{1}{e^x\sqrt{2x}} = 0$

Chapter 6. More Applications of Differentiation

▶ **Example**

1. $\lim_{x \to 0^+} (\sin x)^x =$
2. $\lim_{x \to \infty} x^{\frac{1}{x}} =$
3. $\lim_{x \to \infty} \left(x \sin \frac{1}{x} \right) =$

---

1. Form 0^0:

$y = \lim_{x \to 0^+} (\sin x)^x \Rightarrow \ln y = \lim_{x \to 0^+} x \ln(\sin x) \Rightarrow \ln y = \lim_{x \to 0^+} \frac{\ln(\sin x)}{\frac{1}{x}}$ (Still indeterminate form $\frac{-\infty}{\infty}$)

$\ln y = \lim_{x \to 0^+} \frac{\frac{\cos x}{\sin x}}{-x^{-2}} = \lim_{x \to 0^+} \frac{-x^2}{\tan x} = \lim_{x \to 0^+} \frac{-2x}{\sec^2 x} = \frac{0}{1} = 0 \;\to\; y = e^{\ln y} = e^0 \;\to\; y = 1$

2. Form ∞^0: Take the natural logarithm of both sides.

$\ln y = \lim_{x \to \infty} \frac{1}{x} \ln x \Rightarrow \ln y = \lim_{x \to \infty} \frac{\ln x}{x} = \lim_{x \to \infty} \frac{\frac{1}{x}}{1} = 0$

Therefore, $y = e^{\ln y} = e^0 \;\to\; y = 1$

3. Form $\frac{0}{0}$: $\lim_{x \to \infty} \frac{\sin \frac{1}{x}}{\frac{1}{x}} = \lim_{x \to \infty} \frac{\cos \frac{1}{x} \left(\frac{1}{x}\right)'}{\left(\frac{1}{x}\right)'} = \lim_{x \to \infty} \cos \frac{1}{x} = 1$

Note $\lim_{x \to 0} \frac{\sin x}{x} = 1$ and $\lim_{x \to \infty} \frac{\sin(1/x)}{1/x} = 1$

▶ **Example**

$\lim_{x \to 0} \frac{x^3}{x - \tan x} =$

$\left(\frac{0}{0}\right) \to \lim_{x \to 0} \frac{3x^2}{1 - \sec^2 x} \to \left(\frac{0}{0}\right) \to \lim_{x \to 0} \frac{6x}{-2 \sec x \sec x \tan x} = \lim_{x \to 0} \frac{6x}{-2 \sec^2 x \tan x} = \lim_{x \to 0} \frac{3x}{-\sec^2 x \tan x} \to$

$\left(\frac{0}{0}\right) \to \lim_{x \to 0} \frac{-3}{2 \sec^2 x \tan x + \sec^2 x \sec^2 x} = \lim_{x \to 0} \frac{-3}{0 + 1} = -3$

Chapter 6. More Applications of Differentiation

▶ **Example**

$$\lim_{x \to 0^+} \left(\frac{1}{x} - \frac{1}{1-\cos x} \right) =$$

$$\lim_{x \to 0^+} \left(\frac{1}{x} - \frac{1}{1-\cos x} \right) \to (\infty - \infty) \to \lim_{x \to 0^+} \left(\frac{1-\cos x - x}{x - x\cos x} \right) \to \left(\frac{0}{0} \right) \to \lim_{x \to 0^+} \frac{\sin x - 1}{1 - (\cos x - x \sin x)}$$

$$\lim_{x \to 0^+} \frac{\sin x - 1}{1 - (\cos x - x \sin x)} = \lim_{x \to 0^+} \frac{\sin 0 - 1}{1 - (\cos 0 - \sin 0)} = \frac{-1}{0^+} = +\infty$$

▶ **Example**

$$\lim_{x \to 0^+} \left[\csc x \cdot \ln(1 - \sin x) \right] =$$

$$\lim_{x \to 0^+} \left[\csc x \cdot \ln(1 - \sin x) \right] \to (0 \cdot \infty) \to \lim_{x \to 0^+} \frac{\ln(1 - \sin x)}{\sin x} \to \left(\frac{0}{0} \right) \to \lim_{x \to 0^+} \frac{\frac{-\cos x}{1 - \sin x}}{\cos x} = \lim_{x \to 0^+} \frac{-1}{1 - \sin x} = -1$$

▶ **Example**

$$\lim_{x \to 0^+} x^x =$$

$$y = \lim_{x \to 0^+} x^x \to (0^0) \to e^{\ln y} = \lim_{x \to 0^+} e^{x \ln x} = e^{\lim_{x \to 0} x \ln x} \to$$

$$\ln y = \lim_{x \to 0^+} x \ln x = \lim_{x \to 0^+} \frac{\ln x}{\frac{1}{x}} \to \left(\frac{-\infty}{+\infty} \right) \to$$

$$\lim_{x \to 0^+} \frac{1/x}{-1/x^2} = \lim_{x \to 0^+} (-x) = 0 \quad \text{Therefore, } y = e^{\ln y} = e^0 = 1.$$

Chapter 6. More Applications of Differentiation

G. EVALUATING THE DERIVATIVE OF AN INVERSE FUNCTION

THEOREM. If f and its inverse g are differentiable on an interval I, then
$$g'(y) = \frac{1}{f'(x)}$$

Point $A(x,y)$ and point $B(y,x)$ are corresponding points of f and f^{-1} respectively.
The slope at point $A = \dfrac{dy}{dx}$ and the slope at point $B = \dfrac{dx}{dy}$

Therefore, $\dfrac{dy}{dx} = \dfrac{1}{dx/dy}$ or $f'(x) = \dfrac{1}{g'(y)}$

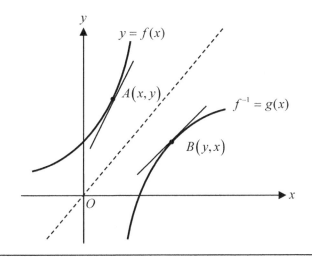

▶ **Example**

If $f(x) = \dfrac{1}{3}x^3 + 5$, what is the value of $\left(f^{-1}\right)'(14)$?

Find the corresponding points: $A(k,14) \leftrightarrow B(14,k)$ and find k. $\quad \dfrac{1}{3}k^3 + 5 = 14 \rightarrow k = 3$

$f'(x) = x^2 \rightarrow f'(3) = 9 \qquad$ So, $g(14) = \dfrac{1}{f'(3)} = \dfrac{1}{9}$

Chapter 6. More Applications of Differentiation

▶ **Example**

If $f(x) = \sqrt{x-4}$ and $g(x)$ is the inverse function of f, what is the value of $g'(8)$?

$$A(k,8) \leftrightarrow B(8,k) \to \sqrt{k-4} = 8 \to k = 68$$

$$f'(x) = \frac{1}{2\sqrt{x-4}} \to f'(68) = \frac{1}{16} \quad \text{So, } g'(8) = \frac{1}{f'(68)} = \frac{1}{1/16} = 16$$

▶ **Example**

Evaluate the derivative of the inverse of $f(x) = 5x^5 + 2x^3 + x - 1$ at $x = -1$.

$$A(k,-1) \leftrightarrow B(-1,k) \to 5k^5 + 2k^3 + k - 1 = -1 \to 5k^5 + 2k^3 + k = 0 \to k(5k^4 + 2k^2 + 1) = 0 \to k = 0$$

$$f'(x) = 25x^4 + 6x^2 + 1 \to f'(0) = 1 \quad \text{So, } g'(-1) = \frac{1}{f'(0)} = \frac{1}{1} = 1.$$

Chapter 7. Integration

A. ANTIDERIVATIVES

Notation for the antiderivative

$$\int f(x)dx = F(x) + C$$

means "The antiderivative of f with respect to x is $F(x) + C$."

Note 1) The antiderivative is also called "the indefinite integral".

2) The process of finding antiderivatives is called antidifferentiation or integration.

3) $\int f(x)dx$ is called an indefinite integral.

Basic Integration

$\int 0 \, dx = C$	$\int k \, dx = kx + C$
$\int x^n \, dx = \dfrac{x^{n+1}}{n+1} + C,\ n \neq -1$	$\int \cos x \, dx = \sin x + C$
$\int \sin x \, dx = -\cos x + C$	$\int \sec^2 x \, dx = \tan x + C$
$\int \sec x \tan x \, dx = \sec x + C$	$\int \csc^2 x \, dx = -\cot x + C$
$\int \csc x \cot x \, dx = -\csc x + C$	

▶ **Example**

Solve.

1. $\int \sqrt[3]{x} \, dx$ 2. $\int \dfrac{1}{x^3} dx$ 3. $\int (\sqrt{x} + 2x - 3) dx$ 4. $\int x^2 \sqrt{x} \, dx$

---

1. $\int \sqrt[3]{x} \, dx = \int x^{\frac{1}{3}} dx = \dfrac{x^{\frac{1}{3}+1}}{\frac{1}{3}+1} = \dfrac{x^{\frac{4}{3}}}{\frac{4}{3}} = \dfrac{3}{4} x^{\frac{4}{3}} + C$ (Power rule)   2. $\int \dfrac{1}{x^3} dx = \int x^{-3} dx = \dfrac{x^{-3+1}}{-3+1} = -\dfrac{1}{2x^2} + C$

3. $\int (\sqrt{x} + 2x - 3) dx = \int \left(x^{\frac{1}{2}} + 2x - 3 \right) dx = \dfrac{2}{3} x^{\frac{3}{2}} + x^2 - 3x + C$ 4. $\int x^2 \sqrt{x} \, dx = \int x^2 x^{\frac{1}{2}} dx = \int x^{\frac{5}{2}} dx = \dfrac{2}{7} x^{\frac{7}{2}} + C$

Chapter 7. Integration

▶ **Example**

1. $\int (2\sin x + \cos x)\,dx$ 2. $\int (x - \csc x \cot x)\,dx$ 3. $\int (\tan^2 x + 1)\,dx$ 4. $\int \sec x (\tan x - \sec x)\,dx$

---

1. $\int (2\sin x + \cos x)\,dx = -2\cos x + \sin x + C$ 2. $\int (x - \csc x \cot x)\,dx = \dfrac{x^2}{2} + \csc x + C$

3. $\int (\tan^2 x + 1)\,dx = \int \sec^2 x\,dx = \tan x + C$

4. $\int \sec x (\tan x - \sec x)\,dx = \int (\sec x \tan x - \sec^2 x)\,dx = \sec x - \tan x + C$

▶ **Example**

1. $\int \left(\dfrac{\sin x}{1 - \sin^2 x}\right)dx$ 2. $\int \dfrac{x^4 - 2x^2 + 1}{x^2}\,dx$ 3. $\int \tan^2 x\,dx$ 4. $\int \left(\dfrac{3}{\cos^2 x} - x\right)dx$

---

1. $\int \left(\dfrac{\sin x}{1 - \sin^2 x}\right)dx = \int \dfrac{\sin x}{\cos^2 x}\,dx = \int \dfrac{1}{\cos x}\dfrac{\sin x}{\cos x}\,dx = \int \sec x \tan x\,dx = \sec x + C$

2. $\int \dfrac{x^4 - 2x^2 + 1}{x^2}\,dx = \int (x^2 - 2 + x^{-2})\,dx = \dfrac{x^3}{3} - 2x - \dfrac{1}{x} + C$

3. $\int \tan^2 x\,dx = \int (\sec^2 x - 1)\,dx = \tan x - x + C$ 4. $\int \left(\dfrac{3}{\cos^2 x} - x\right)dx = \int (3\sec^2 x - x)\,dx = 3\tan x - \dfrac{x^2}{2} + C$

▶ **Example**

1. $\int \left(x + \dfrac{2}{1 - \cos^2 x}\right)dx$ 2. $\int \cot^2 x\,dx$

---

1. $\int \left(x + \dfrac{2}{1 - \cos^2 x}\right)dx = \int \left(x + \dfrac{2}{\sin^2 x}\right)dx = \int (x + 2\csc^2 x)\,dx = \dfrac{x^2}{2} - 2\cot x + C$

2. Since $1 + \cot^2 x = \csc^2 x$, $\cot^2 x = \csc^2 x - 1$. $\int \cot^2 x\,dx = \int (\csc^2 x - 1)\,dx = -\cot x - x + C$

Chapter 7. Integration

B. *U*-SUBSTITUTION IN ORDER TO USE THE POWER RULE

DEFINITION. Since $\dfrac{d}{dx}[F(g(x))] = F'(g(x))g'(x)$, the antiderivative will be

$$\int F'(g(x))g'(x)\,dx = F(g(x)) + C.$$

If $F'(x) = f(x)$ and $u = g(x)$, then $du = g'(x)dx$.
Therefore,

$$\int F'(g(x))g'(x)\,dx = \int f(u)\,du = F(u) + C.$$

▶ **Example**

Evaluate $\int (x^2 + 4)^2 (2x)\,dx$.

Let $u = x^2 + 4$ and $du = 2x\,dx$.

Therefore, $\int (x^2 + 4)^2 (2x)\,dx = \int u^2\,du = \dfrac{u^3}{3} + C = \dfrac{(x^2 + 4)^3}{3} + C$

Note You must recognize the $f(g(x))g'(x)$ pattern.

▶ **Example**

Evaluate the indefinite integral.

1. $\int 4x(x^2 - 1)^2\,dx$ 2. $\int \sqrt{2x+1}\,dx$ 3. $\int x\sqrt{2x+1}\,dx$

---

1. Let $u = x^2 - 1$ and $du = 2x\,dx \Rightarrow x\,dx = \dfrac{du}{2}$.

$$\int 4x(x^2 - 1)^2\,dx = 2\int u^2\,du = \dfrac{2}{3}u^3 + C = \dfrac{2}{3}(x^2 - 1)^3 + C$$

2. Let $u = 2x + 1$ and $du = 2dx \Rightarrow dx = \dfrac{du}{2}$.

$$\int \sqrt{2x+1}\,dx = \int (2x+1)^{\frac{1}{2}}\,dx = \dfrac{1}{2}\int u^{\frac{1}{2}}\,du = \dfrac{1}{2}\left(\dfrac{2}{3}\right)u^{\frac{3}{2}} + C = \dfrac{1}{3}(2x+1)^{\frac{3}{2}} + C \text{ or } \dfrac{1}{3}\sqrt{(2x+1)^3} + C$$

3. Change of variables:

Let $2x + 1 = u$ and $x = \dfrac{u-1}{2} \Rightarrow dx = \dfrac{1}{2}du$

$$\int x\sqrt{2x+1}\,dx = \int \dfrac{u-1}{2}\left(u^{\frac{1}{2}}\right)\dfrac{du}{2} = \dfrac{1}{4}\int \left(u^{\frac{3}{2}} - u^{\frac{1}{2}}\right)du = \dfrac{1}{4}\left(\dfrac{2}{5}u^{\frac{5}{2}} - \dfrac{2}{3}u^{\frac{3}{2}}\right) + C = \dfrac{1}{10}(2x+1)^{\frac{5}{2}} - \dfrac{1}{6}(2x+1)^{\frac{3}{2}} + C$$

Chapter 7. Integration

▶ **Example**

1. $\int (\sin^2 3x \cos 3x)\, dx$ 2. $\int \sec 2x \tan 2x\, dx$ 3. $\int \sec^2 x \tan^2 x\, dx$ 4. $\int \left(1+\frac{1}{x}\right)^3 \left(\frac{1}{x^2}\right) dx$

---

1. Let $u = \sin 3x$ and $du = 3\cos 3x\, dx \Rightarrow \cos 3x\, dx = \dfrac{du}{3}$.

$$\int (\sin^2 3x \cos 3x)\, dx = \frac{1}{3}\int u^2\, du = \frac{1}{3}\left(\frac{1}{3}u^3\right) + C = \frac{1}{9}(\sin 3x)^3 + C$$

2. $\int \sec 2x \tan 2x\, dx = \int \dfrac{\sin 2x}{\cos 2x \cos 2x}\, dx$

   Let $u = \cos 2x$ and $du = -2\sin 2x\, dx \Rightarrow \sin 2x\, dx = \dfrac{du}{-2}$.

$$\int \frac{\sin 2x}{\cos 2x \cos 2x}\, dx = -\frac{1}{2}\int \frac{1}{u^2}\, du = \frac{1}{2}u^{-1} + C = \frac{1}{2\cos 2x} + C$$

3. Let $u = \tan x$ and $du = \sec^2 x\, dx$. $\int \sec^2 x \tan^2 x\, dx = \int u^2\, du = \dfrac{u^3}{3} + C = \dfrac{\tan^3 x}{3} + C$

4. Let $u = 1 + \dfrac{1}{x}$ and $du = -\dfrac{1}{x^2}\, dx \Rightarrow \dfrac{1}{x^2}\, dx = -du$.

$$\int \left(1+\frac{1}{x}\right)^3 \left(\frac{1}{x^2}\right) dx = -\int u^3\, du = -\frac{1}{4}u^4 + C = -\frac{1}{4}\left(1+\frac{1}{x}\right)^4 + C$$

▶ **Example**

1. $\int x^2 \sqrt{1-x}\, dx$ 2. $\int \dfrac{x^2 - 1}{\sqrt{2x-1}}\, dx$

Change of variables:

1. Let $u = 1 - x$ and $x = 1 - u \Rightarrow dx = -du$ and $x^2 = 1 - 2u + u^2$.

$$\int x^2 \sqrt{1-x}\, dx = -\int (1 - 2u + u^2)\sqrt{u}\, du = -\int \left(u^{\frac{1}{2}} - 2u^{\frac{3}{2}} + u^{\frac{5}{2}}\right) dx = -\left(\frac{2}{3}u^{\frac{3}{2}} - \frac{4}{5}u^{\frac{5}{2}} + \frac{2}{7}u^{\frac{7}{2}}\right) + C$$

$$= -\frac{2}{3}(1-x)^{\frac{3}{2}} + \frac{4}{5}(1-x)^{\frac{5}{2}} - \frac{2}{7}(1-x)^{\frac{7}{2}} + C$$

2. Let $u = 2x - 1$ and $du = 2\, dx$. $x = \dfrac{u+1}{2} \Rightarrow x^2 - 1 = \dfrac{u^2 + 2u + 1}{4} - 1 = \dfrac{u^2 + 2u - 3}{4}$

$$\int \frac{x^2 - 1}{\sqrt{2x-1}}\, dx = \int \frac{u^2 + 2u - 3}{4u^{\frac{1}{2}}} \cdot \frac{du}{2} = \frac{1}{8}\int u^{\frac{3}{2}} + 2u^{\frac{1}{2}} - 3u^{-\frac{1}{2}}\, du = \frac{1}{8}\left(\frac{2}{5}u^{\frac{5}{2}} + \frac{4}{3}u^{\frac{3}{2}} - 6u^{\frac{1}{2}}\right) + C$$

$$= \frac{1}{120}(2x-1)^{\frac{5}{2}} + \frac{1}{6}(2x-1)^{\frac{3}{2}} - \frac{3}{4}(2x-1)^{\frac{1}{2}} + C$$

Chapter 7. Integration

C. DEFINITE INTEGRALS

A definite integral is an integral $\int_a^b f(x)dx$ with upper and lower limits, a and b respectively.

If $F(x)$ is the infinite integral for a continuous function $f(x)$, then

$$\int_a^b f(x)dx = F(b) - F(a).$$

This idea is called the Fundamental Theorem of Calculus. A definite integral is used to find the area under a curve.

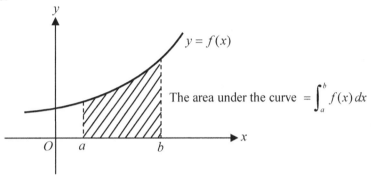

The area under the curve $= \int_a^b f(x)\,dx$

Integration rules for definite integrals:

1. $\int_a^a f(x) = 0$

2. $\int_a^b f(x)dx = \int_a^c f(x)dx + \int_c^b f(x)dx$ for $c \in (a,b)$

3. $\int_a^b f(x)dx = -\int_b^a f(x)dx$

4. $\int_{-a}^a f(x)dx = 2\int_0^a f(x)dx$, where f is an even function.

5. $\int_{-a}^a f(x)dx = 0$, where f is an odd function.

▶ **Example**

Evaluate the definite integral.

1. $\int_{-1}^{1} x(x^2 + 2)^3 dx$ 2. $\int_{-\frac{\pi}{2}}^{\frac{\pi}{2}} \cos x\, dx$ 3. $\int_0^{\pi} (2\sin x + \sin 2x)dx$

---

1. 0 (Odd function)

2. $\int_{-\frac{\pi}{2}}^{\frac{\pi}{2}} \cos x\, dx = 2\int_0^{\frac{\pi}{2}} \cos x\, dx = 2[\sin x]_0^{\frac{\pi}{2}} = 2[1 - 0] = 2$, because $f(x) = \cos x$ is an even function.

3. 4

Chapter 7. Integration

▶ **Example**

If $f(x)$ is an even function and $\int_0^2 f(x)\,dx = 10$, what is the value of the following definite integral?

1. $\int_{-2}^0 f(x)\,dx$

2. $\int_{-2}^2 f(x)\,dx$

3. $\int_0^2 -f(x)\,dx$

4. $\int_{-2}^0 3f(x)\,dx$

---

1. Since $f(x)$ is an even function, $\int_{-2}^0 f(x)\,dx = \int_0^2 f(x)\,dx = 10$.

2. $\int_{-2}^2 f(x)\,dx = 2\int_0^2 f(x)\,dx = 2(10) = 20$

3. $\int_0^2 -f(x)\,dx = -\int_0^2 f(x)\,dx = -10$

4. $\int_{-2}^0 3f(x)\,dx = 3\int_{-2}^0 f(x)\,dx = 3(10) = 30$

Chapter 7. Integration

D. APPROXIMATING INTEGRALS (Finding the Area under a Curve)

If we can find an antiderivative of $f(x)$, then we can compute the area under the curve exactly using the fundamental theorem of calculus. Sometimes, it is hard to find antiderivatives. However, we can approximate $\int_a^b f(x)\,dx$ using Riemann sums as follows.

1) Right endpoints approximation (R_n) = Right Riemann Sum
2) Left endpoints approximation (L_n) = Left Riemann Sum
3) Midpoints approximation (M_n) = Midpoint Riemann Sum
4) Trapezoid approximation (T_n) = Trapezoidal Sum

Typically, the midpoints and trapezoidal approximations are much better than the right endpoints or left endpoints approximations.

Example

Find the approximation of the area of the region between the graph of $f(x) = -x^2 + 6$ and the x-axis between $x = 0$ and $x = 2$ using 4 rectangles.

1. Right Endpoints Approximation

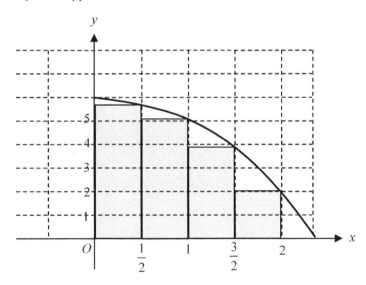

The width of each rectangle: $\dfrac{2-0}{4} = \dfrac{1}{2}$

The height of each rectangle: $f\left(\dfrac{1}{2}\right) = -\left(\dfrac{1}{2}\right)^2 + 6 = \dfrac{23}{4}$, $f(1) = 5 = \dfrac{20}{4}$, $f\left(\dfrac{3}{2}\right) = -\left(\dfrac{3}{2}\right)^2 + 6 = \dfrac{15}{4}$ and $f(2) = 2 = \dfrac{8}{4}$

Therefore, the sum of the areas: $R_4 = \dfrac{1}{2}\left(\dfrac{23}{4} + \dfrac{20}{4} + \dfrac{15}{4} + \dfrac{8}{4}\right) = \dfrac{1}{2}\left(\dfrac{66}{4}\right) = \dfrac{33}{4} = 8.25$

Note: The sum of the areas of the inscribed rectangles is called a **lower sum**.

Chapter 7. Integration

2. Left Endpoints Approximation

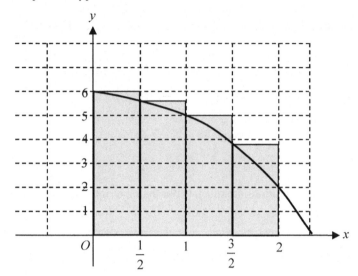

The width of each rectangle $= \dfrac{1}{2}$

The height of each rectangle: $f(0) = 6 = \dfrac{24}{4}$, $f\left(\dfrac{1}{2}\right) = \dfrac{23}{4}$, $f(1) = 5 = \dfrac{20}{4}$, $f\left(\dfrac{3}{2}\right) = \dfrac{15}{4}$

The sum of the areas: $L_4 = \dfrac{1}{2}\left(\dfrac{24}{4} + \dfrac{23}{4} + \dfrac{20}{4} + \dfrac{15}{4}\right) = \dfrac{1}{2}\left(\dfrac{82}{4}\right) = \dfrac{41}{4} = 10.25$

Note: The sum of the areas of the circumscribed rectangles is called an **upper sum**.

Chapter 7. Integration

3. Midpoints Approximation

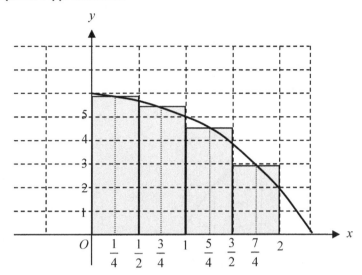

The width of each rectangle = $\dfrac{1}{2}$

The height of each rectangle: $f\left(\dfrac{1}{4}\right)=\dfrac{95}{16}$, $f\left(\dfrac{3}{4}\right)=\dfrac{87}{16}$, $f\left(\dfrac{5}{4}\right)=\dfrac{71}{16}$, $f\left(\dfrac{7}{4}\right)=\dfrac{47}{16}$

The sum of the areas: $M_4 = \dfrac{1}{2}\left(\dfrac{95}{16}+\dfrac{87}{16}+\dfrac{71}{16}+\dfrac{47}{16}\right) = \dfrac{1}{2}\left(\dfrac{300}{16}\right) = \dfrac{75}{8} = 9.375$

Note: Actual area $= \int_0^2 \left(-x^2+6\right)dx = \left[-\dfrac{x^3}{3}+6x\right]_0^2 = \left(-\dfrac{8}{3}+12\right)-0 = \dfrac{28}{3} \approx 9.333$

You notice that the midpoints approximation is **better than** the right end points and left end points approximation.

Chapter 7. Integration

4. Trapezoidal Approximation

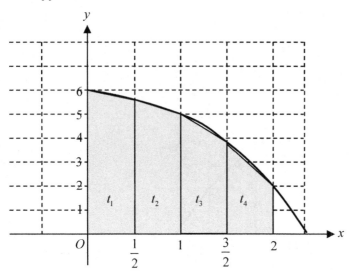

The area of trapezoid $t_1 = \dfrac{\frac{1}{2}\left(f(0)+f\left(\frac{1}{2}\right)\right)}{2}$

" $t_2 = \dfrac{\frac{1}{2}\left(f\left(\frac{1}{2}\right)+f(1)\right)}{2}$

" $t_3 = \dfrac{\frac{1}{2}\left(f(1)+f\left(\frac{3}{2}\right)\right)}{2}$

" $t_4 = \dfrac{\frac{1}{2}\left(f\left(\frac{3}{2}\right)+f(2)\right)}{2}$

Therefore, the sum of the areas:

$$T_4 = \dfrac{\frac{1}{2}\left(f(0)+2f\left(\frac{1}{2}\right)+2f(1)+2f\left(\frac{3}{2}\right)+f(2)\right)}{2}$$

You know that $f(0)=6$, $f\left(\dfrac{1}{2}\right)=\dfrac{23}{4}$, $f(1)=5$, $f\left(\dfrac{3}{2}\right)=\dfrac{15}{4}$, and $f(2)=2$.

$T_4 = \dfrac{1}{4}\left(6+2\cdot\dfrac{23}{4}+2\cdot 5+2\cdot\dfrac{15}{4}+2\right)=\dfrac{37}{4}=9.25$

Chapter 7. Integration

FORMULAS

If the interval $[a, b]$ is partitioned into n subintervals, then each length
$$\Delta x = \frac{b-a}{n}.$$

Therefore, here is how you find the sum of the areas

1. Right Riemann Sum $\quad R_n = \dfrac{b-a}{n}\left(y_1 + y_2 + y_3 \cdots + y_{n-1} + y_n\right)$

2. Left Riemann Sum $\quad L_n = \dfrac{b-a}{n}\left(y_0 + y_1 + y_2 + \cdots + y_{n-1}\right)$

3. Midpoint Riemann Sum $\quad M_n = \dfrac{b-a}{n}\left(y_{\frac{1}{2}} + y_{\frac{3}{2}} + y_{\frac{5}{2}} + \cdots + y_{\frac{2n-1}{2}}\right)$

4. Trapezoidal Sum $\quad T_n = \dfrac{1}{2}\left(\dfrac{b-a}{n}\right)\left(y_0 + 2y_1 + 2y_2 + 2y_3 + \cdots + 2y_{n-1} + y_n\right)$

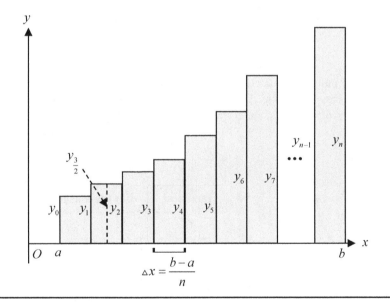

Dr. John Chung's AP Calculus AB/BC

Chapter 7. Integration

▶ **Example**

Approximate the area under the curve $y = 6x - x^2$ from $x = 0$ to $x = 6$ using 6 subintervals and

a) Right Riemann Sum
b) Left Riemann Sum
c) Midpoint Riemann Sum
d) Trapezoidal Sum

a) $R_6 = \dfrac{b-a}{n}(y_1 + y_2 + y_3 + y_4 + y_5 + y_6) = 1(5+8+9+8+5+0) = 35$

b) $L_6 = \dfrac{b-a}{n}(y_0 + y_1 + y_2 + y_3 + y_4 + y_5) = 1(0+5+8+9+8+5) = 35$

c) $M_6 = \dfrac{b-a}{n}(y_{0.5} + y_{1.5} + y_{2.5} + y_{3.5} + y_{4.5} + y_{5.5}) = 1(2.75+6.75+8.75+8.75+6.75+2.75) = 36.5$

d) $T_6 = \dfrac{1}{2}\left(\dfrac{b-a}{n}\right)(y_0 + 2y_1 + 2y_2 + 2y_3 + 2y_4 + 2y_5 + y_6) = \dfrac{1}{2}(1)(0+2\cdot5+2\cdot8+2\cdot9+2\cdot8+2\cdot5+0) = 35$

▶ **Example**

The rate of fuel consumption, in gallons per minute, recorded during a train ride is given by a twice differentiable and increasing function R with respect to time. The graph of the function R and a table with selected values for $R(t)$, for the interval $0 \le t \le 60$ minutes, are shown as follows.

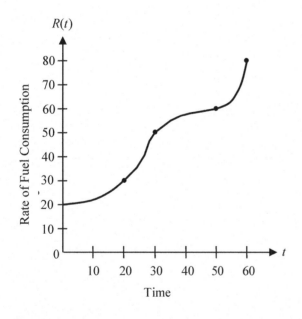

t(minutes)	$R(t)$ (gallons per minute)
0	20
20	30
30	50
50	60
60	80

Chapter 7. Integration

a) Find an approximation for $R'(40)$.

b) Approximate the value of $\int_0^{60} R(t)dt$ using a left Riemann sum with the four subintervals indicated by the data in the table.

c) Is this left Riemann sum less than the value of $\int_0^{60} R(t)\,dt$? Explain.

a) $R'(40) \approx \dfrac{60-50}{20} = 0.5$ gallons per square minute

b)

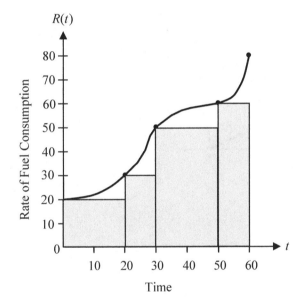

$y_0 = R(0) = 20$
$y_1 = R(20) = 30$
$y_2 = R(30) = 50$
$y_4 = R(50) = 60$

$\int_0^{60} R(t)\,dt \approx 20\cdot 20 + 10\cdot 30 + 20\cdot 50 + 10\cdot 60 = 2300$ Gallons

c) Yes. The approximation is less because the graph of $R(t)$ is increasing on the interval.

Chapter 7. Integration

▶ **Example**

The graph of a differentiable function f on the interval $[0, 18]$ is shown. The graph of f has a horizontal tangent line at $x = 9$. Find a trapezoidal approximation of $\int_0^{18} f(t)\,dt$ using six subintervals of length $\Delta t = 3$.

Graph of f

Graph of f

$$\int_0^{18} f(t)\,dt \approx \frac{1}{2}\left(\frac{18-0}{6}\right)(0 + 2\cdot 1 + 2\cdot 3 + 2\cdot 4 + 2\cdot 3 + 2\cdot 1 + 0) = 36$$

Chapter 7. Integration

E. THEOREM OF CALCULUS

THEOREM. The Fundamental Theorem of Calculus

If a function f is continuous on the interval $[a, b]$ and F is an antiderivative on the interval, then

1. $\int_a^b f(x)\,dx = F(b) - F(a)$

2. $F'(x) = f(x)$ or $\dfrac{d}{dx}\left[\int_a^x f(t)\,dt\right] = f(x)$

3. $\dfrac{d}{dx}\int_a^{u(x)} f(t)\,dt = f(u(x))u'(x)$

▶ **Example**

Evaluate.

1. $\dfrac{d}{dx}\int_x^{2x^2} \sqrt{2t-1}\,dt$

2. $\dfrac{d}{dx}\int_0^{\sin x} \sqrt{t}\,dt$

3. $\dfrac{d}{dx}\int_0^{x^2} \sin^2 t\,dt$

4. $\dfrac{d}{dx}\int_0^{e^x} \cos t\,dt$

5. $\dfrac{d}{dx}\int_0^{\cos^2 2x} \sqrt{t}\,dt$

6. $\dfrac{d}{dx}\int_1^{\ln x} e^t\,dt$

---

1. $\dfrac{d}{dx}\int_x^{2x^2} \sqrt{2t-1}\,dt = \sqrt{2(2x^2)-1}\cdot(2x)' - \sqrt{2(x)-1}(x)' = 2\sqrt{4x^2-1} - \sqrt{2x-1}$

2. $\dfrac{d}{dx}\int_0^{\sin x} \sqrt{t}\,dt = \sqrt{\sin x}\cdot(\sin x)' = \cos x\sqrt{\sin x}$

3. $\dfrac{d}{dx}\int_0^{x^2} \sin^2 t\,dt = \sin^2(x^2)\cdot(x^2)' = 2x\sin^2(x^2)$

4. $\dfrac{d}{dx}\int_0^{e^x} \cos t\,dt = \cos(e^x)\cdot(e^x)' = e^x \cos(e^x)$

5. $\dfrac{d}{dx}\int_0^{\cos^2 2x} \sqrt{t}\,dt = \sqrt{\cos^2(2x)}\cdot(\cos^2 2x)' = 2\cos(2x)(-\sin 2x)(2)\sqrt{\cos^2(2x)}$

   $= -4\sin(2x)\cos(2x)\sqrt{\cos^2(2x)}$

6. $\dfrac{d}{dx}\int_e^{\ln x} e^t\,dt = e^{\ln x}\cdot(\ln x)' = \dfrac{1}{x}e^{\ln x} = \dfrac{x}{x} = 1$

Chapter 7. Integration

THEOREM. Mean-Value Theorem for Integrals

If a function f is continuous on the interval $[a, b]$, then there exists at least one point c in $[a, b]$ such that

$$\int_a^b f(x)dx = f(c)(b-a)$$

Let f be a continuous nonnegative on $[a, b]$, and let m and M be minimum and maximum values of $f(x)$ on this interval.

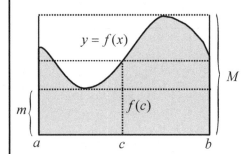

Since $(b-a)m < \int_a^b f(x)dx < (b-a)M$, there is a rectangle over the interval $[a, b]$ of some appropriate height $f(c)$ between m and M whose area is $\int_a^b f(x)dx = f(c)(b-a)$.

THEOREM. The Average Value of a Function (from mean value theorem)

If a function f is continuous on the interval $[a, b]$, then, the average value of a function f over the interval $[a, b]$ is given by

$$f_{\text{average}} = \frac{1}{b-a}\int_a^b f(x)dx$$

▶ **Example**

Let $f(x) = 10\pi x^2$ and $g(x) = k^2 \sin\dfrac{\pi x}{2k}$ for $k > 0$.

a) What is the average value of f on $[1, 4]$?

b) For what value of k will the average value of g on $[0, k]$ be equal to the average value of f on $[1, 4]$?

Chapter 7. Integration

a) $f_{avg} = \dfrac{1}{4-1}\displaystyle\int_1^4 10\pi x^2\,dx = \dfrac{10\pi}{3}\int_1^4 x^2\,dx = \dfrac{10\pi}{3}\left[\dfrac{x^3}{3}\right]_1^4 = \dfrac{10\pi}{3}\left(\dfrac{64}{3}-\dfrac{1}{3}\right) = 70\pi$

b) $\dfrac{1}{k-0}\displaystyle\int_0^k k^2 \sin\left(\dfrac{\pi x}{2k}\right)dx = \dfrac{k^2}{k}\int_0^k \sin\left(\dfrac{\pi x}{2k}\right)dx = k\left[\dfrac{-\cos\left(\dfrac{\pi x}{2k}\right)}{\dfrac{\pi}{2k}}\right]_0^k = \dfrac{-2k^2}{\pi}\left[\cos\left(\dfrac{\pi x}{2k}\right)\right]_0^k$

$= \dfrac{-2k^2}{\pi}\left[\cos\left(\dfrac{\pi k}{2k}\right) - \cos 0\right] = \dfrac{-2k^2}{\pi}\left[\cos\left(\dfrac{\pi}{2}\right) - \cos 0\right] = \dfrac{-2k^2}{\pi}(0-1) = \dfrac{2k^2}{\pi}$

Since $\dfrac{2k^2}{\pi} = 98\pi$, $2k^2 = 98\pi^2 \Rightarrow k^2 = 49\pi^2 \Rightarrow k = 7\pi$.

▶ **Example**

Let f be the function that is defined for all real number x. f has a following properties.:
(i) $f''(x) = 24x - 18$ (ii) $f'(1) = -6$ (iii) $f(2) = 0$

a) Find each x such that the line tangent to the graph of f at $(x, f(x))$ is horizontal.
b) Find $f(x)$.
c) Find the average value of f on the interval $1 \leq x \leq 3$.

a) $f'(x) = \displaystyle\int (24x-18)\,dx = 12x^2 - 18x + C$ and $f'(1) = -6$

$f'(1) = 12 - 18 + C = -6 \Rightarrow C = 0$ Therefore, $f'(x) = 12x^2 - 18x$.

Horizontal tangent line $\Rightarrow f'(x) = 0 \Rightarrow 12x^2 - 18x = 0 \Rightarrow 6x(2x-3) = 0 \to x = 0,\ 3/2$

b) $f(x) = \displaystyle\int (12x^2 - 18x)\,dx = 4x^3 - 9x^2 + C$ and $f(2) = 0$

$f(2) = 32 - 36 + C = 0 \Rightarrow C = 4$ Therefore, $f(x) = 4x^3 - 9x^2 + 4$.

c) $f_{avg} = \dfrac{1}{3-1}\displaystyle\int_1^3 (4x^3 - 9x^2 + 4)\,dx = \dfrac{1}{2}\left[x^4 - 3x^3 + 4x\right]_1^3 = \dfrac{1}{2}\left[(81-81+12)-(1-3+4)\right]$

$= \dfrac{1}{2}(12-2) = 5$

Chapter 7. Integration

F. EXPONENTIAL AND LOGARITHMIC FUNCTIONS

FORMULA

1. $\int \dfrac{1}{x} dx = \ln|x| + C$

2. $\int \dfrac{u'}{u} du = \ln|u| + C$, where u is a differentiable function of x.

3. $\int \dfrac{1}{x-a} dx = \ln|x-a| + C$, where a = constant.

4. $\int a^x dx = \dfrac{a^x}{\ln a} + C$

5. $\int e^x dx = \dfrac{e^x}{\ln e} + C = e^x + C$

6. $\int a^{u(x)} dx = \dfrac{a^{u(x)}}{(\ln a)u'} + C$

7. $\int e^{u(x)} dx = \dfrac{e^{u(x)}}{u'} + C$

▶ **Example**

Find the indefinite integral.

1. $\int \dfrac{1}{x+2} dx$
2. $\int \dfrac{3}{x-5} dx$
3. $\int \dfrac{1}{2-3x} dx$
4. $\int \dfrac{x}{x^2+1} dx$

5. $\int \dfrac{x^2-4}{x} dx$
6. $\int \dfrac{x+2}{x^2+4x+6} dx$
7. $\int \dfrac{(\ln x)^2}{x} dx$
8. $\int \dfrac{1}{x \ln x} dx$

9. $\int \dfrac{1}{x \ln(x^2)} dx$

---

1. $\ln|x+2| + C$ 2. $3\ln|x-5| + C$ 3. $-\dfrac{1}{3}\ln|2-3x| + C$ 4. $\dfrac{1}{2}\ln(x^2+1) + C$ 5. $\dfrac{x^2}{2} - 4\ln|x| + C$

6. $\dfrac{1}{2}\ln|x^2+4x+6| + C$ 7. $\dfrac{(\ln x)^3}{3} + C$ 8. $\ln|\ln x| + C$ 9. $\dfrac{1}{2}\ln|\ln|x|| + C$

▶ **Example**

Solve.

1. $\int \dfrac{x}{(x-1)^2} dx$
2. $\int \dfrac{x(x-2)}{(x-1)^3} dx$

Chapter 7. Integration

1. $\int \dfrac{x}{(x-1)^2} dx \;\to\; u = x - 1$ and $du = dx$, also $x = u + 1$

   $\int \dfrac{x}{(x-1)^2} dx = \int \dfrac{u+1}{u^2} du = \int \left(\dfrac{1}{u} + \dfrac{1}{u^2}\right) du = \ln|u| - \dfrac{1}{u} + C$ Therefore, $\ln|x-1| - \dfrac{1}{x-1} + C$

   Also, we can use **partial fraction**.

   $\dfrac{x}{(x-1)^2} = \dfrac{1}{(x-1)} + \dfrac{1}{(x-1)^2}$

   $\int \dfrac{x}{(x-1)^2} dx = \int \dfrac{1}{(x-1)} + \dfrac{1}{(x-1)^2} dx = \ln|x-1| + \dfrac{1}{x-1} + C$

2. $\int \dfrac{x(x-2)}{(x-1)^3} dx \;\to\; u = x - 1$ and $du = dx$, also $x = u+1$, $x - 2 = u - 1$

   $\int \dfrac{x(x-2)}{(x-1)^3} dx = \int \dfrac{(u+1)(u-1)}{u^3} du = \int \dfrac{u^2-1}{u^3} du = \int \dfrac{1}{u} - \dfrac{1}{u^3} du = \ln|u| + \dfrac{1}{2u^2} + C$

   Therefore, $\int \dfrac{x(x-2)}{(x-1)^3} dx = \ln|x-1| + \dfrac{1}{2(x-1)^2} + C$.

   Or, using partial fraction,

   $\dfrac{x(x-2)}{(x-1)^3} = \dfrac{A}{(x-1)} + \dfrac{B}{(x-1)^2} + \dfrac{C}{(x-1)^3} \;\to\; A = 1, B = 0,$ and $C = -1$

   $\int \dfrac{x(x-2)}{(x-1)^3} dx = \int \dfrac{1}{x-1} - \dfrac{1}{(x-1)^3} dx = \ln|x-1| + \dfrac{1}{2(x-1)^2} + C$

▶ **Example**

Evaluate.

1. $\int_0^4 \dfrac{4}{3x+1} dx$ 2. $\int_{-1}^1 \dfrac{1}{4-2x} dx$ 3. $\int_1^e \dfrac{(1+\ln x)^2}{x} dx$

4. $\int_e^{e^2} \dfrac{1}{x \ln x} dx$ 5. $\int_0^2 \dfrac{x^2-2}{x+1} dx$ 6. $\int_0^1 \dfrac{x-1}{x+1} dx$

---

1. $\int_0^4 \dfrac{4}{3x+1} dx = \dfrac{4}{3}\Big[\ln|3x+1|\Big]_0^4 = \dfrac{4}{3} \ln 13$

2. $\int_{-1}^1 \dfrac{1}{4-2x} dx = -\dfrac{1}{2}\Big[\ln|4-2x|\Big]_{-1}^1 = -\dfrac{1}{2}(\ln 2 - \ln 6) = \dfrac{1}{2} \ln 3$ or $\ln \sqrt{3}$

3. $\dfrac{7}{3}$ 4. $\ln 2$ 5. $-\ln 3$ 6. $1 - 2\ln 2$ 7. $-\ln 3$ 8. $1 - 2\ln 2$

Dr. John Chung's AP Calculus AB/BC

Chapter 7. Integration

Example

Find the indefinite integral.

1. $\int 5^x \, dx$
2. $\int 7^{2x} \, dx$
3. $\int e^{5x} \, dx$
4. $\int 3^{\cos x} \sin x \, dx$

---

1. $\dfrac{5^x}{\ln 5} + C$ 2. $\dfrac{7^{2x}}{2\ln 7} + C$ 3. $\dfrac{e^{5x}}{5} + C$ 4. $-\dfrac{3^{\cos x}}{\ln 3} + C$

G. THE INTEGRALS OF TRIGONOMETRIC FUNCTIONS

1. $\int \cos x \, dx = \sin x + c$
2. $\int \sin x \, dx = -\cos x + C$
3. $\int \sec^2 x \, dx = \tan x + C$
4. $\int \csc^2 x \, dx = -\cot x + C$
5. $\int \sec x \tan x \, dx = \sec x + C$
6. $\int \csc x \cot x \, dx = -\csc x + C$
7. $\int \dfrac{1}{\sqrt{1-x^2}} \, dx = \arcsin x + C$
8. $\int \dfrac{-1}{\sqrt{1-x^2}} \, dx = \arccos x + C$
9. $\int \dfrac{1}{1+x^2} \, dx = \arctan x + C$
10. $\int \dfrac{-1}{1+x^2} \, dx = \text{arc}\cot x + C$
11. $\int \dfrac{1}{|x|\sqrt{x^2-1}} \, dx = \text{arc}\sec x + C$
12. $\int \dfrac{-1}{|x|\sqrt{x^2-1}} \, dx = \text{arc}\csc x + C$

If u is a differentiable function of x, then

1. $\int \dfrac{u'}{u} \, du = \ln|u| + C$
2. $\int \dfrac{u'}{\sqrt{1-u^2}} \, du = \arcsin u + C$
3. $\int \dfrac{u'}{1+u^2} \, du = \arctan u + C$.

Example

Find the indefinite integral.

1. $\int \tan x \, dx$
2. $\int \cot x \, dx$
3. $\int \sec x \, dx$
4. $\int \csc x \, dx$
5. $\int \tan 5x \, dx$
6. $\int \csc 2x \, dx$
7. $\int \dfrac{\cos x}{1+\sin x} \, dx$
8. $\int \dfrac{\sin x}{1+\cos x} \, dx$
9. $\int (\sec x + \tan x) \, dx$
10. $\int \dfrac{\sec x \tan x}{\sec x - 1} \, dx$

Chapter 7. Integration

1. $\int \tan x \, dx = \int \dfrac{\sin x}{\cos x} dx = -\ln|\cos x| + C$ or, $\ln|\sec x| + c$ 2. $\ln|\sin x| + C$

3. $\int \sec x \, dx = \int \dfrac{\sec x(\sec x + \tan x)}{\sec x + \tan x} dx = \int \dfrac{\sec^2 x + \sec x \tan x}{\sec x + \tan x} dx = \ln|\sec x + \tan x| + C$

4. $\int \csc x \, dx = \int \dfrac{\csc x(\csc x + \cot x)}{\csc x + \cot x} dx = -\ln|\csc x + \cot x| + C$

5. $\int \tan 5x \, dx = -\dfrac{1}{5} \int \dfrac{\sin 5x}{\cos 5x} dx = -\dfrac{1}{5} \ln|\cos 5x| + c$

6. $\int \csc 2x \, dx = \int \dfrac{\csc 2x \cdot (\csc 2x + \cot 2x)}{\csc 2x + \cot 2x} dx = -\dfrac{1}{2} \ln|\csc 2x + \cot 2x| + C$

7. $\int \dfrac{\cos x}{1 + \sin x} dx = \ln|1 + \sin x| + C = \ln(1 + \sin x) + C \to (1 + \sin x \geq 0)$

8. $\int \dfrac{\sin x}{1 + \cos x} dx = -\ln(1 + \cos x) + C$ 9. $\int (\sec x + \tan x) dx = \ln|\sec x + \tan x| - \ln|\cos x| + C$

10. $\int \dfrac{\sec x \tan x}{\sec x - 1} dx = \ln|\sec x - 1| + C$

▶ **Example**

Evaluate.

1. $\displaystyle\int_0^{1/4} \dfrac{1}{\sqrt{1 - 4x^2}} dx$ 2. $\displaystyle\int_0^{\sqrt{3}/2} \dfrac{1}{1 + 4x^2} dx$ 3. $\displaystyle\int_0^e \dfrac{x^3}{x^2 + 1} dx$

4. $\displaystyle\int_0^{1/\sqrt{2}} \dfrac{\arcsin x}{\sqrt{1 - x^2}} dx$ 5. $\displaystyle\int_0^{\ln(\sqrt{3})} \dfrac{e^x}{1 + e^{2x}} dx$ 6. $\displaystyle\int_0^2 \dfrac{1}{x^2 - 2x + 2} dx$

---

1. $\dfrac{1}{2} \left[\arcsin(2x) \right]_0^{1/4} = \dfrac{\pi}{12}$ 2. $\dfrac{1}{2} \left[\arctan(x) \right]_0^{\sqrt{3}/2} = \dfrac{\pi}{6}$

3. $\displaystyle\int_0^e \dfrac{x^3}{x^2 + 1} dx = \int_0^e \left(x - \dfrac{x}{x^2 + 1} \right) dx = \left[\dfrac{x^2}{2} - \dfrac{1}{2} \ln(x^2 + 1) \right]_0^e = \dfrac{e^2 - \ln(e^2 + 1)}{2}$

4. $\displaystyle\int_0^{1/\sqrt{2}} \dfrac{\arcsin x}{\sqrt{1 - x^2}} dx = \left[\dfrac{(\arcsin x)^2}{2} \right]_0^{1/\sqrt{2}} = \dfrac{\pi^2}{32}$ 5. $\left[\arctan e^x \right]_0^{\ln\sqrt{3}} = \dfrac{\pi}{12}$

6. $\displaystyle\int_0^2 \dfrac{1}{x^2 - 2x + 2} dx = \int_0^2 \dfrac{1}{(x - 1)^2 + 1} dx = \left[\arctan(x - 1) \right]_0^2 = \dfrac{\pi}{2}$

Chapter 7. Integration

H. Integration by Parts

If u and v are functions of x and differentiable, then

$$\int u\,dv = uv - \int v\,du.$$

Tips for solving Integrals using Integration by Parts:
1) $\ln x$ belongs to u.
2) e^x and trigonometric functions belong to dv.
3) If the integration has both exponential function and trigonometric functions, then
   e^x belongs to u and the trigonometric function belongs to dv.

There is another useful strategy for choosing u and dv that can be applied when integrand is a product of the functions in the list (**LIATE**)

Logarithmic – **I**nverse trigonometric---- **A**lgebraic ---- **T**rigonometric ---**E**xponential

Take u to be the function whose category occurs earlier in the list and take dv to the rest of the integrand. The method does not work all the time, but it works often enough to be useful.

▶ **Example**

Evaluate.

1. $\int xe^x\,dx$
2. $\int x^2 \ln x\,dx$
3. $\int x^2 \sin x\,dx$

4. $\int \ln x\,dx$
5. $\int e^x \cos x\,dx$
6. $\int x \ln x\,dx$

7. $\int x^2 e^x\,dx$
8. $\int e^x \cos 2x\,dx$
9. $\int \sec^3 x\,dx$

Chapter 7. Integration

1. $\int xe^x dx = \int (x)(e^x dx) dx \rightarrow \begin{cases} u = x \rightarrow du = dx \\ dv = e^x dx \rightarrow v = e^x \end{cases}$

   Therefore, $\int xe^x dx = uv - \int v du = xe^x - \int e^x dx = xe^x - e^{x+c}$

2. $\int x^2 \ln x \, dx = \int (\ln x) x^2 dx \rightarrow \begin{cases} u = \ln x \rightarrow du = \dfrac{1}{x} dx \\ dv = x^2 dx \rightarrow v = \dfrac{x^3}{3} \end{cases}$

   Therefore, $\int x^2 \ln x \, dx = uv - \int v du = \dfrac{x^3 \ln x}{3} - \int \dfrac{x^3}{3} \cdot \dfrac{1}{x} dx = \dfrac{x^3 \ln x}{3} - \dfrac{x^3}{9} + C$

3. $(2 - x^2)\cos x + 2x \sin x + C$ 4. $x \ln x - x$ 5. $\dfrac{e^x(\cos x + \sin x)}{2} + C$ 6. $\dfrac{x^2 \ln x}{2} - \dfrac{x^2}{4} + C$

7. $(x^2 - 2x + 2)e^x + C$ 8. $\dfrac{e^x \cos 2x}{5} + \dfrac{2e^x \sin 2x}{5} + C$ 9. $\dfrac{\sec x \tan x}{2} + \dfrac{\ln|\sec x + \tan x|}{2} + C$

Chapter 7. Integration

I. TABULAR INTEGRATION BY PARTS (Short Cut)

DEFINITION. When you have a problem involving repeated integration by parts, the Tabular Method is a nifty way to simplify repeated integration by parts especially for

$$\int (\text{polynomial} \cdot \text{exponential})\,dx \quad \text{or} \quad \int (\text{polynomial} \cdot \text{trigometric})\,dx.$$

For Example:
$$\int x^3 e^x\,dx, \quad \int (x^3 \sin x)\,dx, \quad \int (x^3 \cos x)\,dx, \cdots$$

Evaluate $\int x^3 e^x\,dx$.

Tabular Method:

$$\int x^3 \cdot e^x\,dx = ?$$

Alternate Sign	x^3 and its derivative of x^3	e^x and its antiderivative of e^x	Product
+ →	x^3	e^x	
− →	$3x^2$	e^x	$x^3 e^x$
+ →	$6x$	e^x	$-3x^2 e^x$
− →	6	e^x	$6xe^x$
+	0	e^x	$-6e^x$

(When the derivative is 0, you can stop)

Therefore, $\int x^3 \cdot e^x\,dx = x^3 e^x - 3x^2 e^x + 6xe^x - 6e^x + C.$

Remember: Tabular Integration:

Step 1) Differentiate x^3 repeatedly until you obtain 0, and list the results in the first column.

Step 2) Integrate e^x repeatedly and list the results in the second column.

Step 3) Draw an arrow from each entry in the first column to the entry that is one row down in the second column.

Step 4) Label the arrows with alternating + and − signs, starting with a +.

Step 5) For each arrow, form the product of the expressions at its tip and tail and then multiply that product by +1 or −1 in accordance with the sign on the arrow. Add the results to obtain the value of the integral.

Chapter 7. Integration

 Example

Evaluate. (Use Tabular Integration by Parts)
These are the same questions as in part H. Compare your answers.

1. $\int x^2 \sin x \, dx$ 2. $\int x^2 \cos 3x \, dx$ 3. $\int x^3 \cos 2x \, dx$

---

1. $\int x^2 \sin x \, dx$

+	x^2	$\sin x$	
−	$2x$	$-\cos x$	$-x^2 \cos x$
+	2	$-\sin x$	$-2x \sin x$
−	0	$\cos x$	$2\cos x$

$\int x^2 \sin x \, dx = -x^2 \cos x - 2x \sin x + 2\cos x + C$

2. $\int x^2 \cos 3x \, dx$

+	x^2	$\cos 3x$	
−	$2x$	$\dfrac{1}{3}\sin 3x$	$\dfrac{x^2 \sin 3x}{3}$
+	2	$-\dfrac{1}{9}\cos 3x$	$\dfrac{2\cos 3x}{9}$
−	0	$-\dfrac{1}{27}\sin 3x$	$-\dfrac{2\sin 3x}{27}$

$\int x^2 \cos 3x \, dx = \dfrac{x^2 \sin 3x}{3} + \dfrac{2\cos 3x}{9} - \dfrac{2\sin 3x}{27} + C$

3. $\int x^3 \cos 2x \, dx$

+	x^3	$\cos 2x$	
−	$3x^2$	$\dfrac{1}{2}\sin 2x$	$\dfrac{x^3 \sin 2x}{2}$
+	$6x$	$-\dfrac{1}{4}\cos 2x$	$\dfrac{3x^2 \cos 2x}{4}$
−	6	$-\dfrac{1}{8}\sin 2x$	$-\dfrac{3x \sin 2x}{8}$
	0	$\dfrac{1}{16}\cos 2x$	$-\dfrac{3\cos 2x}{8}$

$\int x^3 \cos 2x \, dx = \dfrac{x^3 \sin 2x}{2} + \dfrac{3x^2 \cos 2x}{4} - \dfrac{3x \sin 2x}{4} - \dfrac{3\cos 2x}{8} + C$

Chapter 7. Integration

J. PARTIAL FRACTION

DEFINITION. Sometimes it makes sense to decompose a rational function $R(x)$ into simpler rational functions to which you can apply the basic integration formulas. This procedure is called the method of partial fractions and it works as follows.

$$R(x) = \frac{N(x)}{D(x)}$$

If the degree of a polynomial $N(x)$ is less than the degree of a polynomial $D(x)$, then you can write the partial fractions as follows.

Rule 1. $\dfrac{N(x)}{x^2 - 5x + 6} = \dfrac{A}{x-3} + \dfrac{B}{x-2}$

Rule 2. $\dfrac{N(x)}{(x-1)^2} = \dfrac{A}{(x-1)} + \dfrac{B}{(x-1)^2}$

Rule 3. $\dfrac{N(x)}{x(x^2+1)} = \dfrac{A}{x} + \dfrac{Bx+C}{(x^2+1)}$

Rule 4. $\dfrac{N(x)}{(x-1)(x+1)^2(x^2+1)} = \dfrac{A}{(x-1)} + \dfrac{B}{(x+1)} + \dfrac{C}{(x+1)^2} + \dfrac{Dx+E}{(x^2+1)}$

Rule 5. $\dfrac{N(x)}{\left(x^2+2\right)^2} = \dfrac{Ax+B}{\left(x^2+2\right)} + \dfrac{Cx+D}{\left(x^2+2\right)^2}$

▶ **Example**

Evaluate.

1. $\displaystyle\int \dfrac{1}{x^2-5x+6}\,dx$

2. $\displaystyle\int \dfrac{x+1}{(x-1)^2}\,dx$

3. $\displaystyle\int \dfrac{x+1}{x(x^2+1)}\,dx$

4. $\displaystyle\int \dfrac{5x^2+20x+6}{x^3+2x^2+x}\,dx$

---

1. $\displaystyle\int \dfrac{1}{x^2-5x+6}\,dx = \int\left(\dfrac{1}{x-3} - \dfrac{1}{x-2}\right)dx = \ln|x-3| - \ln|x-2| + C$

2. $\displaystyle\int \dfrac{x+1}{(x-1)^2}\,dx = \int\left[\dfrac{1}{x-1} + \dfrac{2}{(x-1)^2}\right]dx = \ln|x-1| - \dfrac{2}{x-1} + C$

3. $\displaystyle\int \dfrac{x+1}{x(x^2+1)}\,dx = \int\left(\dfrac{1}{x} - \dfrac{x}{x^2+1} + \dfrac{1}{x^2+1}\right)dx = \ln|x| - \dfrac{1}{2}\ln\left(x^2+1\right) + \arctan x + C$

4. $\displaystyle\int \dfrac{5x^2+20x+6}{x^3+2x^2+x}\,dx = \int\dfrac{5x^2+20x+6}{x(x+1)^2}\,dx = \int\dfrac{6}{x} - \dfrac{1}{x+1} + \dfrac{9}{(x+1)^2}\,dx = 6\ln|x| - \ln|x+1| - \dfrac{9}{x+1} + C$

Chapter 7. Integration

▶ **Example**

1. $\int \dfrac{8x^3 + 13x}{(x^2+2)^2}\, dx$

2. $\int \dfrac{x^2 - 1}{x^3 + x}\, dx$

3. $\int \dfrac{x^2 + 12x + 12}{x^3 - 4x}\, dx$

4. $\int \dfrac{x^2}{x^4 - 2x^2 - 8}\, dx$

---

1. $\int \dfrac{8x^3 + 13x}{(x^2+2)^2}\, dx = \int \left[\dfrac{8x}{x^2+2} - \dfrac{3x}{(x^2+2)^2} \right] dx = 4\ln(x^2+2) + \dfrac{3}{2(x^2+2)} + C$

2. $\int \dfrac{x^2-1}{x^3+x}\, dx = \int \dfrac{x^2-1}{x(x^2+1)}\, dx = \int \left(\dfrac{-1}{x} + \dfrac{2x}{x^2+1} \right) dx = -\ln|x| + \ln(x^2+1) + C$

3. $\int \dfrac{x^2+12x+12}{x^3-4x}\, dx = \int \dfrac{x^2+12x+12}{x(x+2)(x-2)}\, dx = \int \left(\dfrac{-3}{x} - \dfrac{1}{x+2} + \dfrac{5}{x-2} \right) dx = -3\ln|x| - \ln|x+2| + 5\ln|x-2| + C$

4. $\int \dfrac{x^2}{x^4 - 2x^2 - 8}\, dx = \int \dfrac{x^2}{(x+2)(x-2)(x^2+2)}\, dx = \int \dfrac{-1/6}{x+2} + \dfrac{1/6}{x-2} + \dfrac{1/3}{x^2+2}\, dx =$

$-\dfrac{1}{6}\ln|x+2| + \dfrac{1}{6}\ln|x-2| + \dfrac{\sqrt{2}}{6}\arctan\left(\dfrac{x}{\sqrt{2}}\right) + C$

Chapter 7. Integration

K. IMPROPER INTEGRALS

DEFINITION. In calculus, an improper integral is characterized as such because either one or both limits of integration are infinite, or f has infinite discontinuity in the interval.

$\int_{a}^{\infty} f(x)dx$, $\int_{-\infty}^{a} f(x)dx$, and $\int_{-\infty}^{\infty} f(x)dx$ are improper integrals.

$\int_{1}^{10} \frac{1}{\sqrt{x-1}} dx$ and $\int_{-2}^{2} \frac{1}{(x+1)^2} dx$ are improper, because the integrands are discontinuous at $x=1$ and $x=-2$.

If $f(x)$ is continuous on the interval $[a, \infty)$, then

1) $\int_{a}^{\infty} f(x)\,dx = \lim_{b \to \infty} \int_{a}^{b} f(x)\,dx$

If $f(x)$ is continuous on the interval $(-\infty, a]$, then

2) $\int_{-\infty}^{a} f(x)\,dx = \lim_{b \to -\infty} \int_{b}^{a} f(x)\,dx$

If $f(x)$ is continuous on the interval $(-\infty, \infty)$, then

3) $\int_{-\infty}^{\infty} f(x)dx = \int_{-\infty}^{c} f(x)dx + \int_{c}^{\infty} f(x)dx$

▶ **Example**

Evaluate.

$\int_{0}^{\infty} \frac{1}{x^2+1} dx$

$\int_{0}^{\infty} \frac{1}{x^2+1} dx = \lim_{b \to \infty} \int_{0}^{b} \frac{1}{x^2+1} dx = \lim_{b \to \infty} [\arctan x]_{0}^{b} = \lim_{b \to \infty} (\arctan b - \arctan 0) = \frac{\pi}{2} - 0 = \frac{\pi}{2}$

Chapter 7. Integration

L. LENGTH OF A CURVE

1. If the function $y = f(x)$ differentiable on the interval $[a, b]$, then the length of the curve between a and b is

$$L = \int_a^b \sqrt{1 + \left(\frac{dy}{dx}\right)^2}\, dx$$
$$= \int_a^b \sqrt{1 + [f'(x)]^2}\, dx$$

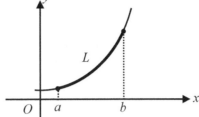

2. If the function $x = f(y)$ differentiable on the interval $[a, b]$, then the length of the curve between a and b is

$$L = \int_a^b \sqrt{1 + \left(\frac{dx}{dy}\right)^2}\, dy$$

3. If a smooth curve C is given by the parametric equations $x = f(t)$ and $y = g(t)$, then the slope of the curve at (x, y) is

$$\frac{dy}{dx} = \frac{dy/dt}{dx/dt},\ \frac{dx}{dt} \neq 0$$

4. Arc length in parametric form
   If a smooth curve C is given by $x = f(t)$ and $y = g(t)$, then the arc length over the interval $a \leq t \leq b$ is given by

$$\ell = \int_a^b \sqrt{\left(\frac{dx}{dt}\right)^2 + \left(\frac{dy}{dt}\right)^2}\, dt$$

▶ **Example**

Find arc length of the function over the indicated interval.

1. $y = \frac{1}{3}(x^2 + 2)^{\frac{3}{2}}$ $[0, 3]$
2. $x = \frac{1}{3}\sqrt{y}(y - 3)$ $[1, 9]$

---

1. $y' = x(x^2 + 2)^{1/2}$ → $L = \int_0^3 \sqrt{1 + x^2(x^2 + 2)}\, dx = \int_0^3 \sqrt{(x^2 + 1)^2}\, dx = \int_0^3 (x^2 + 1)\, dx = \left[\frac{x^3}{3} + x\right]_0^3 = 12$

2. $\frac{dx}{dy} = \frac{1}{2}y^{1/2} - \frac{1}{2}y^{-1/2} = \frac{\sqrt{y}}{2} - \frac{1}{2\sqrt{y}}$

$L = \int_1^9 \sqrt{1 + \left(\frac{dx}{dy}\right)^2}\, dy = \int_1^9 \sqrt{1 + \left(\frac{\sqrt{y}}{2} - \frac{1}{2\sqrt{y}}\right)^2}\, dx = \int_1^9 \frac{y + 1}{2\sqrt{y}}\, dy = \int_0^9 \frac{1}{2}y^{1/2} + \frac{1}{2}y^{-1/2}\, dy =$

$\left[\frac{1}{3}y^{3/2} + y^{1/2}\right]_1^9 = \frac{32}{3}$

Chapter 7. Integration

▶ **Example**

1. $\begin{cases} x = \sin t \\ y = \cos t \end{cases}$ $[0, \pi]$

2. $\begin{cases} x = e^{-t} \cos t \\ y = e^{-t} \sin t \end{cases}$ $\left[0, \dfrac{\pi}{2}\right]$

---

1. $\begin{cases} x = \sin t \\ y = \cos t \end{cases}$ $[0, \pi]$

$\dfrac{dx}{dt} = \cos t, \ \dfrac{dy}{dt} = -\sin t \ \Rightarrow \ \left(\dfrac{dx}{dt}\right)^2 + \left(\dfrac{dy}{dt}\right)^2 = \cos^2 t + \sin^2 t = 1$

$L = \displaystyle\int_0^{\pi} \sqrt{\left(\dfrac{dx}{dt}\right)^2 + \left(\dfrac{dy}{dt}\right)^2} = \int_0^{\pi} \sqrt{1}\, dt = [t]_0^{\pi} = \pi$

2. $\begin{cases} x = e^{-t} \cos t \\ y = e^{-t} \sin t \end{cases}$ $\left[0, \dfrac{\pi}{2}\right]$

$\begin{cases} x' = -e^{-t} \cos t - e^{-t} \sin t \\ y' = -e^{-t} \sin t + e^{-t} \cos t \end{cases}$

$\Rightarrow (x')^2 + (y')^2 = \left(e^{-2t} \cos^2 t + e^{-t}e^{-t} \sin t \cos t + e^{-2t} \sin^2 t\right) + \left(e^{-2t} \sin^2 t - e^{-t}e^{-t} \cos t \sin t + e^{-2t} \cos^2 t\right)$

$= 2e^{-2t}(\cos^2 t + \sin^2 t) = 2e^{-2t}$

$L = \displaystyle\int_0^{\pi/2} \sqrt{2e^{-2t}}\, dt = \sqrt{2}\int_0^{\pi/2} e^{-t}\, dt = \sqrt{2}\left[-e^{-t}\right]_0^{\pi/2} = \sqrt{2}\left(-e^{-\pi/2} + 1\right)$

Chapter 8. Finding the Area between Two Curves

A. FINDING THE AREA BETWEEN TWO CURVES 1

DEFINITION.
Calculating the area of a region between two curves using vertical rectangles

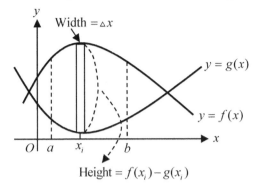

The area of the rectangle $= \left[f(x_i) - g(x_i) \right] \Delta x$

If f and g are continuous on $[a, b]$, then

The area of the given interval $[a, b] = \lim\limits_{n \to \infty} \sum\limits_{i=1}^{n} [f(x_i) - g(x_i)] \Delta x = \int_a^b [f(x) - g(x)] dx$

▶ **Example**

Find the area between two curves.

$$f(x) = -x^2 + 6x \text{ and } g(x) = 0$$

$f(x) = -x^2 + 6x$ and $g(x) = 0$

Point of intersection: $-x^2 + 6x = 0 \Rightarrow -x(x-6) = 0 \Rightarrow x = 0$ and $x = 6$

$$A = \int_0^6 \left(-x^2 + 6x\right) dx = \left[-\frac{x^3}{3} + 3x^2 \right]_0^6 = (-72 + 108) - (0) = 36$$

Dr. John Chung's AP Calculus AB/BC

Chapter 8. Finding the Area between Two Curves

▶ **Example**

Find the area between two curves.

$$f(x) = x^2 - 2x + 1 \text{ and } g(x) = x - 1$$

Point of intersection: $x^2 - 2x + 1 = x - 1 \implies x^2 - 3x + 2 = 0 \implies (x-2)(x-1) = 0 \implies x = 1, 2$

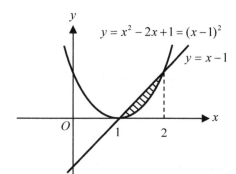

$$A = \int_1^2 \left[(x-1) - (x^2 - 2x + 1)\right] dx = \int_1^2 (3x - 2 - x^2) dx = \left[\frac{3x^2}{2} - 2x - \frac{x^3}{3}\right]_1^2$$

$$= \left(6 - 4 - \frac{8}{3}\right) - \left(\frac{3}{2} - 2 - \frac{1}{3}\right) = \frac{1}{6}$$

Chapter 8. Finding the Area between Two Curves

▶ **Example**

Find the area between two curves.

$$f(x) = (x-1)^3 \text{ and } g(x) = x-1$$

$f(x) = (x-1)^3$ and $g(x) = x-1$

Point of intersection: $(x-1)^3 = x-1 \Rightarrow (x-1)^3 - (x-1) = 0 \Rightarrow (x-1)\left[(x-1)^2 - 1\right] = 0$

$(x-1)x(x-2) = 0 \Rightarrow x = 0, 1, 2$

$$A = \int_0^1 \left[(x-1)^3 - (x-1)\right] dx + \int_1^2 \left[(x-1) - (x-1)^3\right] dx = \left[\frac{(x-1)^4}{4} - \frac{(x-1)^2}{2}\right]_0^1 + \left[\frac{(x-1)^2}{2} - \frac{(x-1)^4}{4}\right]_1^2$$

$= \dfrac{1}{4} + \dfrac{1}{4} = \dfrac{1}{2}$ or $A = 2\int_0^1 \left[(x-1)^3 - (x-1)\right] dx$ ∵ The shaded regions are equal.

Chapter 8. Finding the Area between Two Curves

B. FINDING THE AREA BETWEEN TWO CURVES 2

DEFINITION. Calculating the area of a region between two curves using horizontal rectangles

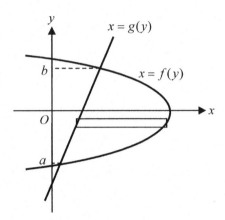

The area of the rectangle $= [f(y) - g(y)] \Delta y$

If f and g are continuous on $[a, b]$, then

The area of the given interval $[a,b] = \lim\limits_{n \to \infty} \sum\limits_{i=1}^{n} [f(y_i) - g(y_i)] \Delta y = \int_{a}^{b} [f(y) - g(y)] dy$

▶ **Example**

Find the area between two curves.

$$f(y) = y^2 \text{ and } g(y) = y + 2$$

$f(y) = y^2$ and $g(y) = y + 2$

Points of intersection: $y^2 = y + 2 \implies y^2 - y - 2 = 0 \implies (y-2)(y+1) = 0 \implies y = 2, -1$

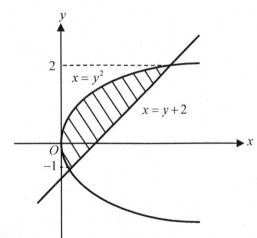

Area $= \int_{-1}^{2} [(y+2) - y^2] dy = \left[\dfrac{y^2}{2} + 2y - \dfrac{y^3}{3} \right]_{-1}^{2} = \left(2 + 4 - \dfrac{8}{3} \right) - \left(\dfrac{1}{2} - 2 + \dfrac{1}{3} \right) = \dfrac{9}{2}$

Chapter 8. Finding the Area between Two Curves

▶ **Example**

Find the area between two curves.

$$x = y^2 - 4y \text{ and } x = y$$

$x = y^2 - 4y$ and $x = y$

Point of intersection: $y^2 - 4y = y \implies y^2 - 5y = 0 \implies y(y-5) = 0 \implies y = 0, 5$

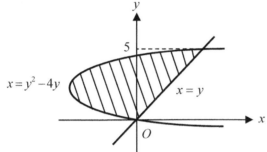

$$\text{Area} = \int_0^5 \left[y - (y^2 - 4y) \right] dy = \int_0^5 (5y - y^2) dy = \left[\frac{5y^2}{2} - \frac{y^3}{3} \right]_0^5 = \left(\frac{125}{2} - \frac{125}{3} \right) - (0) = \frac{125}{6}$$

Chapter 9. Finding the Volume of a Solid

A. The Volume of a Solid of Revolution

DEFINITION. If the region between two curves is rotated about a line, the resulting solid is called a solid of revolution and the line is called the axis of revolution. There are three methods for finding the volume of a solid of revolution.

1) The Disc Method 2) The Washer Method 3) The Cylindrical Shells Method

1. The Disc Method

a) Using the Disc method with a horizontal axis of revolution;

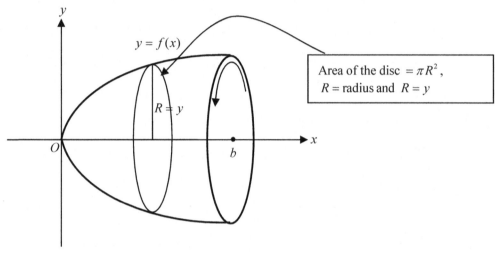

If a region under the graph of $y = f(x)$ is revolved about x-axis, the resulting solid is a solid of revolution. In the figure above, axis of revolution is x-axis.

The volume of the solid $V = \pi \int_a^b R^2 \, dx = \pi \int_a^b y^2 \, dx = \pi \int_a^b [f(x)]^2 \, dx$

b) Using the Disc method with a vertical axis of revolution;

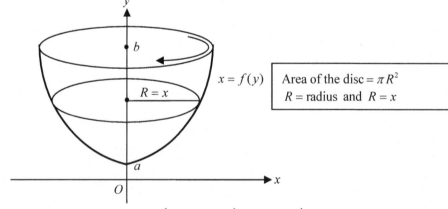

The volume of the solid $V = \pi \int_a^b R^2 \, dy = \pi \int_a^b x^2 \, dy = \pi \int_a^b [f(y)]^2 \, dy$

Chapter 9. Finding the Volume of a Solid

▶ **Example**

1. Let R be the region enclosed by $y = \tan x$, the x-axis, and $x = \dfrac{\pi}{3}$. Find the volume of the solid formed by revolving the region bounded by the graphs about the x-axis.

2. Let R be the region enclosed by $y = \sqrt{x}$, y-axis, and $y = 3$. Find the volume of the solid formed by revolving the region bounded by the graphs about the y-axis.

3. Let R be the region in the first quadrant enclosed by the hyperbola $x^2 - y^2 = 9$, the x-axis, and $x = 5$. Find the volume of the solid generated by revolving R about x-axis.

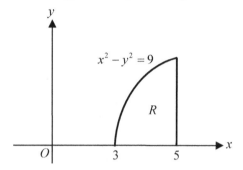

1. $V = \pi \displaystyle\int_0^{\pi/3} \tan^2 x \, dx = \pi \displaystyle\int_0^{\pi/3} (\sec^2 x - 1) \, dx = \pi [\tan x - x]_0^{\pi/3} = \pi \left[\left(\tan \dfrac{\pi}{3} - \dfrac{\pi}{3} \right) - (\tan 0 - 0) \right]$
$= \pi \left(\sqrt{3} - \dfrac{\pi}{3} \right) - (0) = \pi \left(\sqrt{3} - \dfrac{\pi}{3} \right)$

2. $V = \pi \displaystyle\int_0^3 x^2 \, dy = \pi \displaystyle\int_0^3 y^4 \, dy = \pi \left[\dfrac{y^5}{5} \right]_0^3 = \dfrac{243}{5} \pi$

3. $V = \pi \displaystyle\int_3^5 y^2 \, dx = \pi \displaystyle\int_3^5 (x^2 - 9) \, dx = \pi \left[\dfrac{x^3}{3} - 9x \right]_3^5 = \pi \left[\left(\dfrac{125}{3} - 45 \right) - (9 - 27) \right] = \dfrac{44\pi}{3}$

Chapter 9. Finding the Volume of a Solid

B. Finding the Volume of a Solid of Revolution using the Washer Method

DEFINITION. The Washer Method

a) Using the washer method with a horizontal axis of revolution

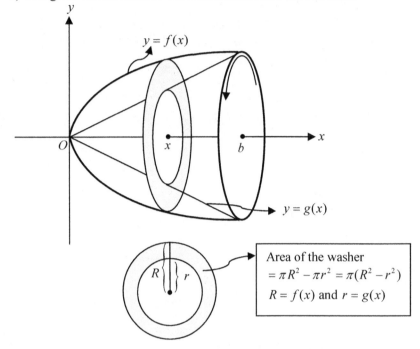

Area of the washer
$= \pi R^2 - \pi r^2 = \pi(R^2 - r^2)$
$R = f(x)$ and $r = g(x)$

The volume of the solid $V = \pi \int_0^b \left[R^2 - r^2 \right] dx = \pi \int_0^b \left[f^2(x) - g^2(x) \right] dx$

b) Using the washer method with a vertical axis of revolution

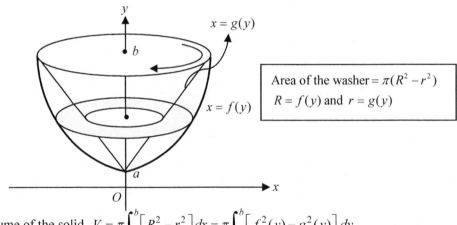

Area of the washer $= \pi(R^2 - r^2)$
$R = f(y)$ and $r = g(y)$

The volume of the solid $V = \pi \int_a^b \left[R^2 - r^2 \right] dx = \pi \int_0^b \left[f^2(y) - g^2(y) \right] dy$

Chapter 9. Finding the Volume of a Solid

▶ **Example**

1. Let R be the region enclosed by the graphs of $y = (27x)^{\frac{1}{4}}$ and $y = x$.
   Find the volume of the solid generated when region R is revolved about the x-axis.

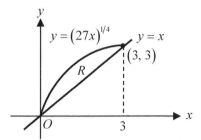

2. Let R be the region in the first quadrant enclosed by the graph of $y = \sqrt{6x+4}$, the line $y = 2x$, and the y-axis. Find the volume of the solid generated when R is revolved about the x-axis.

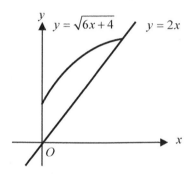

1. $V = \pi \int_0^3 \left(\left[(27x)^{1/4} \right]^2 - x^2 \right) dx = \pi \int_0^3 \left(\left[(27x)^{1/4} \right]^2 - x^2 \right) dx = \pi \int_0^3 \left((27x)^{1/2} - x^2 \right) dx = \pi \left[\frac{\sqrt{27} \cdot x^{\frac{3}{2}}}{3/2} - \frac{x^3}{3} \right]_0^3$

   $= \pi \left[3\sqrt{3} \cdot x^{3/2} - \frac{x^3}{3} \right]_0^3 = \pi \left(3\sqrt{3} \cdot 3^{3/2} - \frac{27}{3} \right) = \pi(18 - 9) = 9\pi$

2. Point of intersection:
   $\sqrt{6x+4} = 2x \Rightarrow 6x + 4 = 4x^2 \Rightarrow 2x^2 - 3x - 2 = (2x+1)(x-2) = 0 \Rightarrow x = -1/2, 2$
   Point $(2, 4)$
   $V = \pi \int_0^2 \left[\left(\sqrt{6x+4} \right)^2 - (2x)^2 \right] dx = \pi \int_0^2 \left(6x + 4 - 4x^2 \right) dx = \pi \left[3x^2 + 4x - \frac{4}{3}x^3 \right]_0^2$

   $= \pi \left[\left(12 + 8 - \frac{32}{3} \right) - (0) \right] = \frac{28\pi}{3}$

Dr. John Chung's AP Calculus AB/BC

Chapter 9. Finding the Volume of a Solid

C. Finding the Volume of a Solid of Revolution using the Cylindrical Shell Method

DEFINITION. The Cylindrical Shell Method

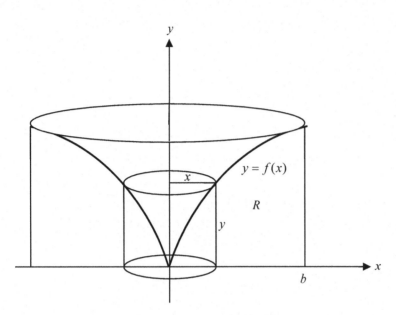

Let R be the region in the first quadrant enclosed by the graph of $y = f(x)$, the line $x = b$, and the x-axis. The volume of the solid generated when R is revolved about the y-axis is

$$V = 2\pi \int_0^b xy\,dx$$

Volume of the cylindrical shell = $2\pi xy \Delta x$

118

Chapter 9. Finding the Volume of a Solid

▶ **Example**

Let R be the region in the first quadrant enclosed by the graph of $y = \sqrt{x}$, the x-axis, and the line $x = 4$. Find the volume of the solid generated by revolving about the y-axis.

(i) Using the washer method

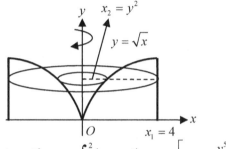

$$V = \pi \int_0^2 \left(x_1^2 - x_2^2\right) dy = \pi \int_0^2 \left(16 - y^4\right) dy = \pi \left[16y - \frac{y^5}{5}\right]_0^2 = \pi \left(32 - \frac{32}{5}\right) = \frac{128\pi}{5}$$

(ii) Using the cylindrical method

$f(4) = \sqrt{4} = 2$

$$V = 2\pi \int_0^4 xy\, dx = 2\pi \int_0^4 x\sqrt{x}\, dx = 2\pi \int_0^4 x^{3/2} dx = 2\pi \left[\frac{2}{5}x^{5/2}\right]_0^4 = 2\pi \left(\frac{2}{5} \cdot 4^{5/2}\right) = \frac{128\pi}{5}$$

Dr. John Chung's AP Calculus AB/BC

Chapter 9. Finding the Volume of a Solid

▶ **Example**

Let R be the region in the first quadrant enclosed by the graph of $y = x - x^3$ and the x-axis. Find the volume of the solid generated by revolving about the y-axis.

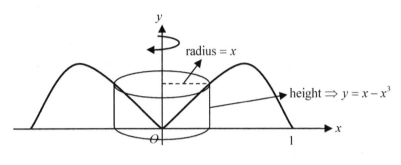

Point of intersection: $x - x^3 = 0 \Rightarrow x(1 - x^2) = 0 \Rightarrow x = 0, x = \pm 1$

$$V = 2\pi \int_0^1 x(x - x^3)\,dx = 2\pi \left[\frac{x^3}{3} - \frac{x^5}{5} \right]_0^1 = 2\pi \left(\frac{1}{3} - \frac{1}{5} \right) = \frac{4\pi}{15}$$

▶ **Example**

Let R be the region in the first quadrant enclosed by the graph of $y = \sqrt{6x + 4}$, the line $y = 2x$, and the y-axis. Set up, but do not integrate, an integral expression in terms of a single variable for the volume of the solid generated when R is revolved about the y-axis.

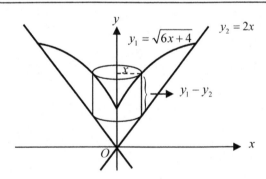

Point of intersection: $\sqrt{6x + 4} = 2x \Rightarrow 6x + 4 = 4x^2 \Rightarrow 2x^2 - 3x - 2 = 0 \Rightarrow (2x + 1)(x - 2) = 0$

$x = 2$

$$V = 2\pi \int_a^b x(y_1 - y_2)\,dx = 2\pi \int_0^2 x\left(\sqrt{6x + 4} - 2x\right)dx$$

Chapter 9. Finding the Volume of a Solid

▶ **Example**

Let R be the region in the first quadrant enclosed by the hyperbola $x^2 - y^2 = 9$, the x-axis, and the line $x = 5$. Setup, but do not integrate, an integral expression in terms of a single variable for the volume of the solid generated when R is revolved about the line $x = -1$.

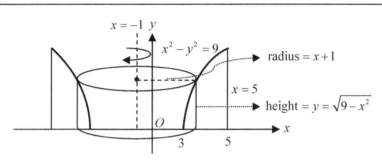

Point of intersection: $x^2 - y^2 = 9$ and $x = 5$ $\Rightarrow$ $25 - y^2 = 9$ $\Rightarrow$ $y = 4$

$y = \sqrt{9 - x^2}$ $\qquad V = 2\pi \int_3^5 (x+1)\sqrt{x^2 - 9}\, dx$

Chapter 9. Finding the Volume of a Solid

D. Volumes of Solids with Known Cross Sections

DEFINITION. We can use the definite integral to find the volume of a solid with specific cross sections on an interval, provided you know a formula for the region determined by each cross section. If the cross sections generated are perpendicular to the x-axis, then their areas will be functions of x, denoted by A. The volume of the solid on the interval $[a, b]$ is

1) For the cross section perpendicular to the x-axis,
$$V = \int_a^b A(x)dx$$

2) For the cross section perpendicular to the y-axis,
$$V = \int_a^b A(y)dy$$

Some common cross sections are squares, rectangles, triangles, semicircles, and trapezoids.

▶ **Example**

The volume of the solid whose base is the region of the circle $x^2 + y^2 = 4$ whose cross sections taken perpendicular to the x-axis are squares.

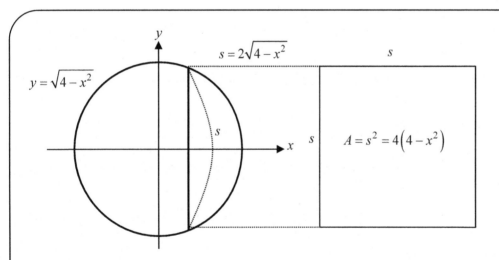

The area of the square $A(x) = s^2 = \left(2\sqrt{4-x^2}\right)^2$

$$V = \int_{-2}^{2} A(x)dx = 2\int_0^2 A(x)dx = 2\int_0^2 \left(2\sqrt{4-x^2}\right)^2 dx = 8\int_0^2 (4-x^2)dx = 8\left[4x - \frac{x^3}{3}\right]_0^2$$

$$= 8\left(8 - \frac{8}{3}\right) = \frac{128}{3}$$

Chapter 9. Finding the Volume of a Solid

▶ Example

Let R be the region in the first quadrant enclosed by $y = \sin x$, $y = \cos x$, and $x = 0$. Find the volume of the solid generated whose base is the region R and whose cross sections, perpendicular to the x-axis, are squares.

Point of intersection: $\sin x = \cos x \implies \tan x = 1 \implies x = \pi/4$

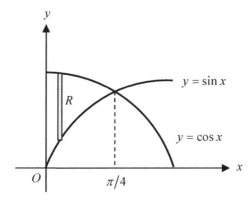

$$V = \int_0^{\pi/4} (\cos x - \sin x)^2 \, dx = \int_0^{\pi/4} (1 - 2\sin x \cos x) \, dx = \int_0^{\pi/4} (1 - \sin 2x) \, dx = \left[x + \frac{\cos 2x}{2} \right]_0^{\pi/4}$$

$$= \left(\frac{\pi}{4}\right) - \left(\frac{1}{2}\right) = \frac{\pi - 2}{4}$$

Chapter 10. Differential Equations

A. Definitions and Basic Concepts

DEFINITION
A differential equation is an equation involving an unknown function and one or more its derivatives.

Example of first order differential equation: $xy' - 2y = 0$, $\dfrac{dy}{dx} = 3x^2$, $\dfrac{dy}{dx} = \dfrac{y}{x}$

Example of second order differential equation: $y'' + 2y' + y = 1$

The **order** is determined by the **highest-order derivative** in the equation.

General Solutions and Particular Solutions:
a) A **general solution** is a solution of a differential equation with its **arbitrary constant** C undetermined.
b) A **particular solution** is a solution of a differential equation with its constant C determined from an initial condition.

▶ **Example** First order differential equation

1. Find the **general solution** of $\dfrac{dy}{dx} = 2x^2$.

2. Find the particular solution of $\dfrac{dy}{dx} = 2x^2$ with the initial condition $y = 10$ when $x = 3$.

Solution

1. $y = \displaystyle\int 2x^2\, dx = \dfrac{2}{3}x^3 + C$

2. For initial value $(3, 10)$, $10 = \dfrac{2}{3}(3^3) + C \;\to\; 10 = 18 + C \;\to\; C = -8$

   Therefore, particular solution is $y = \dfrac{2}{3}x^2 + 4$

▶ **Example**

1. Show that $y = x + 1 + ce^x$ is a solution to $\dfrac{dy}{dx} = y - x$.

2. Find a particular solution for initial value $y(1) = 2$.

Solution

$\dfrac{dy}{dx} = 1 + ce^x$ (substitute in differential equation) $\to 1 + ce^x = \left[y - x = (x + 1 + ce^x) - x\right] = 1 + ce^x$

Therefore, $y = x + 1 + ce^x$ is a solution to $y' = y - x$.

$y(0) = 2$ means $(0, 2) \;\to\; 2 = 0 + 1 + c \;\to\; c = 1$ Therefore, the particular solution is

$y = x + 1 + e^x$. (Arbitary constant c is determined)

Chapter 10. Differential Equations

B. Separable Differential Equations

Definition
One of the popular techniques for solving differential equations is the Separation of Variables.

Note There are many types of differential equations, but only differential equations using the separation of variables will be on the test. I will skip putting homogeneous and non-homogeneous differential equations in this book.

Note Separable differential equations are usually written as a product of two functions or two term of each variable as follows.

$$1)\ \frac{dy}{dx} = f(x)g(y) \qquad 2)\ f(x)dx + g(y)dy = 0$$

Example of separable differential equations:

$$1)\ (x^2+1)dx + \frac{1}{y}dy = 0 \qquad 2)\ \frac{dy}{dx} = ky \qquad 3)\ \frac{dy}{dx} = xy^2$$

▶ **Example**

Find the general solution to differential equation $(x^2+1)\frac{dy}{dx} = xy$.

Solution
Step 1) Separate the Variables, putting all the y's on one side and the x's on the other side.

$$(x^2+1)\frac{dy}{dx} = xy \;\to\; \frac{1}{y}\frac{dy}{dx} = \frac{x}{(x^2+1)}$$

Step 2) Take the integral.

$$\int\left(\frac{1}{y}\frac{dy}{dx}\right)dx = \int\left(\frac{x}{x^2+1}\right)dx \;\to\; \int\left(\frac{1}{y}\frac{dy}{dx}\right)dx = \int\left(\frac{x}{x^2+1}\right)dx \;\to\; \int\frac{1}{y}dy = \int\left(\frac{x}{x^2+1}\right)dx$$

Step 3) Solve.

$$\ln|y| = \frac{1}{2}\ln(x^2+1) + C_1 = \ln\sqrt{x^2+1} + C_1 = \ln\sqrt{x^2+1} + \ln e^{C_1} = \ln\left(e^{C_1}\sqrt{x^2+1}\right)$$

$$\ln|y| = \ln\left(e^{C_1}\sqrt{x^2+1}\right)$$

$$\therefore y = \pm e^{C_1}\sqrt{x^2+1} \;\to\; \left(C = \pm e^{C_1}\right)$$

Because $\pm e^{C_1}$ is a constant, you can write a general solution as

$$y = C\sqrt{x^2+1}$$

Chapter 10. Differential Equations

▶ **Example**

Solve the differential equation
$$\frac{dy}{dx} = \frac{y}{x}$$

Solution

$\dfrac{1}{y}\dfrac{dy}{dx} = \dfrac{1}{x} \;\to\; \int\left(\dfrac{1}{y}\dfrac{dy}{dx}\right)dx = \int\left(\dfrac{1}{x}\right)dx \;\to\; \int\dfrac{1}{y}dy = \int\dfrac{1}{x}dx$

$\ln|y| = \ln|x| + C_1 \;\to\; \ln|y| = \ln|x| + \ln e^{C_1} = \ln|C_2 x|, \;\;(C_2 = e^{C_1})$ Therefore,

$\ln|y| = \ln|C_2 x| \;\to\; |y| = |C_2 x| \;\to\; y = \pm C_2 x \;\to\; y = Cx, \;\;(C = \pm C_2)$

▶ **Example**

Solve the differential equation
$$\begin{cases} \dfrac{dy}{dx} = \left(\dfrac{x}{x^2+1}\right) y \\ y(0) = 5 \end{cases}$$

Solution

$\dfrac{dy}{dx} = \left(\dfrac{x}{x^2+1}\right)y \;\to\; \dfrac{1}{y}\dfrac{dy}{dx} = \left(\dfrac{x}{x^2+1}\right) \;\to\; \int\dfrac{1}{y}dy = \int\left(\dfrac{x}{x^2+1}\right)dx \;\to\; \ln|y| = \dfrac{1}{2}\ln|x^2+1| + C_1$

$\ln|y| = \ln(x^2+1)^{1/2} + \ln e^{C_1} = \ln\left(e^{C_1}\sqrt{x^2+1}\right)$ So,

$y = \pm e^{C_1}\sqrt{x^2+1} \;\to\; y = C\sqrt{x^2+1}$ (Initial value) $5 = C\sqrt{1} \;\to\; C = 5$

Therefore, $y = 5\sqrt{x^2+1}$.

Chapter 10. Differential Equations

▶ **Example**

Solve the differential equation

$$\begin{cases} ye^x dx - (1-y^2)dy = 0 \\ y(0) = 1 \end{cases}$$

Solution

$ye^x dx - (1-y^2)dy = 0 \;\rightarrow\; (1-y^2)\dfrac{dy}{dx} = ye^x \;\rightarrow\; \int \left(\dfrac{1-y^2}{y}\right)dy = \int e^x dx$

$\int \left(\dfrac{1-y^2}{y}\right)dy = \int e^x dx \;\rightarrow\; \int \left(\dfrac{1}{y} - y\right)dy = \int e^x dx \;\rightarrow\; \ln|y| - \dfrac{y^2}{2} = e^x + C_1$

(For better looking, multiply by 2)

$2\ln|y| - y^2 = 2e^x + 2C_1 \;\rightarrow\; \ln y^2 - y^2 = 2e^x + C$, $(C = 2C_1)$ (General solution)

Initial value: $\ln 1 - 1 = 2 + C \;\rightarrow\; C = -3$

Therefore, $\ln y^2 - y^2 = 2e^x - 3$. (Particular solution)

Chapter 10. Differential Equations

▶ **Example**

Solve. $\begin{cases} xy' = y\ln x \\ y(1) = e \end{cases}$

Solution

$x\dfrac{dy}{dx} = y\ln x \;\rightarrow\; \dfrac{1}{y}\dfrac{dy}{dx} = \dfrac{\ln x}{x} \;\rightarrow\; \displaystyle\int \dfrac{1}{y}dy = \int \dfrac{\ln x}{x}dx$

$\displaystyle\int \dfrac{1}{y}dy = \ln|y|$

For $\displaystyle\int \dfrac{\ln x}{x}dx$, $u = \ln x \;\rightarrow\; du = \dfrac{1}{x}dx$ Therefore, $\displaystyle\int \dfrac{\ln x}{x}dx = \int u\,du = \dfrac{u^2}{2} = \dfrac{(\ln x)^2}{2}$

$\ln|y| = \dfrac{(\ln x)^2}{2} + C_1 \;\rightarrow\; 2\ln|y| = (\ln x)^2 + 2C_1 \;\rightarrow\; \ln y^2 = (\ln x)^2 + C$

Initial value: $\ln e^2 = (\ln 1)^2 + C \;\rightarrow\; C = 2$

Therefore, $\ln y^2 = (\ln x)^2 + 2$.

Also, you can change the form as follows.

$\ln y^2 = (\ln x)^2 + C_2 \;\rightarrow\; e^{\ln y^2} = e^{(\ln x)^2 + C_2} = e^{C_2}e^{(\ln x)^2} = C_1 e^{(\ln x)^2}$

$y^2 = C_1 e^{(\ln x)^2} \;\rightarrow\; y = \pm\sqrt{C_1}\left(e^{(\ln x)^2}\right)^{1/2} = Ce^{\frac{(\ln x)^2}{2}}$

Initial value: $2 = Ce^0 \;\rightarrow\; C = 2$

Therefore, another expression is $y = 2e^{\frac{(\ln x)^2}{2}}$.

Chapter 10. Differential Equations

C. Exponential Growth and Decay

Definition

Usually y represents a "**population**."

When the rate of change of a variable y with respect to time is proportional to the size of y.

$$\begin{cases} \dfrac{dy}{dt} = ky \\ y(0) = y_0 \quad \text{(Initial population)} \end{cases}$$

The general solution is

$$\dfrac{dy}{dt} = ky \;\to\; \dfrac{1}{y}\dfrac{dy}{dt} = k \;\to\; \int\left(\dfrac{1}{y}\dfrac{dy}{dt}\right)dt = \int k\,dt \;\to\; \int \dfrac{1}{y}dy = \int k\,dt$$

$$\ln|y| = kt + C_1 \;\to\; e^{\ln|y|} = e^{kt+C_1} = e^{C_1}e^{kt} = C_2 e^{kt} \;\to\; |y| = C_2 e^{kt}$$

$$y = \pm C_2 e^{kt} \;\to\; y = Ce^{kt} \quad (C = \pm C_2)$$

Initial Value: $y_0 = Ce^{k(0)} \;\to\; y_0 = C$

Therefore, $y = y_0 e^{kt}$ is the general solution.

a) Exponential growth occurs: $\dfrac{dy}{dt} = ky$ $(k > 0)$ $\to$ Solution is $y = Ce^{kt}$

b) Exponential decay occurs: $\dfrac{dy}{dt} = -ky$ $(k > 0)$ $\to$ Solution is $y = Ce^{-kt}$

▶ **Example**

The rate of growth of population of flies is proportional to the size of population. In an experiment, it was observed that there were 200 flies after the second day and 1000 flies after the fourth day. How many flies were there in the original population?

Solution

Exponential growth: $y = y_0 e^{kt}$ (at t days)

At $t = 2$, $200 = y_0 e^{2k}$

At $t = 4$, $1000 = y_0 e^{4t}$

$$\dfrac{1000}{200} = \dfrac{y_0 e^{4k}}{y_0 e^{2k}} = e^{2k} \;\to\; 5 = e^{2k} \;\to\; (5)^{1/2} = (e^{2k})^{1/2} \;\to\; \sqrt{5} = e^k \quad \text{(Put in the equation)}$$

$y = y_0 e^{kt} \;\to\; y = y_0 (\sqrt{5})^t$

Now, find y_o from $200 = y_0(\sqrt{5})^2 \;\to\; 200 = 5y_0 \;\to\; y_0 = 40$

Chapter 10. Differential Equations

▶ **Example**

The rate of growth of bacteria is proportional to its population. Initially there are 400 bacteria, and 10,000 bacteria after 4 hours. Find the equation of the population with respect to time T (hours).

Solution

We start with $y = y_0 e^{kT}$ and $y_0 = 400$. $\rightarrow$ $y = 400 e^{kT}$

At $T = 4$, $y = 10,000$ $\rightarrow$ $10000 = 400 e^{4k}$ $\rightarrow$ $25 = e^{4k}$ $\rightarrow$ $e^k = 25^{1/4}$

Therefore, $y = 400 \left(25^{1/4}\right)^T$ $\rightarrow$ $y = 400 (25)^{T/4}$

Alternately,

$25 = e^{4k}$ $\rightarrow$ $4k = \ln 25$ $\rightarrow$ $k = \dfrac{\ln 25}{4} = \ln 25^{1/4}$

Therefore, $y = 400 e^{kT} = 400 e^{\left(\ln 25^{1/4}\right)T} = 400 e^{\ln(25)^{T/4}} = 400(25)^{T/4}$ (same as before)

D. Radioactive Decay (Half-Life)

Definition

Radioactive Decay is the process by which an atomic nucleus of an unstable atom loses energy by emitting ionizing particles. Radioactive Decay is measured in terms of Half-Life.

Define: y = Amount of a radioactive material, y_0 = Initial value, and half-life = H

In Radioactive decay,

"The rate of decay is proportional to y."

Therefore, the general solution is $y = y_0 e^{kt}$. Because the number years (H) is required for the half of the radioactive material to decay.

$\dfrac{1}{2} y_0 = y_0 e^{kH}$ $\rightarrow$ $\dfrac{1}{2} = e^{kH}$ $\rightarrow$ $\left(e^{kH}\right)^{1/He} = \left(\dfrac{1}{2}\right)^{1/H}$ $\rightarrow$ $e^k = \left(\dfrac{1}{2}\right)^{1/H}$

Substitute into the general solution.

$y = y_0 \left[\left(\dfrac{1}{2}\right)^{1/H}\right]^t$ $\rightarrow$ $y = y_0 \left(\dfrac{1}{2}\right)^{\frac{t}{H}}$

▶ **Example**

A radioactive material has a half-life of 1000 years. How long will it take for 100 grams of the material to decay to 10 grams?

Chapter 10. Differential Equations

> Solution
> We know that $y = y_0 \left(\dfrac{1}{2}\right)^{\frac{t}{H}}$. So,
>
> $10 = 100\left(\dfrac{1}{2}\right)^{\frac{t}{1000}} \;\to\; \dfrac{1}{10} = \left(\dfrac{1}{2}\right)^{\frac{t}{1000}} \;\to\; \ln\dfrac{1}{10} = \dfrac{t}{1000}\ln\dfrac{1}{2} \;\to\; t = \dfrac{1000\ln\left(\dfrac{1}{100}\right)}{\ln\left(\dfrac{1}{2}\right)} = 6643.86$ years

E. Newton's Law of Cooling

Definition

Let y be the temperature, in ^oF, of an object in a room whose temperature is kept constant at p. Newton's Law of Cooling states that the rate of change in y is proportional to the difference between y and p. This can be written as follows.

$$\dfrac{dy}{dt} = k(y-p), \text{ where } y > p.$$

The general solution is

$$\dfrac{1}{y-p}\dfrac{dy}{dt} = k \;\to\; \int\left(\dfrac{1}{y-p}\dfrac{dy}{dt}\right)dt = \int k\, dt \;\to\; \ln(y-p) = kt + C_1 \;\to\; e^{\ln(y-p)} = e^{kt+C_1}$$

$$y - p = e^{C_1}e^{kt} \;\to\; y = p + Ce^{kt}, \; (C = e^{C_1})$$

▶ **Example**

When an object is removed from a furnace and placed in a room with a constant temperature of $60^o F$, its core temperature is $1000^o F$. One hour after it is removed, the core temperature is $800^o F$. Find the core temperature 10 hours after it is removed from the furnace.

> Solution
> We can see $y - p = Ce^{kt}$ and $p = 60 \;\to\; y - 60 = Ce^{kt}$
>
> $y(0) = 1000 \;\to\; 1000 - 60 = Ce^0 \;\to\; C = 940$ Then $y = 60 + 940e^{kt}$
>
> At $t = 1$, $800 - 60 = 940e^{kt} \;\to\; 740 = 940e^k \;\to\; \dfrac{740}{940} = e^k \;\to\; e^k = \dfrac{37}{47}$
>
> Therefore, $y = 60 + 940\left(\dfrac{37}{47}\right)^t$. So, $y(10) = 60 + 940\left(\dfrac{37}{47}\right)^{10} \equiv 145.9$

Dr. John Chung's AP Calculus AB/BC

Chapter 10. Differential Equations

F. Logistic Growth

Definition

In population models, the size of population approaches a positive constant M, called the **carrying capacity** of the system. One model with this property is provided by the logistic differential equation.

$$\frac{dP}{dt} = kP\left(1 - \frac{P}{M}\right) \qquad \text{As } P \to M, \ \frac{dP}{dt} = 0$$

Where P = population $\quad k$ = constant $\quad M$ = carrying capacity $\quad t$ = time

Constrained natural growth is called **logistic growth**.

The solution to the **logistic differential equation** is

$$P = \frac{M}{1 + Ae^{-kt}}$$

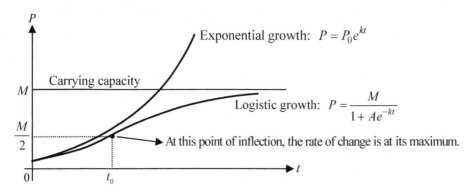

At this point of inflection, the rate of change is at its maximum.

Remember

Logistic differential equation: $\quad \dfrac{dP}{dt} = kP\left(1 - \dfrac{P}{M}\right)$

Solution to the differential equation: $\quad P = \dfrac{M}{1 + Ae^{-kt}}$

Remember

1. At time $t = 0$ the initial value of P is
$$P(0) = \frac{M}{1+A}$$

2. At time $t_0 = \dfrac{\ln A}{k}$ the population is growing the fastest, where $P = \dfrac{M}{2}$

3. Point $\left(t_0, \dfrac{M}{2}\right)$ is the inflection point.

4. $\lim\limits_{t \to \infty} P(t) = m$ (Carrying capacity).

5. $\lim\limits_{t \to \infty} \dfrac{dP}{dt} = 0$

Chapter 10. Differential Equations

▶ **Example**

Solve the logistic differential equation.

$$\frac{dP}{dt} = kP\left(1 - \frac{P}{M}\right), \text{ where } k \text{ is a constant and } M \text{ is carrying capacity.}$$

Solution

$\frac{dP}{dt} = kP\left(1 - \frac{P}{M}\right) = kP\left(\frac{M-P}{M}\right) = k\left(\frac{P(M-P)}{M}\right)$, We got $\frac{M}{P(M-P)} \frac{dP}{dt} = k$

$\int \left[\frac{M}{P(M-P)} \frac{dP}{dt}\right] dt = \int k \, dt \rightarrow \int \left(\frac{1}{P} + \frac{1}{M-P}\right) dP = \int k \, dt$ So,

$\ln P - \ln(M-P) = kt$, (P is positive and $M > P$) We don't need absolute.

$\ln\left(\frac{P}{M-P}\right) = kt + C_1 \rightarrow e^{\ln\left(\frac{P}{M-P}\right)} = e^{kt+C_1} \rightarrow \frac{P}{M-P} = e^{C_1} e^{kt} \rightarrow \frac{P}{M-P} = Ce^{kt}$

Now

$\frac{P}{M-P} = Ce^{kt} \rightarrow P = MCe^{kt} - PCe^{kt} \rightarrow P + PCe^{kt} = MCe^{kt} \rightarrow P(1 + Ce^{kt}) = MCe^{kt}$

$P = \frac{MCe^{kt}}{1 + Ce^{kt}}$ (We divide by e^{kt}) $P = \frac{MCe^{kt}/Ce^{kt}}{(1+Ce^{kt})/Ce^{kt}} = \frac{M}{1 + \frac{1}{Ce^{kt}}} = \frac{M}{1 + \left(\frac{1}{C}\right)e^{-kt}}$

Now let $\frac{1}{C} = A$, therefore, $P = \frac{M}{1 + Ae^{-kt}}$.

1) Initial population: $P(0) = \frac{M}{1 + A}$

2) Inflection point: $\frac{dP}{dt} = kP\left(1 - \frac{P}{M}\right) \rightarrow \frac{d^2P}{dt^2} = k\frac{dP}{dt}\left(1 - \frac{P}{M}\right) + kP\left(-\frac{1}{M}\right) = 0$

$\cancel{k}\frac{dP}{dt}\left(1 - \frac{P}{M}\right) + \cancel{k}P\left(-\frac{1}{M}\right) = 0 \rightarrow \frac{dP}{dt}\left(1 - \frac{P}{M} - \frac{P}{M}\right) = 0 \rightarrow \frac{dP}{dt}\left(1 - \frac{2P}{M}\right) = 0$

$\frac{dP}{dt} \neq 0, \rightarrow 1 - \frac{2P}{M} = 0 \rightarrow 1 = \frac{2P}{M} \leftarrow P = \frac{M}{2}$

It means "Population is growing the fastest at $P = \frac{M}{2}$."

At $P = \frac{M}{2}$, $\frac{M}{2} = \frac{M}{1 + Ae^{-kt}} \rightarrow 2 = 1 + Ae^{-kt} \rightarrow 1 = Ae^{-kt} \rightarrow A = e^{kt} \rightarrow e^{\ln A} = e^{kt}$

Therefore, $kt = \ln A \rightarrow t = \frac{\ln A}{k}$

So, Inflection point: $\left(\frac{\ln A}{k}, \frac{M}{2}\right)$.

Chapter 10. Differential Equations

▶ **Example**

A bacterial culture is growing at the rate of $y = \dfrac{2.5}{1+0.25e^{-0.5t}}$, where y is the weight of the culture in grams and t is the time in hours. Find the weight of the culture at;

(a) $t = 0$ hour.   (b) $t = 1$ hour   (c) $t = 5$ hours   (d) $t = 10$ hours   (e) $t = \infty$

Solution

a) 2   b) 2.17   c) 2.45   d) 2.50   e) $t = \infty$, $\displaystyle\lim_{t \to \infty} \dfrac{2.5}{1+0.25e^{-\infty}} = 2.5$: Carrying capacity

---

▶ **Example**

A lake is stocked with 500 fish. If the population increases according to the logistic curve $y = \dfrac{10,000}{1+19e^{-t/5}}$, where y is the fish population and t is measured in months.

1. At what rate is the fish population changing at the end of one month?
2. After how many months is the population increasing the most rapidly?

Solution

1. We know that logistic growth arise from the equation.

$$y(0) = \dfrac{10000}{1+A} \to 500 = \dfrac{10000}{1+A} \to 1+A = 20 \to A = 19$$

$$\dfrac{dy}{dt} = \dfrac{-10,000\left[(-19/5)e^{-t/5}\right]}{\left(1+19e^{-t/5}\right)^2} = \dfrac{38000e^{-t/5}}{\left(1+19e^{-t/5}\right)^2}, \text{ so } \left.\dfrac{dy}{dx}\right|_{t=5} = \dfrac{38000e^{-1}}{\left(1+19e^{-1}\right)^2} = 219$$

2. At $t = \dfrac{\ln A}{k}$ the population is growing the fastest.

We got $A = 19$ and $k = \dfrac{1}{5}$, so $t = \dfrac{\ln 19}{1/5} = 5\ln 19 \approx 14.72$ months

At that time, $P(5\ln 19) = \dfrac{10000}{2} = 5000$, (Half of Carrying capacity)

Chapter 10. Differential Equations

▶ **Example**

Solve the differential equation
$$\frac{dy}{dt} = \frac{8}{25}y\left(\frac{5}{4} - y\right), \quad y(0) = 1.$$

Solution

We know that the differential equation represents a Logistic growth
$$\frac{dy}{dt} = ky\left(1 - \frac{y}{m}\right)$$

$$\frac{dy}{dt} = \frac{8}{25}y\left(\frac{5}{4} - y\right) \rightarrow \frac{dy}{dt} = \frac{8}{25}\left(\frac{5}{4}\right)y\left(1 - \frac{4}{5}y\right) \rightarrow \frac{dy}{dt} = \frac{2}{5}y\left(1 - \frac{y}{5/4}\right)$$

It follows that
$$ky\left(1 - \frac{y}{m}\right) = \frac{2}{5}y\left(1 - \frac{y}{5/4}\right)$$

Now we obtain
$$k = \frac{2}{5} \text{ and } M = \frac{5}{4}$$

The solution is in this form
$$y = \frac{M}{1 + Ae^{-kt}} \rightarrow y = \frac{5/4}{1 + Ae^{-\frac{2t}{5}}}$$

From the initial value $y(0) = 1$,
$$y(0) = \frac{5/4}{1 + A} = 1 \rightarrow A = \frac{1}{4}$$

The solution of the differential equation is
$$y = \frac{5/4}{1 + \frac{1}{4}e^{-\frac{2t}{5}}}$$

Chapter 10. Differential Equations

▶ **Example**

Growing of a population is modeled by the following function: (P increases according to the logistic differential equation)

$$\frac{dP}{dt} = \frac{4}{5}P\left(1 - \frac{P}{20}\right)$$

1. Find $P(t)$ if $P(0) = 5$.
2. What is $\lim_{t \to \infty} P(t)$?
3. For what value of P is the population growing the fastest?
4. For what value of t is the population growing the fastest?

Solution

1. From the equation
$$k = \frac{4}{5} \text{ and } M = 20$$

We have a solution form
$$P = \frac{M}{1 + Ae^{-kt}} \to P = \frac{20}{1 + Ae^{-\frac{4}{5}t}} \text{ and } P(0) = \frac{20}{1 + A}$$

From $P(0) = 5$
$$\frac{20}{1 + A} = 5 \to A = 3$$

Thus
$$P(t) = \frac{20}{1 + 3e^{-\frac{4t}{5}}}$$

2. $\lim_{t \to \infty} P(t) = 20$ Carrying capacity

3. At $P = \frac{M}{2}$, the population is growing the fastest.
$$P = \frac{M}{2} = \frac{20}{2} = 10$$

4. At $t = t_0 = \frac{\ln A}{k}$, the population is growing the fastest
$$t = \frac{\ln 3}{4/5} = \frac{5}{4}\ln 3 \approx 1.37$$

Chapter 10. Differential Equations

▶ **Example**

Let $y = f(t)$ be the particular solution to the logistic differential equation $\dfrac{dy}{dt} = \dfrac{y}{12}(6-y)$ with $f(0) = 10$.

a) Find $y = f(t)$.

b) Sketch possible solution curve through the point $(0, 10)$.

c) Find $\lim\limits_{t \to \infty} f(t)$ and $\lim\limits_{t \to \infty} f'(t)$.

d) What is the range of f for $t \geq 0$?

e) For what value of y does the graph of f have a point of inflection?

Solution
Logistic Decay

a) $\dfrac{dy}{dt} = \dfrac{y}{12}(6-y) \;\rightarrow\; \dfrac{dy}{dt} = \dfrac{y}{12}(6)\left(1 - \dfrac{y}{6}\right) \;\rightarrow\; \dfrac{dy}{dt} = \dfrac{1}{2}y\left(1 - \dfrac{y}{6}\right) \;\rightarrow\; k = \dfrac{1}{2},\; m = 6$

$y = \dfrac{6}{1 + Ae^{-\frac{1}{2}t}}$ and $y(0) = \dfrac{6}{1+A} = 10 \;\rightarrow\; A = -\dfrac{2}{5}$

Thus
$$y = \dfrac{6}{1 - \dfrac{2}{5}e^{-\frac{1}{2}t}}$$

b) The graph of f is a Logistic decay, since the initial value is greater than the carrying capacity that results negative value of A.

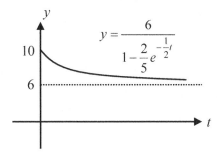

c) $\lim\limits_{t \to \infty} f(t) = 6$ and $\lim\limits_{t \to \infty} f'(t) = 0$

d) $6 \leq y \leq 10$

e) No inflection point

$\dfrac{d}{dt}\left(\dfrac{dy}{dt}\right) = \dfrac{d}{dt}\left(\dfrac{1}{2}y - \dfrac{1}{12}y^2\right) = \dfrac{1}{2}\dfrac{dy}{dt} - \dfrac{1}{6}y\dfrac{dy}{dt} \;\rightarrow\; \dfrac{dy}{dt}\left(\dfrac{1}{2} - \dfrac{y}{6}\right) = 0$

But $6 \leq y \leq 10$ and $\dfrac{dy}{dt} < 0$ (Graph is decreasing), $y'' > 0$

means the graph of y is concave up over entire x.

Chapter 10. Differential Equations

G. Euler's Method for Approximating the Solution of a Differential Equation

THEOREM. Euler's Method uses a linear approximation with increments, h, for solving differential equation with a given initial value.

Define: (1) step size $h = x_n - x_{n-1}$ (2) $f'(x_n) = y'_n$ (3) $(x_0, y_0) = (x_0, f(x_0))$ Initial value

First approximation $y_1 \approx y_0 + hy'_0$
Second approximation $y_2 \approx y_1 + hy'_1$
Third approximation $y_3 \approx y_2 + hy'_2$

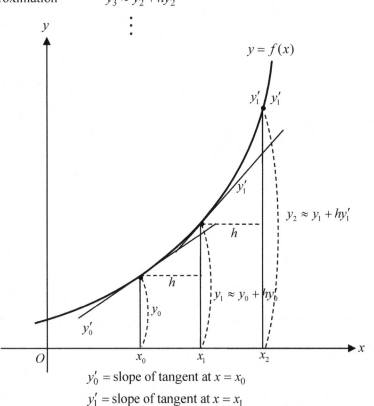

y'_0 = slope of tangent at $x = x_0$
y'_1 = slope of tangent at $x = x_1$

Euler method is just a procedure of repeating a linear approximation.

Note $h = \Delta x$

▶ **Example**

1. Use Euler's method with a step size of $h = 1/4$ (or $\Delta x = 1/4$) to approximate $y(2)$ if $\dfrac{dy}{dx} = y + 1$ and point (1, 1) belongs to the graph of the solution of the differential equation.

Chapter 10. Differential Equations

Solution
Better use organized table.

$y_{n+1} = y_n + \triangle x (y'_n)$

Points	$\triangle x = h$ (step size)	$\dfrac{dy}{dx} = y + 1$	$\triangle y = \triangle x \cdot \left(\dfrac{dy}{dx}\right)$	$y + \triangle y$
$(1, 1)$	0.25	2	$0.25(2) = 0.5$	1.5
$(1.25, 1.5)$	0.25	2.5	0.625	2.125
$(1.5, 2.125)$	0.25	3.125	0.78125	2.90625
$(1.75, 2.90625)$	0.25	3.90625	0.9765625	3.8828125
$(2, 3.8828125)$				

Therefore, $y(2) \approx 3.8828125$

▶ **Example**

Use Euler's method with step size $h = 0.2$ to approximate $y(1.6)$ if $\dfrac{dy}{dx} = \dfrac{x+y}{x}$ and $y(1) = 2$.

Solution

Points	$\triangle x = h$ (step size)	$\dfrac{dy}{dx} = \dfrac{x+y}{x}$	$\triangle y = \triangle x \cdot \left(\dfrac{dy}{dx}\right)$	$y + \triangle y$
$(1, 2)$	0.2	3	0.6	2.6
$(1.2, 2.6)$	0.2	3.17	0.634	3.234
$(1.4, 3.234)$	0.2	3.31	0.662	3.896
$(1.6, 3.896)$				

Therefore, $y(1.6) \approx 3.896$

Chapter 10. Differential Equations

H. Slope Field

Definition The derivative of a function gives its slope.

$$\text{If } \frac{dy}{dx} = f(x,y), \text{ then slope } y' = f(x,y).$$

A slope field is a graphical representation of the solutions of a first-order differential equation. It is achieved without solving the differential equation analytically, and thus it is useful. The representation may be used to qualitatively visualize solutions, or to numerically approximate them.

How do you sketch Slope Field?

Example: Sketch the slope field of the function $\frac{dy}{dx} = \frac{x}{y}$.

    Substitute x- and y-coordinates into the derivative function.

    At $(0,0)$, the slope is $\frac{0}{0}$ (undefined).

    At $(0,\pm1)\ (0,\pm2)(0,\pm3)\cdots$, the slope is 0.

    At $(1,2)$, the slope is $1/2$. And so on.

The slope is drawn as follows.

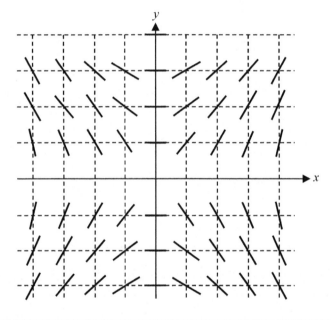

Chapter 10. Differential Equations

▶ **PRACTICE**

1. Sketch the slope field of the differential equation $\dfrac{dy}{dx} = x^2(y-2)$ on the axis provided and find the particular solution to the given differential equation with the initial condition $y = 3$ when $x = 0$.

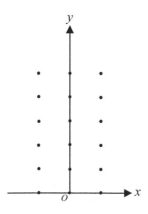

Solution

1. At each point $(0, y)$ and $(x, 2)$: Zero slope

	$x = -1$	$x = 0$	$x = 1$
$y = 0$	-2	0	-2
$y = 1$	-1	0	-1
$y = 2$	0	0	0
$y = 3$	1	0	1
$y = 4$	2	0	2
$y = 5$	3	0	3

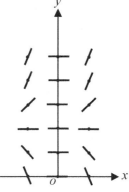

$\dfrac{dy}{dx} = x^2(y-2) \;\Rightarrow\; \dfrac{dy}{(y-2)} = x^2 dx \;\Rightarrow\; \ln|y-2| = \dfrac{x^3}{3} + C$

$|y-2| = e^{C_1} e^{x^3/3} \;\Rightarrow\; y - 2 = \pm e^{C_1} e^{x^3/3} \;\Rightarrow\; y - 2 = C e^{x^3/3}$

Initial condition: $3 - 2 = Ce^0 \;\Rightarrow\; C = 1$

The particular solution is $y = e^{x^3/3} + 2$

Chapter 10. Differential Equations

▶ **PRACTICE**

2. a) Sketch the slope field for the given differential equation $\dfrac{dy}{dx} = -\dfrac{x}{y}$ on the axis provided.

   b) If $y = f(x)$ is the particular solution to the differential equation with initial condition $f(1) = -2$, Find the particular solution.

   c) Write an equation for the line tangent to the graph of f at $(1, -2)$ and use it to approximate $f(1.05)$.

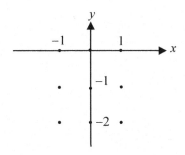

---

2. a) $\dfrac{dy}{dx} = -\dfrac{x}{y}$

	$x = -1$	$x = 0$	$x = 1$
$y = 0$	Undefined	Undefined	Undefined
$y = -1$	-1	0	1
$y = -2$	$-1/2$	0	$1/2$

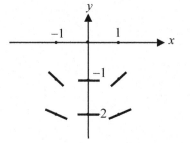

b) $\dfrac{dy}{dx} = -\dfrac{x}{y} \Rightarrow y\,dy = -x\,dx \Rightarrow \dfrac{y^2}{2} = -\dfrac{x^2}{2} + C$

Initial condition: $\dfrac{(-2)^2}{2} = -\dfrac{1^2}{2} + C \Rightarrow C = \dfrac{5}{2}$

$y^2 = -x^2 + 5 \Rightarrow y = \pm\sqrt{5 - x^2}$ ($y < 0$ at $x = 1$)

Therefore, the particular solution is $y = -\sqrt{5 - x^2}$.

c) $\dfrac{dy}{dx} = \dfrac{1}{2}$ at point $(1, -2)$

The line tangent to f at the point is $y + 2 = \dfrac{1}{2}(x - 1) \Rightarrow y = \dfrac{1}{2}x - \dfrac{5}{2}$

Therefore, $f(1.05) \approx \dfrac{1}{2}(1.05) - \dfrac{5}{2} = -1.975$.

Chapter 10. Differential Equations

▶ **PRACTICE**

3. Consider the differential equation $\dfrac{dy}{dx} = -3x^2 y$.

   a) Sketch a slope field for the given differential equation on the axis provided.

   b) If $y = f(x)$ is the particular solution to this differential equation with the initial condition $f(-1) = 1$.

   c) Write an equation for the line tangent to the graph of f at $x = -1$, and use it to approximate $f(-1.1)$.

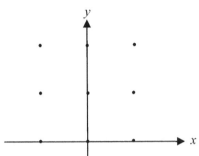

---

3. a)

	$x = -1$	$x = 0$	$x = 1$
$y = 0$	0	0	0
$y = 1$	-3	0	-3
$y = 2$	-6	0	-6

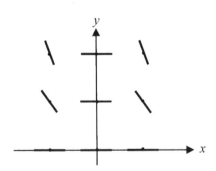

b) $\dfrac{dy}{dx} = -3x^2 y \Rightarrow \dfrac{dy}{y} = -3x^2 dx \Rightarrow \ln|y| = -x^3 + C_1 \Rightarrow y = Ce^{-x^3}$

Initial condition: $1 = Ce \Rightarrow C = e^{-1}$

The particular solution is $y = e^{-1} \cdot e^{-x^3} = e^{-(1+x^3)}$.

c) At $(-1, 1)$, slope $= -3$. $y - 1 = -3(x+1)$ or $y = -3x - 2$

$f(-1.1) \approx -3(-1.1) - 2 = 1.3$

Dr. John Chung's AP Calculus AB/BC

Chapter 11. Sequences and Infinite Series

11.1 SEQUNCES

A. SEQUENCES

1. A sequence is an ordered list of numbers that follow a pattern.
   a_1 = The first term a_2 = The second term a_3 = The third term $\cdots$ a_n = The nth term (The general term)

2. Brace notation of a sequence: $\{a_n\}$ A sequence can be written as follows;
   $\{a_n\}_{n=1}^{\infty} = a_1, a_2, a_3, a_4, \cdots a_n, a_{n+1}, \cdots,$ $\{a_n\}_{n=5}^{n=8} = a_5, a_6, a_7, a_8$
   The numbers $a_1, a_2, a_3, \cdots$ are called the terms of the sequence.

   $\left\{\dfrac{(-1)^n 2^n}{n}\right\}_{n=1}^{\infty} = \overset{a_1}{-2}, \overset{a_2}{2}, \overset{a_3}{-8/3}, \overset{a_4}{4}, \cdots$ $\left\{\dfrac{(-1)^n 2^n}{n}\right\}_{2}^{\infty} = \overset{a_2}{2}, \overset{a_3}{-8/3}, \overset{a_4}{4}, \cdots$

3. A sequence is an alternative notation for the function $f(n) = a_n, n = 1, 2, 3, \cdots$

   $$\{a_n\}_{n=1}^{\infty} \equiv f(n) = a_n, n = 1, 2, 3, \cdots$$

4. Limit of a sequence
   If $\lim_{n \to \infty} a_n = L$ and L is a finite real number, then the limit of the sequence is L.

5. Sequence that has a finite limit is said to "**converge**".
   Sequence that does not have some finite limit is said to "**diverge**".

6. How to find the nth term (general term) of a sequence

   Example:
   Find the general term of the sequence starting with $n = 1$.
   $$1, -\frac{1}{3}, \frac{1}{9}, -\frac{1}{27}, \cdots$$
   Solution
   $1, -\dfrac{1}{3}, \dfrac{1}{9}, -\dfrac{1}{27}, \cdots \rightarrow \dfrac{1}{3^0}, (-1)\dfrac{1}{3^1}, \dfrac{1}{3^2}, (-1)\dfrac{1}{3^3}, \cdots \rightarrow a_n = (-1)^{n-1}\dfrac{1}{3^{n-1}}$

   Brace notation: $\left\{(-1)^{n-1}\dfrac{1}{3^{n-1}}\right\}_{n=1}^{\infty}$ Function notation: $f(n) = (-1)^{n-1}\dfrac{1}{3^{n-1}}, n = 1, 2, 3, \cdots$

7. $f(n)$ are values of $f(x)$ taken at positive integers.
   If $\lim_{x \to \infty} f(x) = L$, then $\lim_{n \to \infty} a_n = L$.

Chapter 11. Sequences and Infinite Series

▶ **Example**

If $f(x) = \dfrac{1}{1+x^2}$, then $a_1 = f(1) = \dfrac{1}{2}$, $a_2 = f(2) = \dfrac{1}{5}$, $a_3 = f(3) = \dfrac{1}{10}$, $\cdots$, $a_n = \dfrac{1}{1+n^2}$, $\cdots$.

What is the value of $\displaystyle\lim_{n \to +\infty} \dfrac{1}{1+n^2}$?

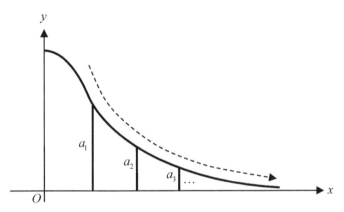

$$\lim_{x \to \infty} \frac{1}{1+x^2} = \lim_{n \to \infty} \frac{1}{1+n^2} = 0$$

- $f(x) = \dfrac{1}{1+x^2}$ is a continuous curve and $f(n) = \dfrac{1}{1+n^2}$ is a succession of isolated points.
- If $f(x)$ approaches a limit L, then the sequence converges to the same limit L.

Note If $f(x)$ approaches a limit L, then the sequence converges to the same limit L.

▶ **Example**

Find the limit of the sequence $f(n) = \dfrac{n}{e^n}$, $n = 1, 2, 3, \cdots$.

Dr. John Chung's AP Calculus AB/BC

Chapter 11. Sequences and Infinite Series

The expression $\dfrac{n}{e^n}$ is an indeterminate form of type ∞/∞ as $n \to +\infty$, so L'Hôpital's rule is indicated.

However, we cannot apply this rule directly to $f(n)$ because the function $f(n)$ has been defined only at the positive integers, and hence are not differentiable function. To circumvent this problem, we extend the domains of this function to all real numbers, here implied by replacing n by x, and apply L'Hôpital's rule to the limit of the quotient $\dfrac{x}{e^x}$. This yields $f(x) = \dfrac{x}{e^x}$ and

$$\lim_{x\to+\infty} f(x) = \lim_{x\to+\infty} \frac{x}{e^x} = \lim_{x\to+\infty} \frac{(x)'}{(e^x)'} = \lim_{x\to+\infty} \frac{1}{e^x} = 0$$

Now we can conclude that

$$\lim_{n\to+\infty} \frac{n}{e^n} = 0$$

Note You can consider n as "a real variable".

B. SEQUENCE CONVERGENCE

THEOREM If $\lim\limits_{n\to+\infty} a_n$ exists, then sequence $\{a_n\}$ is convergent.

If the limit does not exist, sequence $\{a_n\}$ is divergent.

PROPERTIES

1. $\lim\limits_{n\to\infty} c = c$
2. $\lim\limits_{n\to\infty} ca_n = c \lim\limits_{n\to\infty} a_n$
3. $\lim\limits_{n\to\infty} (a_n + b_n) = \lim\limits_{n\to\infty} a_n + \lim\limits_{n\to\infty} b_n$
4. $\lim\limits_{n\to\infty} (a_n - b_n) = \lim\limits_{n\to\infty} a_n - \lim\limits_{n\to\infty} b_n$
5. $\lim\limits_{n\to\infty} (a_n b_n) = \lim\limits_{n\to\infty} a_n \cdot \lim\limits_{n\to\infty} b_n$
6. $\lim\limits_{n\to\infty} \left(\dfrac{a_n}{b_n}\right) = \dfrac{\lim\limits_{n\to\infty} a_n}{\lim\limits_{n\to\infty} b_n}, \left(\text{if } \lim\limits_{n\to\infty} b_n \ne 0\right)$
7. $\lim\limits_{n\to\infty} [a_n]^p = \left[\lim\limits_{n\to\infty} a_n\right]^p$

▶ **Example**

1. $\{(-1)^n\} = -1, 1, -1, 1, \cdots \;\to\; \lim\limits_{n\to\infty}(-1)^n$ is 1 or -1 $\to \{a_n\}$ diverges

2. $\left\{\dfrac{2n}{n+2}\right\} \;\to\; \lim\limits_{n\to\infty} \dfrac{2n}{n+2} = 2 \;\to\; \{a_n\}$ converges.

3. $\left\{\dfrac{2+(-1)^n}{n+1}\right\} \;\to\; \lim\limits_{n\to\infty} \dfrac{n+(-1)^n}{n+1} = \lim\limits_{n\to\infty} \dfrac{n}{n+1} + \lim\limits_{n\to\infty} \dfrac{(-1)^n}{n+1} = 1+0 = 1 \;\to\; \{a_n\}$ converges.

Chapter 11. Sequences and Infinite Series

4. $\left\{\dfrac{\ln n}{n}\right\}$ $\to$ $\lim\limits_{n\to\infty}\dfrac{\ln n}{n} = \lim\limits_{n\to\infty}\dfrac{(\ln n)'}{(n)'} = \lim\limits_{n\to\infty}\dfrac{1/n}{1} = 0$ $\to$ $\{a_n\}$ converges. (L'Hôpital's rule applied)

THEOREM A sequence converges to a limit L if and only if the sequences of even-numbered terms and odd numbered terms both converge to L.

▶ **Example**

Determine whether the sequence converges or diverges. If it converges, find the limit.

$$\dfrac{1}{2},\dfrac{1}{3},\dfrac{1}{2^2},\dfrac{1}{3^2},\dfrac{1}{2^3},\dfrac{1}{3^3}\cdots$$

> The sequence converges to 0, since both even-numbered terms and odd-numbered terms both converge to 0.

▶ **Example**

Determine whether the sequence converges or diverges. If it converges, find the limit.

$$1,\dfrac{1}{2},1,\dfrac{1}{2^2},1,\dfrac{1}{2^3}\cdots$$

> The sequence diverges, since the odd-numbered terms converges to 1 and the even-numbered terms converges to 0.

THEOREM If $\lim\limits_{n\to+\infty}|a_n|=0$, then $\lim\limits_{n\to+\infty}a_n=0$

▶ **Example**

Determine whether the sequence converges or diverges.

$$1,-\dfrac{1}{2},\dfrac{1}{2^2},-\dfrac{1}{2^3},\cdots(-1)^n\dfrac{1}{2^n},\cdots$$

> The sequence converges to 0, since $\lim\limits_{n\to\infty}|a_n|=0$.
>
> Because $\lim\limits_{n\to\infty}\left|\dfrac{(-1)^n}{2^n}\right| = \lim\limits_{n\to\infty}\dfrac{1}{2^n}=0$, $\lim\limits_{n\to\infty}(-1)^n\dfrac{1}{2n}$ is also 0.
>
> or, $\lim\limits_{n\to\infty}\dfrac{(-1)^n}{2^n}=0$ because $(-1)^2$ is 1 or -1 and $2^\infty=$ infinite $\to\dfrac{\text{constant}}{\infty}=0$

Dr. John Chung's AP Calculus AB/BC

Chapter 11. Sequences and Infinite Series

THEOREM If $f(x)$ is continuous and $\lim_{n\to\infty} a_n = L$, then $\lim_{x\to\infty} f(a_n) = f\left(\lim_{x\to\infty} a_n\right)$.

▶ **Example**

Find $\lim_{n\to\infty} e^{n\sin(1/n)}$.

$f(x) = e^x$ is continuous and $\lim_{n\to\infty} e^{n\sin(1/n)} = \lim_{n\to\infty} e^{\left[\frac{\sin(1/n)}{1/n}\right]} = e^{\left[\lim_{n\to\infty} \frac{\sin(1/n)}{1/n}\right]} = e^1 = e$.

SQUEEZE THEOREM If $a_n \leq b_n \leq c_n$ and $\lim_{n\to\infty} a_n = \lim_{n\to\infty} c_n = L$, then $\lim_{n\to\infty} b_n = L$.

▶ **Example**

Determine the convergence or divergence of the sequence with the given nth term $a_n = \dfrac{n!}{n^n}$.

$$\frac{n!}{n^n} = \frac{(n)(n-1)(n-2)\cdots 1}{n^n} = \frac{\overbrace{(n)(n-1)(n-2)\cdots 2}^{n-1 \text{ terms}}}{n^n} \leq \frac{n\cdot n\cdot n\cdots n}{n^n} = \frac{n^{n-1}}{n^n}$$

We can see $\dfrac{n!}{n^n} \leq \dfrac{n^{n-1}}{n^n}$ and $\lim_{n\to\infty} \dfrac{n^{n-1}}{n^n} = \lim_{n\to\infty} \dfrac{1}{n} = 0$. We also can see $\lim_{n\to\infty} \dfrac{n!}{n^n} \geq 0$. (cannot be negative)

Since $0 \leq \lim_{n\to\infty} \dfrac{n!}{n^n} \leq 0$, then $\lim_{n\to\infty} \dfrac{n!}{n^n} = 0$ by the squeeze theorem.

For $\{a_n\}$, if $a_1 > a_2 > a_3 > \cdots$, then $\{a_n\}$ is always decreasing. $\qquad a_n > a_{n+1}$ for all n
For $\{a_n\}$, if $a_1 < a_2 < a_3 < \cdots$, then $\{a_n\}$ is always increasing. $\qquad a_n < a_{n+1}$ for all n
then $\{a_n\}$ is called a **monotonic sequence**.

▶ **Example**

Show that sequence $\left\{\dfrac{n}{n+1}\right\}$ is increasing.

Chapter 11. Sequences and Infinite Series

We need show $a_n < a_{n+1}$ → $\dfrac{n}{n+1} < \dfrac{n+1}{(n+1)+1}$ → $\dfrac{n}{n+1} < \dfrac{n+1}{n+2}$ → $n(n+2) < (n+1)(n+1)$

$n^2 + 2n < n^2 + 2n + 1$ → $0 < 1$ means "always true."

Or, we can use first derivative function.

Let $f(x) = \dfrac{x}{x+1}$. Then $f'(x) = \dfrac{1}{(x+1)^2} \geq 0$ for $x \geq 1$. Since $f(x)$ is increasing, the $\left\{\dfrac{n}{n+1}\right\}$ is also increasing.

▶ **PRACTICE**

Determine the convergence or divergence of the sequence with the given nth term. If the sequence converges, find its limit.

1. $a_n = \dfrac{n+2}{n-1}$

2. $a_n = \dfrac{1+(-1)^n}{n}$

3. $a_n = \dfrac{n^2}{2^n - 1}$

4. $a_n = \left(1 + \dfrac{1}{n}\right)^{2n}$

5. $a_n = \dfrac{n!}{(n+1)!}$

6. $a_n = \left(1 + \dfrac{2}{n}\right)^{3n}$

7. Find $\lim\limits_{n \to +\infty} \sqrt[n]{n} =$

8. $a_n = \sqrt{n+1} - \sqrt{n}$

9. Determine whether $\left\{\dfrac{n}{e^n}\right\}_{n=1}^{\infty}$ is increasing or decreasing.

Chapter 11. Sequences and Infinite Series

Solution

1. $a_n = \dfrac{n+2}{n-1}$ $\Rightarrow$ $\lim\limits_{n\to\infty} \dfrac{n+2}{n-1} = 1$ (Convergence)

   Note: $\lim\limits_{n\to\infty} \dfrac{n+2}{n-1} = \lim\limits_{x\to\infty} \dfrac{x+2}{x-1}$ (Function of a real variable). You can consider n as a real variable.

2. $a_n = \dfrac{1+(-1)^n}{n}$ $\Rightarrow$ $\lim\limits_{n\to\infty} \dfrac{1+(-1)^n}{n} = 0$ $\to$ $0, \dfrac{2}{2}, 0, \dfrac{2}{4}, 0, \dfrac{2}{6}, 0, \dfrac{2}{6}, \cdots$ Converges to 0, since the odd-numbered and the even-numbered terms both converge to 0.

   or, $\lim\limits_{n\to\infty}\left[\dfrac{1+(-1)^n}{n}\right] = \lim\limits_{n\to\infty}\dfrac{1}{n} + \lim\limits_{n\to\infty}\dfrac{(-1)^n}{n} = 0+0 = 0$ $\to$ $\{a_n\}$ converges.

3. $\lim\limits_{n\to\infty} \dfrac{n^2}{2^n-1} = \lim\limits_{n\to\infty} \dfrac{(n^2)'}{(2^n-1)'} = \lim\limits_{n\to\infty} \dfrac{2n}{2^n \ln 2} = \lim\limits_{n\to\infty} \dfrac{(2n)'}{(2^n \ln 2)'} = \lim\limits_{n\to\infty} \dfrac{2}{2^n (\ln 2)^2} = 0$

   $\{a_n\}$ converges. $\left(\text{L'Hôpital's rule applied, } \dfrac{\infty}{\infty}\right)$

4. $\lim\limits_{n\to\infty}\left(1+\dfrac{1}{n}\right)^{2n} = \left[\lim\limits_{n\to\infty}\left(1+\dfrac{1}{n}\right)^n\right]^2 = e^2$ $\to$ $\{a_n\}$ converges.

   Remind: $\lim\limits_{n\to 0}(1+n)^{1/n} = e$, $\lim\limits_{n\to\infty}\left(1+\dfrac{1}{n}\right)^n = e$, and $\lim\limits_{n\to\infty}\left(1+\dfrac{c}{n}\right)^{n/c} = e$ (c is a constant.)

5. $\lim\limits_{n\to\infty} \dfrac{n!}{(n+1)!} = \lim\limits_{n\to\infty} \dfrac{n!}{(n+1)n!} = \lim\limits_{n\to\infty} \dfrac{1}{(n+1)} = 0$ $\to$ $\{a_n\}$ converges.

6. $\lim\limits_{n\to\infty}\left(1+\dfrac{2}{n}\right)^{3n} = \left[\lim\limits_{n\to\infty}\left(1+\dfrac{2}{n}\right)^{\frac{n}{2}}\right]^6 = e^6$ $\to$ $\{a_n\}$ converges.

7. $\lim\limits_{n\to\infty} \sqrt[n]{n} = \lim\limits_{n\to\infty} n^{1/n} = \lim\limits_{n\to\infty} e^{\ln n^{1/n}} = \lim\limits_{n\to\infty} e^{(1/n)\ln n} = \lim\limits_{n\to\infty} e^{\frac{\ln n}{n}} = \lim\limits_{n\to\infty} e^{\left[\lim\limits_{n\to\infty}\frac{\ln n}{n}\right]} = \lim\limits_{n\to\infty} e^{\left[\lim\limits_{n\to\infty}\left(\frac{1/n}{1}\right)\right]} = e^0 = 1$

   $\left(\text{L'Hôpital's rule applied to } \dfrac{\ln n}{n}\right)$

8. $\lim\limits_{n\to\infty} \dfrac{(\sqrt{n+1}-\sqrt{n})}{1} \cdot \dfrac{(\sqrt{n+1}+\sqrt{n})}{(\sqrt{n+1}+\sqrt{n})} = \lim\limits_{n\to\infty} \dfrac{1}{\sqrt{n+1}+\sqrt{n}} = 0$ $\to$ $\{a_n\}$ converges.

9. $f(x) = \dfrac{x}{e^x}$ $\to$ $f'(x) = \dfrac{1-x}{e^x} \le 0$ for $x \ge 1$ $\to$ Therefore, $\{a_n\}$ is decreasing.

Chapter 11. Sequences and Infinite Series

11.2 INFINITE SERIES $\left(\sum_{k=1}^{\infty} a_k \text{ or } \sum_{n=1}^{\infty} a_n\right)$

An infinite series is an expression that can be written in the form
$\sum_{k=1}^{\infty} a_k = a_1 + a_2 + a_3 + \cdots a_n + \cdots$ is a series. The numbers $a_1, a_2, a_3, \cdots$ are called the terms of the series.

The nth "**partial sum**" of the series $s_n = a_1 + a_2 + a_3 + \cdots + a_n = \sum_{k=1}^{n} a_k$

THEOREM Let $\{S_n\} = S_1, S_2, S_{3,\ldots}$ be the **sequence of partial sums** of the series.

If $\{s_n\}$ converges to a limit S, then the series $\sum a_n$ is said to "converged to S."

$$\sum_{k=1}^{\infty} a_k = S \quad \text{or} \quad \lim_{n \to \infty} \sum_{k=1}^{n} a_k = \lim_{n \to \infty} S_n = S$$

If $\{S_n\}$ diverges, $\sum a_n$ diverges.

▶ **Example**

Determine whether the series $\sum_{n=1}^{\infty} n$ converges or diverges.

Step 1) Find the partial sum and find the formula. $S_n = 1 + 2 + 3 + \cdots + n = \dfrac{n(n+1)}{2}$

Step 2) Take a limit as $n \to \infty$. $\lim_{n \to \infty} \dfrac{n(n+1)}{2} = \infty \quad \to \quad \sum a_n$ diverges.

▶ **Example**

Determine whether the series $2 - 2 + 2 - 2 + 2 - 2 \cdots$ converges or diverges.

The partial sums are $s_1 = 2, \ s_2 = 0, \ s_3 = 2, \ s_4 = 0, S_5 = 2, \cdots$ The series diverges. Since the sequence of partial sums is $2, 0, 2, 0, \cdots$.

Chapter 11. Sequences and Infinite Series

A. TELESCOPING SERIES

The series in the form $\sum_{k=1}^{\infty} \dfrac{1}{k(k+a)} = \sum_{k=1}^{\infty} \dfrac{1}{a}\left(\dfrac{1}{k} - \dfrac{1}{k+a}\right)$ is called telescoping series.

▶ **Example**

Determine whether the series $\sum_{k=1}^{\infty} \dfrac{1}{k(k+1)}$ converges or diverges.

The nth partial sum

$$s_n = \sum_{n=1}^{\infty} \dfrac{1}{n(n+1)} = \sum_{n=1}^{\infty}\left(\dfrac{1}{n} - \dfrac{1}{n+1}\right) = \left(1 - \dfrac{1}{2}\right) + \left(\dfrac{1}{2} - \dfrac{1}{3}\right) + \left(\dfrac{1}{3} - \dfrac{1}{4}\right) + \cdots + \left(\dfrac{1}{n} - \dfrac{1}{n+1}\right) = 1 - \dfrac{1}{n+1}$$

So, the sum of the series is $s = \sum_{n=1}^{\infty} \dfrac{1}{n(n+1)} = \lim_{n\to\infty}\left(1 - \dfrac{1}{n+1}\right) = 1$. The series $\sum a_n$ converges.

B. GEOMETRIC SERIES

A geometric series $\sum_{n=1}^{\infty} ar^{n-1} = a + ar + ar^2 + \cdots + ar^{n-1} + \cdots$ $(a \neq 0)$ converges if $|r| < 1$ and diverges if $|r| \geq 1$.

$$S_n = a + ar + ar^2 + \cdots + ar^{n-1} = \dfrac{a(1-r^n)}{1-r}$$

If $|r| < 1$, $\lim_{n\to\infty} r^n = 0$ and the series converges and the sum is $S = \lim_{n\to\infty} S_n = \dfrac{a}{1-r}$.

▶ **Example**

Determine whether the series converges, and if so find the sum of the series $\sum_{k=1}^{\infty} 3^k 5^{1-k}$.

$$\sum_{k=1}^{\infty} 3^k 5^{1-k} = \sum_{k=1}^{\infty} 3^k \dfrac{5}{5^k} = 5\sum_{k=1}^{\infty}\left(\dfrac{3}{5}\right)^k \ \to\ a_1 = 5\left(\dfrac{3}{5}\right) = 3 \text{ and } r = \dfrac{3}{5} \quad \text{(Geometric series)}$$

Since $r = \dfrac{3}{5} < 1$, the series converges and its sum is $\dfrac{a}{1-r} = \dfrac{3}{1-3/5} = \dfrac{15}{2}$

Remind: $\sum_{n=1}^{\infty} 5\left(\dfrac{3}{5}\right)^n = \sum_{n=1}^{\infty} 5\left(\dfrac{3}{5}\right)\left(\dfrac{3}{5}\right)^{n-1} = \sum_{n=1}^{\infty} 3\left(\dfrac{3}{5}\right)^{n-1} = \sum_{n=1}^{\infty} ar^{n-1}$ You can see $a = 3$ and $r = \dfrac{3}{5}$

Chapter 11. Sequences and Infinite Series

▶ **Example**

Determine whether the series converges, and if so find the sum of the series $\sum_{n=1}^{\infty} \dfrac{e^n}{4^{n+1}}$.

$\sum_{n=1}^{\infty} \dfrac{e^n}{4^{n+1}} = \sum_{n=1}^{\infty} \dfrac{e \cdot e^{n-1}}{4^2 \cdot 4^{n-1}} = \sum_{n=1}^{\infty} \left(\dfrac{e}{16}\right)\left(\dfrac{e}{4}\right)^{n-1}$ We can see $a = \dfrac{e}{16}$ and $r = \dfrac{e}{4} < 1 \to \sum a_n$ converges.

$\sum_{n=1}^{\infty} \dfrac{e^n}{4^{n+1}} = \dfrac{a}{1-r} = \dfrac{e/16}{1-e/4} = \dfrac{e}{16-4e}$

C. HARMONIC SERIES

The terms in the harmonic series

$\sum_{k=1}^{\infty} \dfrac{1}{k} = 1 + \dfrac{1}{2} + \dfrac{1}{3} + \dfrac{1}{4} + \cdots$ are all positive and the series diverges, since $s_1 < s_2 < s_3 < s_4 < \cdots s_n < s_{n+1} < \cdots$

These partial sums **strictly increasing**.

▶ **Example**

Determine whether the series $\sum_{k=2}^{\infty} \dfrac{2}{k-1}$ converges or diverges.

$\sum_{k=2}^{\infty} \dfrac{2}{k-1} = 2 \sum_{k=2}^{\infty} \dfrac{1}{k-1} = 2\left(1 + \dfrac{1}{2} + \dfrac{1}{3} + \cdots\right)$ diverges, since the series is a harmonic series.

Chapter 11. Sequences and Infinite Series

11.3 CONVERGENCE TESTS

> To find the sum of a series by finding a closed form for the nth partial sum and taking its limit.
> $$S = \lim_{n\to\infty} S_n \quad \text{or} \quad S = \sum a_k = \lim_{n\to\infty} \sum_1^n a_k, \text{ where } S_n = \sum_1^n a_k \text{ is a partial sum.}$$
> However, it is relatively rare that one can find a closed form for the nth partial sum of a series, so alternative methods are needed for finding the sum of a series. We have various tests that can be used to determine whether a given series converges or diverges.

A. THE CONVERGENCE AND DIVERGENCE TEST

THEOREM
1. If $\lim_{n\to\infty} a_n \neq 0$, then the series $\sum a_n$ diverges. (Useful for only **divergence test**)
2. If $\lim_{n\to\infty} a_n = 0$, then the series $\sum a_n$ may **either converges or diverges.** (Use another test)
3. If the series $\sum a_n$ converges, then $\lim_{k\to\infty} a_k = 0$.
4. If $\sum a_k$ and $\sum b_k$ are convergent series, then $\sum a_k + \sum b_k$ and $\sum a_k - \sum b_k$ are convergent series.
5. If the series $\sum a_k$ converges, the $c\sum a_k$ converges. (c is a nonzero constant.)
6. If the series $\sum a_k$ diverges, the $c\sum a_k$ diverges. (c is a nonzero constant.)
7. Convergence or divergence is **unaffected by deleting a finite number of terms from a series**

▶ **Example** Determine whether the following series converge or diverge.

1) $\sum_{n=1}^{\infty} \dfrac{n}{n+1}$ 2) $\sum_{k=1}^{\infty} \left(\dfrac{2}{3^k} - \dfrac{1}{5^{k-1}} \right)$

3) $\sum_{n=1}^{\infty} \dfrac{10}{n}$ 4) $\sum_{k=5}^{\infty} \dfrac{2}{k}$

---

1) The series diverges since $\lim_{n\to+\infty} \dfrac{n}{n+1} = \lim_{n\to+\infty} \dfrac{1}{1+1/n} = 1 \neq 0$. (**Divergence test applied**)

2) $\sum_{k=1}^{\infty} \left(\dfrac{2}{3^k} - \dfrac{1}{5^{k-1}} \right) = \sum_{k=1}^{\infty} \dfrac{2}{3^k} - \sum_{k=1}^{\infty} \dfrac{1}{5^{k-1}} = \left(\dfrac{\frac{2}{3}}{1-\frac{1}{3}} \right) - \left(\dfrac{1}{1-\frac{1}{5}} \right) = 1 - \dfrac{5}{4} = -\dfrac{1}{4}$ (**Geometric convergent series**)

3) $\sum_{n=1}^{\infty} \dfrac{10}{n} = 10 \sum_{n=1}^{\infty} \dfrac{1}{n}$ The series diverges. (**A constant times the divergent harmonic series**)

4) $\sum_{k=5}^{\infty} \dfrac{2}{k} = 2\left(\dfrac{1}{5} + \dfrac{1}{6} + \dfrac{1}{7} + \cdots \right)$ The series diverges (**Divergent harmonic series**)

(Convergence or divergence is unaffected by deleting a finite number of terms from a series)

Chapter 11. Sequences and Infinite Series

B. THE INTEGRAL TEST

THEOREM $\sum a_n$ is a series with **positive terms** and f is a function that is **decreasing** and **continuous** on an interval $[c, +\infty)$. If $a_n = f(n)$ for all $n \geq c$, then

$$\sum_{n=1}^{\infty} a_n \text{ and } \int_c^{\infty} f(x)\,dx$$

both converge or diverge.

Note In order to use integral test, we need to make sure that $f(x)$ is **positive, decreasing, and continuous on** $[c, \infty)$.

Note Integral Test only tells us that $\sum_{n=1}^{\infty} a_n$ and $\int_c^{\infty} f(x)\,dx$ both converge or diverge.

It doesn't tell us $\sum_{n=1}^{\infty} a_n = \int_c^{\infty} f(x)\,dx$.

▶ **Example**

Use the integral test to determine whether the following series converges or diverges.

1) $\sum_{n=1}^{\infty} \dfrac{1}{n}$ 2) $\sum_{n=1}^{\infty} \dfrac{1}{n^2}$

1) The series is the divergent harmonic series. The integral test is simply another way of establishing the divergence. We can see 1) the series have positive terms, 2) $f(x) = \dfrac{1}{x}$ is decreasing and 3) continuous.

$\int_1^{+\infty} \dfrac{1}{x}\,dx = \lim_{b \to +\infty} \int_1^b \dfrac{1}{x}\,dx = \lim_{b \to +\infty} [\ln x]_1^b = \lim_{b \to +\infty} [\ln b - \ln 1] = +\infty$ The integral diverges.

$\int_1^{\infty} \dfrac{1}{x}\,dx$ diverges. Hence $\sum_1^{\infty} \dfrac{1}{n}$ also diverges.

2) The series: positive terms. We know that $f(x) = \dfrac{1}{x^2}$ is obviously decreasing and continuous on $[1, \infty)$. Now we can use integral test.

$\sum_{k=1}^{\infty} \dfrac{1}{k^2} \to \int_1^{+\infty} \dfrac{1}{x^2} = \lim_{b \to \infty} \left[-\dfrac{1}{x}\right]_1^b = \lim_{b \to \infty} \left(1 - \dfrac{1}{b}\right) = 1$ The integral converges to 1.

Therefore, $\sum_{n=1}^{\infty} \dfrac{1}{n^2}$ also converges.

Note It doesn't mean the sum of the series converges to 1. Just tells "convergence or divergence."

Chapter 11. Sequences and Infinite Series

▶ **Example**

Use the integral test to determine whether the following series converges or diverges.

1) $\sum_{n=1}^{\infty} \dfrac{\ln n}{n}$

2) $\sum_{n=1}^{\infty} \dfrac{e^{1/n}}{n^2}$

---

1) The series (positive terms), $f(x) = \dfrac{\ln x}{x}$ is continuous on $[1, \infty)$.

But we are **not sure** $f(x)$ is decreasing. We need to check.

$f(x) = \dfrac{\ln x}{x} \;\rightarrow\; f'(x) = \dfrac{1-\ln x}{x^2} \leq 0 \;\rightarrow\; 1-\ln x \leq 0 \;\rightarrow\; 1 \leq \ln x \;\rightarrow\; x \geq e$

That means the function is decreasing on $[3, \infty)$. So, apply the integral test.

$$\int_3^{\infty} \dfrac{\ln x}{x} dx = \lim_{b \to \infty} \int_3^b \dfrac{\ln x}{x} dx$$

Now, $\int \dfrac{\ln x}{x} dx \;\rightarrow\; u = \ln x \;\rightarrow\; du = \dfrac{1}{x} dx$ so, $\int \dfrac{\ln x}{x} dx = \int u\, du = \dfrac{u^2}{2} \;\rightarrow\; \dfrac{(\ln x)^2}{2}$

Therefore, $\int_3^{\infty} \dfrac{\ln x}{x} dx = \lim_{b \to \infty} \left[\dfrac{(\ln x)^2}{2} \right]_3^b = \lim_{b \to \infty} \dfrac{1}{2}\left[(\ln b)^2 - (\ln 3)^2 \right] = \infty \;\rightarrow\;$ diverges

So, $\sum_{n=1}^{\infty} \dfrac{\ln n}{n}$ also diverges.

2) We can see $f(x) = \dfrac{e^{1/x}}{x^2}$ is positive, continuous, and decreasing (?) on $[1, \infty)$.

But, we don't have to check it's decreasing every time. If $\int_1^{\infty} f(x) dx$ has a finite number, then the series also converges. So,

$\int_1^{\infty} \dfrac{e^{1/x}}{x^2} dx \;\rightarrow\; u = \dfrac{1}{x}$ and $du = -\dfrac{1}{x^2} dx \;\rightarrow\; \int \dfrac{e^{1/x}}{x^2} dx = -\int e^u du = -e^u = -e^{1/x}$

$\int_1^{\infty} \dfrac{e^{1/x}}{x^2} dx = \lim_{b \to \infty} \left[-e^{1/x} \right]_1^b = \lim_{b \to \infty} \left[-e^{1/b} + e \right] = -1 + e$ (finite number), Therefore,

$\int_1^{\infty} \dfrac{e^{1/x}}{x^2} dx$ converges, and $\sum_{n=1}^{\infty} \dfrac{e^{1/n}}{n^2}$ also converges.

Chapter 11. Sequences and Infinite Series

C. *P*-SERIES TEST

THEOREM $\sum_{n=1}^{\infty} \frac{1}{n^p} = 1 + \frac{1}{2^p} + \frac{1}{3^p} + \cdots$ converges if $p > 1$ and diverges if $0 < p \leq 1$.

Proof

We use the integral test with $p \neq 1$. (If $p = 1$, it is divergent harmonic series)

We can see that $f(x) = \frac{1}{x^p}$ is positive, decreasing, and continuous on $[1, \infty)$.

$$\int_1^{\infty} \frac{1}{x^p} dx = \lim_{b \to \infty} \int_1^b x^{-p} dx = \lim_{b \to \infty} \left[\frac{x^{1-p}}{1-p} \right]_1^b = \lim_{b \to \infty} \left[\left(\frac{b^{1-p}}{1-p} \right) - \left(\frac{1}{1-p} \right) \right]$$

1. If $p > 1$,

$$\lim_{b \to \infty} \left[\frac{b^{1-p}}{1-p} - \frac{1}{1-p} \right] = \lim_{b \to \infty} \left[\frac{1}{b^{p-1}(1-p)} - \frac{1}{1-p} \right] = 0 - \frac{1}{1-p} = \frac{1}{p-1}. \text{ (Finite number)}$$

$\int_1^{\infty} \frac{1}{x^p} dx$ converges and also the series $\sum_{n=1}^{\infty} \frac{1}{n}$ converges.

2. If $0 < p < 1$,

$$\lim_{b \to \infty} \left[\frac{x^{1-p}}{1-p} \right]_1^b = +\infty. \text{ The series diverges.}$$

3. If $p = 1$, $\sum_{k=1}^{\infty} \frac{1}{k}$ is the divergent harmonic series. The series diverges.

▶ **Example** Determine whether the series converges or diverges.

1. $\sum_{k=1}^{\infty} \frac{1}{\sqrt[3]{k}}$

2. $\sum_{k=1}^{\infty} 2k^{-3}$

1. $\sum_{k=1}^{\infty} \frac{1}{\sqrt[3]{k}} = \sum_{k=1}^{\infty} \frac{1}{k^{\frac{1}{3}}}$ The series diverges, since it is a *p*-series with $p = \frac{1}{3} < 1$.

2. $\sum_{k=1}^{\infty} 2k^{-3} = 2\sum_{k=1}^{\infty} \frac{1}{k^3}$ The series converges, since it is a *p*-series with $p = 3 > 1$

Dr. John Chung's AP Calculus AB/BC

Chapter 11. Sequences and Infinite Series

D. THE COMPARISON TEST

THEOREM Let $\sum a_k$ and $\sum b_k$ be series with **nonnegative** terms and
$$a_1 \leq b_1,\ a_2 \leq b_2,\ a_3 \leq b_3, \cdots a_k \leq b_k, \cdots$$

1. If the bigger series $\sum b_k$ converges, then the smaller series $\sum a_k$ also converges.
2. If the smaller series $\sum b_k$ diverges, then the bigger series $\sum a_k$ also diverges.

Note 1) We must find a **convergent series** where we already known whose terms are bigger than the corresponding terms of $\sum a_k$.
2) We must find a **divergent series** where we already known whose terms are smaller than the corresponding terms of $\sum a_k$.

▶ **Example** Use the comparison test to determine whether the series converges or diverges.

1. $\sum_{k=1}^{\infty} \dfrac{1}{\sqrt{k}-2}$
2. $\sum_{k=1}^{\infty} \dfrac{1}{2k^2+k}$
3. $\sum_{n=1}^{\infty} \dfrac{1}{n^2+2}$
4. $\sum_{n=1}^{\infty} \dfrac{1}{1+3^n}$

---

1. We must find a divergent p-series. We can see $\sum_{k=1}^{\infty} \dfrac{1}{\sqrt{k}-2} > \sum_{k=1}^{\infty} \dfrac{1}{\sqrt{k}}$.

   $\sum_{k=1}^{\infty} \dfrac{1}{\sqrt{k}} = \sum_{k=1}^{\infty} \dfrac{1}{k^{1/2}}\ \left(p=\dfrac{1}{2}\right)$. The series diverges.

   Since the smaller series diverges, then the bigger series $\sum_{k=1}^{\infty} \dfrac{1}{\sqrt{k}-2}$ also diverges.

2. $\sum_{k=1}^{\infty} \dfrac{1}{2k^2+k} < \sum_{k=1}^{\infty} \dfrac{1}{2k^2}$ and $2\sum_{k=1}^{\infty} \dfrac{1}{k^2}$ is a constant times a convergent p-series $(p=2)$. The series converges.

   Therefore, $\sum_{k=1}^{\infty} \dfrac{1}{2k^2+k}$ also converges.

3. We can see $\sum_{n=1}^{\infty} \dfrac{1}{n^2+1} < \sum_{n=1}^{\infty} \dfrac{1}{n^2}$ and $\sum_{n=1}^{\infty} \dfrac{1}{n^2}$ is convergent p-series. $(p=2>1)$

   Therefore, the series converges.

4. We can see $\sum_{n=1}^{\infty} \dfrac{1}{1+3^n} < \sum_{n=1}^{\infty} \dfrac{1}{3^n}$ and $\sum_{n=1}^{\infty} \dfrac{1}{3^n}$ is convergent geometric series. $\left(a=\dfrac{1}{3}\ \text{and}\ r=\dfrac{1}{3}<1\right)$

   Therefore, $\sum_{n=1}^{\infty} \dfrac{1}{1+3^n}$ also converges.

Chapter 11. Sequences and Infinite Series

E. THE LIMIT COMPARISON TEST

THEOREM Let $\sum a_k$ and $\sum b_k$ be series with **positive terms** and
$$\rho = \lim_{k \to \infty} \frac{a_k}{b_k}$$
If ρ is finite and $\rho > 0$, then the series both converges or diverges.

▶ **Example**

Use the limit comparison test to determine whether the series converges or diverges.

1. $\sum_{k=1}^{\infty} \frac{1}{\sqrt{k}+2}$

2. $\sum_{k=1}^{\infty} \frac{1}{2k^2+k}$

---

1. The series is likely to behave *the divergent p-series*. We can see $\sum_{k=1}^{\infty} \frac{1}{\sqrt{k}+1} < \sum_{k=1}^{\infty} \frac{1}{\sqrt{k}}$.

   $\sum_{k=1}^{\infty} \frac{1}{\sqrt{k}} = \sum_{k=1}^{\infty} \frac{1}{k^{1/2}} \to$ divergent p-series $(p = 1/2)$. We cannot see that the series $\sum_{k=1}^{\infty} \frac{1}{\sqrt{k}+1}$ diverges or converges. Now we apply **the limit comparison test**. Define $a_k = \frac{1}{\sqrt{k}+1}$ and $b_k = \frac{1}{\sqrt{k}}$. We obtain

   $\rho = \lim_{k \to \infty} \frac{a_k}{b_k} = \lim_{k \to \infty} \frac{a_k}{b_k} = \lim_{k \to \infty} \frac{1/(\sqrt{k}+1)}{(1/\sqrt{k})} = \lim_{k \to \infty} \frac{\sqrt{k}}{\sqrt{k}+1} = 1$. Since $\rho = 1$ is **finite and positive**, the given

   series $\sum_{k=1}^{\infty} \frac{1}{\sqrt{k}+2}$ diverges.

   Note When comparison test fails, this limit comparison test will be recommended.

2. The series is likely to behave *the convergent p-series*. $a_k = \frac{1}{2k^2+k}$ and $b_k = \frac{1}{2k^2}$ (convergent p-series,

   $p = 2$). We obtain $\rho = \lim_{k \to \infty} \frac{a_k}{b_k} = \lim_{k \to \infty} \frac{a_k}{b_k} = \lim_{k \to \infty} \frac{2k^2}{2k^2+k} = 1$. Since $\rho = 1$ is finite and positive, the given

   series converges.

   Note $\sum_{k=1}^{\infty} \frac{1}{2k^2+k} < \sum_{k=1}^{\infty} \frac{1}{2k^2} \to$ convergent p-series $(p = 2)$. Therefore, $\sum_{k=1}^{\infty} \frac{1}{2k^2+k}$ converges by comparison test.

Chapter 11. Sequences and Infinite Series

▶ **Example**

Use the limit comparison test to determine whether the series converges or diverges.

1. $\sum_{k=1}^{\infty} \dfrac{2k^3 - k^2 + 4}{k^6 - k^3 + 2}$

2. $\sum_{n=1}^{\infty} \dfrac{n^2}{\sqrt{n^7 + 1}}$

---

1. The series is likely to behave *the convergent p-series*.
   Note When you try to find the corresponding series, **End-Behavior Series** will be useful as follows.

   $\sum_{k=1}^{\infty} \dfrac{2k^3 - k^2 + 4}{k^6 - k^3 + 2} \rightarrow$ When k goes to infinite, $\sum_{k=1}^{\infty} \dfrac{2k^3 - k^2 + 4}{k^6 - k^3 + 2} \rightarrow \sum_{k=1}^{\infty} \dfrac{2k^3 - \cancel{k^2} + \cancel{4}}{k^6 - \cancel{k^3} + \cancel{2}} = \sum_{k=1}^{\infty} \dfrac{2k^3}{k^6}$

   So, $\sum_{k=1}^{\infty} \dfrac{2k^3}{k^6} = \sum_{k=1}^{\infty} \dfrac{2}{k^3} = 2\sum_{k=1}^{\infty} \dfrac{1}{k^3} \rightarrow$ convergent $(p = 3)$

   $\rho = \lim_{k \to \infty} \dfrac{\dfrac{2k^3 - k^2 + 4}{k^6 - k^3 + 2}}{\dfrac{2}{k^3}} = \lim_{k \to \infty} \dfrac{2k^6 - 2k^3 + 4}{2k^6 - k^5 + 4k^3} = 1$

   Since $\rho = 1$ is finite and positive, the given series converges.

2. $\sum_{n=1}^{\infty} \dfrac{n^2}{\sqrt{n^7 + 1}}$ By end-behavior series $\sum_{n=1}^{\infty} \dfrac{n^2}{\sqrt{n^7 + 1}} \rightarrow \sum_{n=1}^{\infty} \dfrac{n^2}{\sqrt{n^7 \cancel{+1}}} = \sum_{n=1}^{\infty} \dfrac{n^2}{n^{7/2}} = \sum_{n=1}^{\infty} \dfrac{1}{n^{3/2}}$

   $\sum_{n=1}^{\infty} \dfrac{1}{n^{3/2}}$ is convergent p-series $\left(p = \dfrac{3}{2}\right)$. $\sum_{n=1}^{\infty} \dfrac{1}{n^{3/2}}$ converges.

   Now, $\lim_{n \to \infty} \dfrac{\dfrac{n^2}{\sqrt{n^7+1}}}{\dfrac{1}{n^{3/2}}} = \lim_{n \to \infty} \dfrac{n^{7/2}}{\sqrt{n^7+1}} = \lim_{n \to \infty} \dfrac{1}{\sqrt{1 + 1/n^7}} = 1$

   Since $\rho = 1$ is finite and positive, the given series converges.

Chapter 11. Sequences and Infinite Series

F. THE ALTERNATING SERIES TEST

THEOREM An alternating series $\sum_{k=1}^{\infty}(-1)^{k+1}a_k$ or $\sum_{k=1}^{\infty}(-1)^{k}a_k$ converges if the following two conditions are satisfied.

1. $a_1 \geq a_2 \geq a_3 \geq \cdots \geq a_k \geq \cdots$ 2. $\lim_{k\to\infty} a_k = 0$

▶ **Example** Use the alternating series test to determine whether the following series converge.

1. $\sum_{k=1}^{\infty}(-1)^k \dfrac{1}{k}$ (Alternating harmonic series) 2. $\sum_{k=1}^{\infty}(-1)^{k+1}\dfrac{k+2}{k(k+1)}$

3. $\sum_{n=1}^{\infty}(-1)^n \dfrac{2n}{4n-1}$ 4. $\sum_{n=2}^{\infty}(-1)^n \dfrac{\sqrt{n+1}}{n-1}$

---

1. $\sum_{k=1}^{\infty}\dfrac{1}{k} = \dfrac{1}{1}+\dfrac{1}{2}+\dfrac{1}{3}+\cdots$ We can see terms are obviously decreasing and $\lim_{k\to\infty} a_k = \lim_{k\to\infty}\dfrac{1}{k} = 0$.

   Therefore, the alternating series converges, since the two conditions are satisfied.

   Note If you are not sure its decreasing, you can check the ratio $\dfrac{a_{k+1}}{a_k}$.

   $\dfrac{a_{k+1}}{a_k} = \dfrac{1/(k+1)}{1/k} = \dfrac{k}{k+1} < 1$ (for all $k \geq 1$) show that term are decreasing.

   Remember Alternating harmonic series $\sum_{k=1}^{\infty}(-1)^k \dfrac{1}{k}$ converges.

2. $\dfrac{a_{k+1}}{a_k} = \dfrac{(k+3)}{\cancel{(k+1)}(k+2)} \cdot \dfrac{k\cancel{(k+1)}}{(k+2)} = \dfrac{k^2+3k}{k^2+4k+4} = \dfrac{(k^2+3k)}{(k^2+3k)+k+4} < 1$

   Terms are decreasing and $\lim_{k\to\infty} a_k = \lim_{k\to\infty}\dfrac{k+2}{k(k+1)} = \lim_{k\to\infty}\dfrac{k+2}{k^2+k} = 0$

   The alternating series converges, since the two conditions are satisfied.

3. $a_n = \dfrac{2n}{4n-1} \;\to\; \lim_{n\to\infty}\dfrac{2n}{4n-1} = \dfrac{1}{2} \neq 0 \;\to\;$ The alternating series $\sum_{n=1}^{\infty}(-1)^n \dfrac{2n}{4n-1}$ diverges.

   We don't have to check terms are decreasing.

4. $a_n = \dfrac{\sqrt{n+1}}{n-1} \;\to\; \lim_{n\to\infty} a_n = \dfrac{\sqrt{\dfrac{1}{n}+\dfrac{1}{n^2}}}{1-\dfrac{1}{n}} = \dfrac{\sqrt{0+0}}{1-0} = 0$

   In order to check terms' decreasing, we can also use the first derivative.

   $f(x) = \dfrac{\sqrt{x+1}}{x-1} \;\to\; f'(x) = \dfrac{-x-3}{2\sqrt{x+1}\cdot(x-1)^2} < 0$ for $x \geq 2$.

   Therefore, the alternating series converges.

Dr. John Chung's AP Calculus AB/BC

Chapter 11. Sequences and Infinite Series

> **THEOREM** Error of alternating convergent series
>
> If $S_n = \sum_{k=1}^{n}(-1)^{k+1} a_k$ and S is the sum of the series, then
>
> Absolute error $|R_n| = |S - S_n| \leq a_{n+1}$

▶ **Example**

Find the upper bound on the absolute error that results the sum of the series is approximated by the nth partial sum.

1) $\sum_{n=1}^{\infty} \dfrac{(-1)^{n+1}}{n}$; $n = 7$

2) $\sum_{n=1}^{\infty} \dfrac{(-1)^{n+1}}{\sqrt{k}}$; $n = 99$

3) $\sum_{n=0}^{\infty} \dfrac{(-1)^{n}}{n!}$; $n = 6$

4) Find the value of n for which the partial sum is ensured to approximate the sum of the series to the stated accuracy.

$$\sum_{n=0}^{\infty} \dfrac{(-1)^{n}}{n!} \; ; \; |R_n| < 0.0005$$

---

1) We already know that the alternating series converges. ($a_n = \dfrac{1}{n}$, $\lim\limits_{n \to \infty} \dfrac{1}{n} = 0$, terms are decreasing.)

Therefore, $|R_n| = |S - S_n| \leq a_{n+1} = \dfrac{1}{n+1}$ → $|R_7| \leq \dfrac{1}{8}$ or $|R_7| \leq 0.125$

2) The series is an alternating convergent series and $a_n = \dfrac{1}{\sqrt{n}}$.

Therefore, $|R_n| = |S - S_n| \leq a_{n+1} = \dfrac{1}{\sqrt{n+1}}$ → $|R_{99}| \leq \dfrac{1}{\sqrt{99+1}} = \dfrac{1}{10}$ or $R_{99} \leq 0.1$

3) The series is an alternating convergent series and $a_n = \dfrac{1}{n!}$.

So, $|R_n| \leq a_{n+1} = \dfrac{1}{(n+1)!}$ → $|R_6| \leq \dfrac{1}{7!}$ or $|R_6| \leq 0.000198 \cdots$.

4) $|R_n| = \dfrac{1}{(n+1)!} \leq 0.0005 = \dfrac{1}{2000}$ When $n = 6$, $\dfrac{1}{7!} = \dfrac{1}{5040} \leq \dfrac{1}{2000}$.

Chapter 11. Sequences and Infinite Series

G. ABSOLUTE CONVERGENCE

Absolute convergence test can be applied to the series that *has mixed signs.*

> **THEOREM**
> 1. A series $\sum a_k = a_1 + a_2 + \cdots + a_k + \cdots$ is said to *"converge absolutely"* if the series of absolute values $\sum |a_k| = |a_1| + |a_2| + \cdots + |a_k| + \cdots$ converges.
>
> 2. A series $\sum a_k = a_1 + a_2 + \cdots + a_k + \cdots$ is said to *"diverge absolutely"* if the series of absolute values $\sum |a_k| = |a_1| + |a_2| + \cdots + |a_k| + \cdots$ diverges.

▶ **Example** Determine whether the following series converge absolutely.

1. $1 + \dfrac{1}{2} - \dfrac{1}{2^2} - \dfrac{1}{2^3} + \dfrac{1}{2^4} - \dfrac{1}{2^5} - \cdots$

2. $1 - \dfrac{1}{2} + \dfrac{1}{3} - \dfrac{1}{4} + \dfrac{1}{5} - \cdots$

1. The series of **absolute values** $1 + \dfrac{1}{2} + \dfrac{1}{2^2} + \dfrac{1}{2^3} + \dfrac{1}{2^4} + \dfrac{1}{2^5} - \cdots$ is the convergent geometric series.
   The series is **absolutely convergent. (Stronger than convergent)**

2. This is an **alternating convergent harmonic series**. But when we apply absolute convergent test, the series of absolute values is the divergent harmonic series. So, we might think the given series *diverges absolutely*. But it is wrong.
   Remember: The alternating harmonic series converges, *but does not converge absolutely because*
   $1 - \dfrac{1}{2} + \dfrac{1}{3} - \dfrac{1}{4} + \dfrac{1}{5} - \cdots$ converges but its absolute values $1 + \dfrac{1}{2} + \dfrac{1}{3} + \dfrac{1}{4} + \dfrac{1}{5} + \cdots$ diverges.
   Therefore, we can say the alternating harmonic series **converges, but not absolutely converges.**

▶ **Example** Determine whether the following series converge absolutely.

1. $1 - \dfrac{1}{2} - \dfrac{1}{2^2} + \dfrac{1}{2^3} + \dfrac{1}{2^4} - \dfrac{1}{2^5} - \dfrac{1}{2^6} + \cdots$

2. $\sum_{k=1}^{\infty} \dfrac{\sin k}{k^2}$

Dr. John Chung's AP Calculus AB/BC

Chapter 11. Sequences and Infinite Series

1. This is not an alternating series because the alternate in pairs after the first term. But the series converges absolutely (geometric series). The series *converges absolutely* and hence *converges.*

$$\sum_{n=1}^{\infty}|a_n| = \frac{1}{1} + \frac{1}{2} + \frac{1}{2^2} + \frac{1}{2^3} + \cdots \text{ convergent geometric series. Therefore, the given series } absolutely$$

converges.

Remember $1 - \frac{1}{2} - \frac{1}{2^2} + \frac{1}{2^3} + \frac{1}{2^4} - \frac{1}{2^5} - \frac{1}{2^6} + \cdots < \sum_{k=1}^{\infty} \frac{1}{2^{k-1}}$ (convergent geometric series)

2. $\sum_{k=1}^{\infty} \left|\frac{\sin k}{k^2}\right| = \sum_{k=1}^{\infty} \frac{|\sin k|}{k^2}$ We can see $|\sin k| \le 1$ for all k. So,

$\sum_{k=1}^{\infty} \frac{|\sin k|}{k^2} \le \sum_{k=1}^{\infty} \frac{1}{k^2} \to \sum_{k=1}^{\infty} \frac{1}{k^2}$ is a convergent *p*-series.

So the series of absolute values converges by the comparison test. The given series $\sum_{k=1}^{\infty} \frac{\sin k}{k^2}$ converges *absolutely* and hence *converges.*

H. THE RATIO TEST

THEOREM Let $\sum a_k$ be a series with *positive terms* and

$$\rho = \lim_{k \to \infty} \frac{a_{k+1}}{a_k}$$

1. If $\rho < 1$, the series **converges.**
2. If $\rho > 1$ or $\rho = +\infty$, the series diverges.
3. If $\rho = 1$, the series may converge or diverge. (In conclusive. Another test must be tried.)

▶ **Example** Use the ratio test to determine whether the following series converge or diverge.

1. $\sum_{k=1}^{\infty} \frac{2}{k!}$ 2. $\sum_{k=1}^{\infty} \frac{k}{5^k}$

1. $\rho = \lim_{k \to \infty} \frac{a_{k+1}}{a_k} = \lim_{k \to \infty} \frac{\frac{2}{(k+1)!}}{\frac{1}{k!}} = \lim_{k \to \infty} \frac{2k!}{(k+1)!} = \lim_{k \to \infty} \frac{2}{k+1} = 0$ The series converges, since $\rho < 1$.

2. $\rho = \lim_{k \to \infty} \frac{a_{k+1}}{a_k} = \lim_{k \to \infty} \frac{\frac{k+1}{5^{k+1}}}{\frac{k}{5^k}} = \lim_{k \to \infty} \frac{k+1}{5^{k+1}} \cdot \frac{5^k}{k} = \lim_{k \to \infty} \frac{1}{5} \cdot \frac{k+1}{k} = \frac{1}{5}$ The series converges, since $\rho < 1$.

Chapter 11. Sequences and Infinite Series

▶ **Example** Use the ratio test to determine whether the following series converge or diverge.

1. $\sum_{k=1}^{\infty} \dfrac{k^k}{k!}$
2) $\sum_{n=1}^{\infty} \dfrac{n!}{n^n}$

---

1. $\rho = \lim\limits_{k\to\infty} \dfrac{a_{k+1}}{a_k} = \lim\limits_{k\to\infty} \dfrac{(k+1)^{k+1}}{(k+1)!} \cdot \dfrac{k!}{k^k} = \lim\limits_{k\to\infty} \dfrac{(k+1)^{k+1}}{k^k} \cdot \dfrac{k!}{(k+1)!} = \lim\limits_{k\to\infty} \dfrac{(k+1)^{k+1}}{k^k} \cdot \dfrac{1}{k+1}$

$= \lim\limits_{k\to\infty} \dfrac{(k+1)^k}{k^k} = \lim\limits_{k\to\infty}\left(1+\dfrac{1}{k}\right)^k = e$ The series diverges, since $\rho > 1$.

2. $\rho = \lim\limits_{n\to\infty} \dfrac{a_{n+1}}{a_n} = \lim\limits_{n\to\infty} \dfrac{\dfrac{(n+1)!}{(n+1)^{n+1}}}{\dfrac{n!}{n^n}} = \lim\limits_{n\to\infty} \dfrac{(n+1)!}{(n+1)^{n+1}} \cdot \dfrac{n^n}{n!} = \lim\limits_{n\to\infty} \dfrac{(n+1)\,n!}{(n+1)^n(n+1)} \cdot \dfrac{n^n}{n!} = \lim\limits_{n\to\infty} \dfrac{n^n}{(n+1)^n}$

Remember $\dfrac{n^n}{(n+1)^n} = \left(\dfrac{n}{n+1}\right)^n = e^{\ln\left(\dfrac{n}{n+1}\right)^n} = e^{n\ln\left(\dfrac{n}{n+1}\right)} = e^{\dfrac{\ln\left(\dfrac{n}{n+1}\right)}{1/n}}$

So, $\lim\limits_{n\to\infty} \dfrac{n^n}{(n+1)^n} = \lim\limits_{n\to\infty} e^{\ln\left(\dfrac{n}{n+1}\right)^n} = e^{\left[\lim\limits_{n\to\infty}\dfrac{\ln\left(\tfrac{n}{n+1}\right)}{1/n}\right]} = e^{\left[\lim\limits_{n\to\infty}\dfrac{\ln n - \ln(n+1)}{1/n}\right]}$ (Apply L'Hôpital's rule: $\dfrac{0}{0}$)

$e^{\left[\lim\limits_{n\to\infty}\dfrac{\ln n - \ln(n+1)}{1/n}\right]} = e^{\left[\lim\limits_{n\to\infty}\dfrac{\tfrac{1}{n}-\tfrac{1}{n+1}}{-1/n^2}\right]} = e^{\left[\lim\limits_{n\to\infty}\dfrac{\tfrac{1}{n(n+1)}}{-1/n^2}\right]} = e^{\left[\lim\limits_{n\to\infty}\dfrac{-n^2}{n(n+1)}\right]} = e^{-1} = \dfrac{1}{e} < 1$ The series converges.

Chapter 11. Sequences and Infinite Series

I. RATIO TEST (FOR ABSOLUTE CONVERGENCE)

THEOREM If $\sum a_k$ be a series with **alternating terms** and

$$\rho = \lim_{k \to \infty} \frac{|a_{k+1}|}{|a_k|}$$

1. If $\rho < 1$, then the series $\sum a_k$ **converges absolutely** and therefore *converges*.
2. If $\rho > 1$ or $\rho = \infty$, then the series $\sum a_k$ *diverges*.
3. If $\rho = 1$, inconclusive.

Note You can also use this test for the series with positive terms.

▶ **Example** Use the ratio test for absolute convergence to determine whether the series converges.

1. $\sum_{k=1}^{\infty} (-1)^k \dfrac{3^k}{k!}$
2. $\sum_{k=1}^{\infty} (-1)^{k+1} \dfrac{(4k-1)!}{5^k}$

---

1. $|a_k| = \left|(-1)^k \dfrac{3^k}{k!}\right| = \dfrac{3^k}{k!}$ → $\rho = \lim_{k \to \infty} \dfrac{|a_{k+1}|}{|a_k|} = \lim_{k \to \infty} \dfrac{3^{k+1}}{(k+1)!} \cdot \dfrac{k!}{3^k} = \lim_{k \to \infty} \dfrac{3}{k+1} = 0 < 1$ → *The series absolutely converges* and therefore *converges*.

2. $|a_k| = \lim_{k \to \infty} \left|(-1)^{k+1} \dfrac{(4k-1)!}{5^k}\right| = \dfrac{(4k-1)!}{5^k}$ → $\rho = \lim_{k \to \infty} \dfrac{|a_{k+1}|}{|a_k|} = \lim_{k \to \infty} \dfrac{(4(k+1)-1)!}{5^{k+1}} \cdot \dfrac{5^k}{(4k-1)!}$

$= \lim_{k \to \infty} \dfrac{(4k+3)!}{5^{k+1}} \cdot \dfrac{5^k}{(4k-1)!} = \lim_{k \to \infty} \dfrac{(4k+3)(4k+2)(4k+1)4k}{5} = \infty$ → *The series diverges.*

Chapter 11. Sequences and Infinite Series

J. THE ROOT TEST

When it is difficult to find the limit required for the ratio test, the root test is sometimes useful.

THEOREM If $\sum a_k$ is a series with positive terms, then $\rho = \lim\limits_{n \to \infty} \sqrt[n]{a_n}$

If $\sum a_k$ is an alternating terms, then $\rho = \lim\limits_{n \to \infty} \sqrt[n]{|a_n|}$

1. If $\rho < 1$, the series converges.
2. If $\rho > 1$ or $\rho = \infty$, the series diverges.
3. If $\rho = 1$, inconclusive. The series may converge or diverge. (Another test will be applied.)

▶ **Example** Use the root test to determine whether the following series converge or diverge.

1. $\sum\limits_{n=1}^{\infty} \left(\dfrac{5n-3}{3n+2} \right)^n$ 2. $\sum\limits_{n=1}^{\infty} \left[\dfrac{1}{\ln(n+2)} \right]^n$ 3. $\sum\limits_{n=1}^{\infty} (-1)^{n-1} \dfrac{2^{n+3}}{(n+1)^n}$

1. $\rho = \lim\limits_{n \to \infty} \sqrt[n]{a_n} = \lim\limits_{n \to \infty} \sqrt[n]{\left(\dfrac{5n-3}{3n+2} \right)^n} = \lim\limits_{n \to \infty} \dfrac{5n-3}{3n+2} = \dfrac{5}{3} > 1 \rightarrow$ The series diverges, since $\rho > 1$.

2. $\rho = \lim\limits_{n \to \infty} \sqrt[n]{a_n} = \lim\limits_{n \to \infty} \dfrac{1}{\ln(n+2)} = 0 < 1 \rightarrow$ The series converges, since $\rho < 1$.

3. $|a_n| = \dfrac{2^{n+3}}{(n+1)^n} = \dfrac{8(2^n)}{(n+1)^n} = 8\left(\dfrac{2}{n+1}\right)^n$ We apply the root test.

$\lim\limits_{n \to \infty} \sqrt[n]{8\left(\dfrac{2}{n+1}\right)^n} = \lim\limits_{n \to \infty} 8^{1/n} \left(\dfrac{2}{n+1}\right) = 8^0 \times 0 = 0 < 1$

There, the given series $\sum\limits_{n=1}^{\infty} (-1)^{n-1} \dfrac{2^{n+3}}{(n+1)^n}$ is absolutely convergent by root test.

Chapter 11. Sequences and Infinite Series

SUMMARY OF CONVERGENCE TESTS

TEST NAME	THEOREM ($\sum a_k$ and $\sum b_k$ are series with positive terms)
Divergence Test (*n*th-term Test)	If the sequence does not converge to 0, then the series must diverge. ○ If $\lim\limits_{n\to\infty} a_n \neq 0$, then the series $\sum a_k$ diverges. ○ If $\lim\limits_{n\to\infty} a_n = 0$, then the test is inconclusive. ($\sum a_k$ may or may not converge.)
***p*-Series Test**	The form of *p*-series is $$\sum_{k=1}^{\infty} \frac{1}{k^p} = \frac{1}{1^p} + \frac{1}{2^p} + \frac{1}{3^p} + \cdots$$ ○ If $p > 1$, then the series converges. ○ If $0 < p \leq 1$, then the series diverges
Harmonic Series Test	For $p = 1$, the series is the harmonic series. $$\sum_{k=1}^{\infty} \frac{1}{k} = \frac{1}{1} + \frac{1}{2} + \frac{1}{3} + \cdots$$ The harmonic series diverges. The integral test is convenient for establishing the convergence or divergence of *p*-series. $$\sum_{k=1}^{\infty} (-1)^{k+1} \frac{1}{n} = \frac{1}{1} - \frac{1}{2} + \frac{1}{3} - \frac{1}{4} + \cdots$$ The alternating harmonic series converges.
Geometric Series Test	For the geometric series $\sum\limits_{k=0}^{\infty} ar^k = \frac{a}{1-r}$ ○ If $|r| \geq 1$, the series diverges. ○ If $|r| < 1$, the series converges.
Ratio Test	○ If $\lim\limits_{k\to\infty} \left|\frac{a_{k+1}}{a_k}\right| < 1$, then the series $\sum a_n$ converges. ○ If $\lim\limits_{k\to\infty} \left|\frac{a_{k+1}}{a_k}\right| > 1$ or $\lim\limits_{n\to\infty} \left|\frac{a_{k+1}}{a_k}\right| = \infty$, then the series diverges. ○ If $\lim\limits_{k\to\infty} \left|\frac{a_{k+1}}{a_k}\right| = 1$, then the test is inconclusive.
Integral Test	○ If f is positive, continuous, and decreasing and $a_k = f(k)$ for all $k \geq c$ then $$\sum_{k=1}^{\infty} a_k \text{ and } \int_c^{\infty} f(x)dx$$ Either both converge or both diverge.

Chapter 11. Sequences and Infinite Series

TEST NAME	THEOREM
Comparison Test	Let $0 \leq a_k \leq b_k$ for all k, ○ If $\sum_{k=1}^{\infty} b_k$ converges, then $\sum_{k=1}^{\infty} a_k$ converges. ○ If $\sum_{k=1}^{\infty} a_k$ diverges, $\sum_{k=1}^{\infty} b_k$ diverges.
Limit Comparison Test	○ If $a_k > 0, b_k > 0,$ and $$\rho = \lim_{k \to \infty} \frac{a_k}{b_k}$$ ○ If ρ is finite and positive, then the series $\sum a_k$ and $\sum b_k$ either both converge or both diverge.
Root Test	$\sum a_k$ is a series with positive terms and $\rho = \lim_{k \to \infty} \sqrt[k]{a_k} = \lim_{k \to \infty} (a_k)^{1/k}$ ○ If $\rho < 1$, the series converges. ○ If $\rho > 1$ or $\rho = \infty$, the series diverges. ○ If $\rho = 1$, the test is inconclusive.
Alternating Series Test	The series contain both positive and negative terms. Let $a_k > 0$. $\sum_{k=1}^{\infty} (-1)^k a_k$ or $\sum_{k=1}^{\infty} (-1)^{k+1} a_k$ are alternating series. ○ If $a_{k+1} \leq a_k$ for all k and $\lim_{k \to \infty} a_k = 0$, then the series converges.

Chapter 11. Sequences and Infinite Series

More Practice

Verify that the series diverges. (Use the Convergence Test)

1. $\sum_{n=1}^{\infty} \dfrac{n}{n+3}$
2. $\sum_{n=1}^{\infty} \dfrac{n^2+1}{n^3}$
3. $\sum_{n=1}^{\infty} 5\left(\dfrac{3}{2}\right)^n$
4. $\sum_{n=1}^{\infty} \dfrac{2}{3^n}$
5. $\sum_{n=1}^{\infty} \dfrac{n^2}{e^n}$
6. $\sum_{n=1}^{\infty} \dfrac{1}{n(n+1)}$

---

1. Divergent
2. Inconclusive (Another test must be applied) 3. Inconclusive (Another test must be applied)
4. Inconclusive (Another test must be applied)
5. $\lim\limits_{n\to\infty} \dfrac{n^2}{e^n} = \lim\limits_{n\to\infty} \dfrac{2n}{e^n} = \lim\limits_{n\to\infty} \dfrac{2}{e^n} = 0$ Inconclusive (Another test must be applied)
6. Inconclusive (Another test must be applied)

Determine whether the series converges or diverges. (Use p-series Test)

1. $\sum_{n=1}^{\infty} \dfrac{1}{n}$
2. $\sum_{n=1}^{\infty} \dfrac{1}{n^3}$
3. $\sum_{n=1}^{\infty} \dfrac{1}{n+2}$
4. $1 + \dfrac{1}{4} + \dfrac{1}{9} + \dfrac{1}{16} + \cdots$
5. $\sum_{n=1}^{\infty} \dfrac{1}{\sqrt[3]{n}}$
6. $\sum_{n=1}^{\infty} \dfrac{5}{n^{1/2}}$
7. $1 + \dfrac{1}{2\sqrt{2}} + \dfrac{1}{3\sqrt{3}} + \dfrac{1}{4\sqrt{4}} + \cdots$
8. $1 + \dfrac{1}{\sqrt[3]{4}} + \dfrac{1}{\sqrt[3]{9}} + \dfrac{1}{\sqrt[3]{16}} + \cdots$

---

1. $p = 1$, Divergent (Harmonic series)   2. $p = 3$, Convergent
3. Part of harmonic series is also divergent.   4. $p = 2$, Convergent
5. $p = 1/3$, divergent   6. $\sum_{n=1}^{\infty} \dfrac{5}{n^{1/2}} = 5\sum_{n=1}^{\infty} \dfrac{1}{n^{1/2}}$, $p = 1/2$, Divergent
7. $1 + \dfrac{1}{2\sqrt{2}} + \dfrac{1}{3\sqrt{3}} + \dfrac{1}{4\sqrt{4}} + \cdots = \dfrac{1}{1} + \dfrac{1}{2^{3/2}} + \dfrac{1}{3^{3/2}} + \dfrac{1}{4^{3/2}} + \cdots$ $p = 3/2$, Convergent
8. $1 + \dfrac{1}{\sqrt[3]{4}} + \dfrac{1}{\sqrt[3]{9}} + \dfrac{1}{\sqrt[3]{16}} + \cdots = \sum_{n=1}^{\infty} \dfrac{1}{n^{2/3}}$, $p = 2/3$, Divergent

Chapter 11. Sequences and Infinite Series

Determine whether the series converges or diverges. (Use the Geometric Series Test)

1. $\sum_{n=1}^{\infty} 10\left(\dfrac{3}{2}\right)^n$

2. $\sum_{n=1}^{\infty} 1000(0.8)^n$

3. $\sum_{n=1}^{\infty} 2\left(-\dfrac{1}{3}\right)^n$

4. $\sum_{n=1}^{\infty} \dfrac{3}{2^n}$

---

1. Divergent 2. Convergent 3. Convergent 4. Convergent

Determine whether the series converges or diverges. (Use the Ratio Test)

1. $\sum_{n=1}^{\infty} \dfrac{3^n}{n!}$

2. $\sum_{n=1}^{\infty} \dfrac{n^n}{n!}$

3. $\sum_{n=1}^{\infty} \dfrac{2^{n+1} n^2}{3^n}$

4. $\sum_{n=1}^{\infty} \dfrac{4^n}{3^n+1}$

5. $\sum_{n=1}^{\infty} \dfrac{(-1)^n 3^{n-1}}{n!}$

6. $\sum_{n=1}^{\infty} \dfrac{1}{n^3}$

7. $\sum_{n=1}^{\infty} \dfrac{1}{\sqrt{n}}$

8. $\sum_{n=0}^{\infty} \dfrac{(n!)^2}{(3n)!}$

---

1. Convergent 2. Divergent 3. Convergent 4. Divergent 5. Convergent
6. Inconclusive

$$\lim_{n \to \infty} \dfrac{a_{n+1}}{a_n} = \lim_{n \to \infty} \dfrac{1}{(n+1)^3} \cdot \dfrac{n^3}{1} = \lim_{n \to \infty} \left(\dfrac{n}{n+3}\right)^3 = 1$$

But the series is convergent when you apply *p*-series test.

7. Inconclusive. If you use the *p*-series test you can see that the series diverges, because $p = \dfrac{1}{2} < 1$

8. Convergent

$$\sum_{n=0}^{\infty} \dfrac{(n!)^2}{(3n)!} \Rightarrow \lim_{n \to \infty} \left| \dfrac{[(n+1)!]^2}{(3n+3)!} \cdot \dfrac{(3n)!}{(n!)^2} \right| = \lim_{n \to \infty} \left| \dfrac{(n+1)!(n+1)!}{n!n!} \cdot \dfrac{(3n)!}{(3n+3)!} \right|$$

$$= \lim_{n \to \infty} \left| \dfrac{(n+1)^2}{(3n+3)(3n+2)(3n+1)} \right| = 0 < 1$$

Dr. John Chung's AP Calculus AB/BC

Chapter 11. Sequences and Infinite Series

Determine whether the series converges or diverges. (Use the Integral Test)

1. $\sum_{n=1}^{\infty} \dfrac{n}{n^2+1}$

2. $\sum_{n=1}^{\infty} \dfrac{1}{n^2+1}$

3. $\sum_{n=1}^{\infty} e^{-n}$

4. $\sum_{n=1}^{\infty} \dfrac{1}{n^3}$

5. $\sum_{n=1}^{\infty} n e^{n}$

6. $\sum_{n=2}^{\infty} \dfrac{1}{n(\ln n)^3}$

7. $\sum_{n=2}^{\infty} \dfrac{1}{n\sqrt{\ln n}}$

8. $\sum_{n=2}^{\infty} \dfrac{\ln n}{n}$

---

1. Divergent
2. Convergent

$$\sum_{n=1}^{\infty} \dfrac{1}{n^2+1} \Rightarrow \lim_{b \to \infty} \int_1^b \dfrac{1}{x^2+1}\,dx = \lim_{b \to \infty}\left[\arctan x\right]_1^b = \lim_{b \to \infty}(\arctan b - \arctan 1) = \dfrac{\pi}{2} - \dfrac{\pi}{4} = \dfrac{\pi}{4}$$

3. Convergent

$$\sum_{n=1}^{\infty} e^{-n} \Rightarrow \lim_{b \to \infty} \int_1^b e^{-x}\,dx = \lim_{b \to \infty}\left[-e^{-x}\right]_1^b = \lim_{b \to \infty}\left(-e^{-b}\right) - \left(-e^{-1}\right) = 0 + \dfrac{1}{e} = \dfrac{1}{e}$$

4. Convergent
   Note: If you use the *p*-series test, you can see that the series converges. $p = 3 > 1$

5. Divergent

$$\sum_{n=1}^{\infty} n e^{n} \Rightarrow \lim_{b \to \infty} \int_1^b x e^x\,dx = \lim_{b \to \infty}\left[e^x(x-1)\right]_1^b = \lim_{b \to \infty}\left[e^b(b-1) - 0\right] = \infty$$

Integration by parts: $\begin{cases} u = x,\ du = dx \\ dv = e^x dx,\ v = e^x \end{cases}$ $\int xe^x\,dx = xe^x - \int e^x\,dx = xe^x - e^x + C$

6. Convergent

$$\sum_{n=2}^{\infty} \dfrac{1}{n(\ln n)^3} \Rightarrow \lim_{b \to \infty} \int_2^b \dfrac{1}{x(\ln x)^3}\,dx = \lim_{b \to \infty} \int_{\ln 2}^{\ln b} \dfrac{1}{u^3}\,du = \lim_{b \to \infty}\left[-\dfrac{1}{2u^2}\right]_{\ln 2}^{\ln b} = -\dfrac{1}{2}\lim_{b \to \infty}\left[\dfrac{1}{u^2}\right]_{\ln 2}^{\ln b}$$

$$= -\dfrac{1}{2}\lim_{b \to \infty}\left(\dfrac{1}{(\ln b)^2} - \dfrac{1}{(\ln 2)^2}\right) = -\dfrac{1}{2}\left(-\dfrac{1}{(\ln 2)^2}\right) = \dfrac{1}{2(\ln 2)^2}$$

u-substitution: $u = \ln x$, $du = \dfrac{1}{x}\,dx$

Chapter 11. Sequences and Infinite Series

Determine whether the series converges or diverges. (Use the Comparison Test)

1. $\sum_{n=1}^{\infty} \dfrac{1}{3^n+1}$

2. $\sum_{n=1}^{\infty} \dfrac{1}{\sqrt{n}+1}$

3. $\sum_{n=1}^{\infty} \dfrac{1}{n^3+1}$

4. $\sum_{n=1}^{\infty} \dfrac{1}{3n^2+2}$

1.

Given series $\sum a_n$	Comparison series $\sum b_n$
$\sum_{n=1}^{\infty} \dfrac{1}{3^n+1}$	$\sum_{n=1}^{\infty} \dfrac{1}{3^n}$: Convergent Geometric Series Test

Check: $a_n = \dfrac{1}{3^n+1} < \dfrac{1}{3^n} = b_n$ for all $n \geq 1$

Since $\sum_{n=1}^{\infty} \dfrac{1}{3^n}$ converges by the geometric series test, the given series $\sum_{n=1}^{\infty} \dfrac{1}{3^n+1}$ converges.

2.

Given series $\sum a_n$	Comparison series $\sum b_n$
$\sum_{n=1}^{\infty} \dfrac{1}{\sqrt{n}+1}$	$\sum_{n=1}^{\infty} \dfrac{1}{\sqrt{n}} = \sum_{n=1}^{\infty} \dfrac{1}{n^{1/2}}$: Divergent p-Series Test

Check: $a_n < b_n$ for all $n \geq 1$

Because the comparison series diverges, the Direct Comparison Test tells you nothing.

If you expecting the series to diverge, you have to use another comparison series like $\sum \dfrac{1}{n}$.

$a_n = \dfrac{1}{\sqrt{n}+1} > \dfrac{1}{n} = b_n$ for $n \geq 4$

Because the harmonic series diverges, the given series diverges.

3.

Given series $\sum a_n$	Comparison series $\sum b_n$
$\sum_{n=1}^{\infty} \dfrac{1}{n^3+1}$	$\sum_{n=1}^{\infty} \dfrac{1}{n^3}$: Convergent p-Series Test $(p=3)$

Check: $a_n < b_n$ for all $n \geq 1$

Since $\sum_{n=1}^{\infty} \dfrac{1}{n^3}$ converges by p-Series Test, the given series $\sum_{n=1}^{\infty} \dfrac{1}{n^3+1}$ converges.

Dr. John Chung's AP Calculus AB/BC

Chapter 11. Sequences and Infinite Series

4.

Given series $\sum a_n$	Comparison series $\sum b_n$
$\sum_{n=1}^{\infty} \dfrac{1}{3n^2+2}$	$\dfrac{1}{3}\sum_{n=1}^{\infty} \dfrac{1}{n^2}$: Convergent p-Series Test $(p=2)$

Check: $a_n < b_n$ for all $n \geq 1$

Since $\sum_{n=1}^{\infty} \dfrac{1}{3n^2}$ converges by p-series test, the given series $\sum_{n=1}^{\infty} \dfrac{1}{3n^2+2}$ converges.

Determine whether the series converges or diverges. (Use the Limit Comparison Test)

1. $\sum_{n=1}^{\infty} \dfrac{\sqrt{n}}{n^2+1}$

2. $\sum_{n=1}^{\infty} \dfrac{n2^n}{4n^3+1}$

3. $\sum_{n=1}^{\infty} \dfrac{n}{n^2+1}$

4. $\sum_{n=1}^{\infty} \dfrac{1}{\sqrt{n^2+1}}$

5. $\sum_{n=1}^{\infty} \dfrac{1}{n\sqrt{n^2+1}}$

---

1. **Step 1)** Disregard all but the height powers in the numerator and the denominator.

   $\sum_{n=1}^{\infty} \dfrac{\sqrt{n}}{n^2+1} \Rightarrow \sum_{n=1}^{\infty} \dfrac{\sqrt{n}}{n^2} = \sum_{n=1}^{\infty} \dfrac{n^{1/2}}{n^2} = \sum_{n=1}^{\infty} \dfrac{1}{n^{3/2}}$: $p = \dfrac{3}{2} > 1 \Rightarrow$ Convergent p-Series.

   Step 2) Find the limit.

   $\lim_{n \to \infty} \dfrac{a_n}{b_n} = \lim_{n \to \infty} a_n \cdot \dfrac{1}{b_n} = \lim_{n \to \infty} \dfrac{\sqrt{n}}{n^2+1} \cdot \dfrac{n^{3/2}}{1} = \lim_{n \to \infty} \dfrac{n^2}{n^2+1} = 1$

   Since the limit is finite and positive and comparison series converges, the given series converges.

2. **Step 1)**

   $\sum_{n=1}^{\infty} \dfrac{n2^n}{4n^3+1} \Rightarrow \sum_{n=1}^{\infty} \dfrac{n2^n}{n^3} = \sum_{n=1}^{\infty} \dfrac{2^n}{n^2}$

   By the nth Term Test: $\lim_{n \to \infty} \dfrac{2^n}{n^2} = \lim_{n \to \infty} \dfrac{2^n \ln 2}{2n} = \lim_{n \to \infty} \dfrac{2^n (\ln 2)^2}{2} = \infty$ The series diverges.

   Step 2)

   $\lim_{n \to \infty} \dfrac{a_n}{b_n} = \lim_{n \to \infty} \dfrac{n2^n}{4n^3+1} \cdot \dfrac{n^2}{2^n} = \lim_{n \to \infty} \dfrac{n^3}{4n^3} = \dfrac{1}{4}$

   Since the limit is finite and positive and comparison series diverges, the given series diverges.

Chapter 11. Sequences and Infinite Series

3. Step 1) $\sum_{n=1}^{\infty} \frac{n}{n^2+1} \Rightarrow \sum_{n=1}^{\infty} \frac{n}{n^2} = \sum_{n=1}^{\infty} \frac{1}{n}$: Divergent Harmonic Series

   Step 2) $\lim_{n \to \infty} \frac{a_n}{b_n} = \lim_{n \to \infty} \frac{n}{n^2+1} \cdot \frac{n}{1} = \lim_{n \to \infty} \frac{n^2}{n^2+1} = 1$

   Since the limit is finite and positive and comparison series diverges, the given series diverges.

4. Step 1) $\sum_{n=1}^{\infty} \frac{1}{\sqrt{n^2+1}} \Rightarrow \sum_{n=1}^{\infty} \frac{1}{\sqrt{n^2}} = \sum_{n=1}^{\infty} \frac{1}{n}$: Divergent Harmonic Series

   Step 2) $\lim_{n \to \infty} \frac{a_n}{b_n} = \lim_{n \to \infty} \frac{1}{\sqrt{n^2+1}} \cdot \frac{n}{1} = \lim_{n \to \infty} \frac{n}{\sqrt{n^2+1}} = 1$

   Since the limit is finite and positive and comparison series diverges, the given series diverges.

5. Step 1) $\sum_{n=1}^{\infty} \frac{1}{n\sqrt{n^2+1}} \Rightarrow \sum_{n=1}^{\infty} \frac{1}{n\sqrt{n^2}} = \sum_{n=1}^{\infty} \frac{1}{n^2}$: Convergent p-Series

   Step 2) $\lim_{n \to \infty} \frac{1}{n\sqrt{n^2+1}} \cdot \frac{n^2}{1} = \lim_{n \to \infty} \frac{n^2}{n^2\sqrt{1+1/n^2}} = 1$

   Since the limit is finite and positive and comparison series converges, the given series converges.

Determine whether the series converges or diverges. (Use the Alternating Series Test)

1. $\sum_{n=0}^{\infty} \left(-\frac{1}{3}\right)^n$

2. $\sum_{n=1}^{\infty} (-1)^n \frac{1}{n}$

3. $\sum_{n=1}^{\infty} (-1)^{n+1} \frac{n}{2^{n-1}}$

4. $\sum_{n=1}^{\infty} (-1)^{n+1} \frac{(n+1)}{n}$

5. $\sum_{n=1}^{\infty} \frac{(-1)^{n+1}}{2n-1}$

6. $\sum_{n=1}^{\infty} \frac{(-1)^n}{n!}$

---

1. $\sum_{n=0}^{\infty} \left(-\frac{1}{3}\right)^n = \sum_{n=0}^{\infty} (-1)^n \left(\frac{1}{3}\right)^n \Rightarrow a_n = \left(\frac{1}{3}\right)^n$

   Step 1) Check $a_{n+1} \leq a_n$ (The sequence is decreasing)

   Step 2) $\lim_{n \to \infty} \left(\frac{1}{3^n}\right) = 0$

   The series converges, because the two conditions are met.

2. $\sum_{n=1}^{\infty} (-1)^n \frac{1}{n} \Rightarrow a_n = \frac{1}{n}$

   Step 1) $a_{n+1} \leq a_n$ Step 2) $\lim_{n \to \infty} \left(\frac{1}{n}\right) = 0$

   The series converges, because the two conditions are met.

Dr. John Chung's AP Calculus AB/BC

Chapter 11. Sequences and Infinite Series

3. $\sum_{n=1}^{\infty}(-1)^{n+1}\dfrac{n}{2^{n-1}}\;\Rightarrow\;a_n=\dfrac{n}{2^{n-1}}$

   Step 1) $a_{n+1}\le a_n\;\Rightarrow\;\dfrac{a_{n+1}}{a_n}\le 1\;\Rightarrow\;\dfrac{n+1}{2^n}\cdot\dfrac{2^{n-1}}{n}=\dfrac{n+1}{n}\cdot 2^{-1}=\dfrac{n+1}{2n}\le 1$ for all $n\ge 1$

   Step 2) $\lim\limits_{n\to\infty}\dfrac{n}{2^{n-1}}=\lim\limits_{n\to\infty}\dfrac{(n)'}{(2^{n-1})'}=\lim\limits_{n\to\infty}\dfrac{1}{2^{n-1}\ln 2}=0$

   The series converges, because the two conditions are met.

4. $\sum_{n=1}^{\infty}(-1)^{n+1}\dfrac{(n+1)}{n}\;\Rightarrow\;a_n=\dfrac{n+1}{n}$

   Step 1) $a_{n+1}\le a_n\;\Rightarrow\;\dfrac{a_{n+1}}{a_n}=\dfrac{n+2}{n+1}\cdot\dfrac{n}{n+1}=\dfrac{n^2+2n}{n^2+2n+1}\le 1$ for all $n\ge 1$

   Step 2) $\lim\limits_{n\to\infty}\dfrac{n+1}{n}=1\ne 0$

   Since the series passes the first condition, but the series doesn't pass the second condition, you cannot apply the Alternating Series Test.
   Note: The series diverges by the nth-Term Test.

5. $\sum_{n=1}^{\infty}\dfrac{(-1)^{n+1}}{2n-1}\;\Rightarrow\;a_n=\dfrac{1}{2n-1}$

   Step 1) $a_{n+1}\le a_n\;\Rightarrow\;\dfrac{1}{2+1}\le\dfrac{1}{2n-1}$ for all $n\ge 1$

   Step 2) $\lim\limits_{n\to\infty}\dfrac{1}{2n-1}=0$

   The series converges, because the two conditions are met.

6. $\sum_{n=1}^{\infty}\dfrac{(-1)^n}{n!}\;\Rightarrow\;a_n=\dfrac{1}{n!}$

   Step 1) $a_{n+1}\le a_n\;\Rightarrow\;\dfrac{1}{(n+1)!}\le\dfrac{1}{n!}$ for all $n\ge 1$

   Step 2) $\lim\limits_{n\to\infty}\dfrac{1}{n!}=0$

   The series converges, because the two conditions are met.

Chapter 11. Sequences and Infinite Series

K. EXPRESSING THE RIEMANN SUM AS A DEFINITE INTEGRAL

For the Riemann sum of an infinite series

$$S = \lim_{n \to \infty} \frac{1^2 + 2^2 + 3^2 + \cdots + n^2}{n^3} = \lim_{n \to \infty} \frac{\sum_{k=1}^{n} k^2}{n^3}$$

The sum can be obtained from the formula $\sum_{k=1}^{n} k^2 = 1^2 + 2^2 + 3^2 + \cdots + n^2 = \frac{n(n+1)(2n+1)}{6}$.

Therefore, $S = \lim_{n \to \infty} \frac{n(n+1)(2n+1)}{6(n^3)} = \frac{1}{3}$.

The sum also can be obtained from a definite integral.

$$S = \lim_{n \to \infty} \frac{1^2 + 2^2 + 3^2 + \cdots + n^2}{n^3} = \lim_{n \to \infty} \frac{1^2 + 2^2 + 3^2 + \cdots + n^2}{n^2} \cdot \frac{1}{n}$$

$$= \lim_{n \to \infty} \left\{ \left(\frac{1}{n}\right)^2 \cdot \frac{1}{n} + \left(\frac{2}{n}\right)^2 \cdot \frac{1}{n} + \left(\frac{3}{n}\right)^2 \cdot \frac{1}{n} + \left(\frac{4}{n}\right)^2 \cdot \frac{1}{n} + \cdots + \left(\frac{n-1}{n}\right)^2 \cdot \frac{1}{n} + \left(\frac{n}{n}\right)^2 \cdot \frac{1}{n} \right\}$$

$$= \lim_{n \to \infty} \sum_{k=1}^{n} \left(\frac{k}{n}\right)^2 \cdot \left(\frac{1}{n}\right) = \lim_{n \to \infty} \sum_{k=1}^{n} f(x_k) \cdot \left(\frac{1-0}{n}\right) = \int_0^1 x^2 \, dx = \left[\frac{x^3}{3}\right]_0^1 = \frac{1}{3}$$

The Process of Converting a Riemann Sum to a definite integral

$$\lim_{n \to \infty} \sum_{k=1}^{n} f\left(a + \frac{pk}{n}\right) \cdot \frac{p}{n} = \int_a^b f(x) \, dx, \text{ where } b = a + p$$

For example,

$$S = \lim_{n \to \infty} \sum_{k=1}^{n} \left(\frac{k}{n}\right)^2 \cdot \left(\frac{1}{n}\right) \Rightarrow p = 1, a = 0, \text{ and } f(x) = x^2$$

Therefore,

$$S = \lim_{n \to \infty} \sum_{k=1}^{n} \left(\frac{k}{n}\right)^2 \cdot \left(\frac{1}{n}\right) = \int_0^1 x^2 \, dx$$

Chapter 11. Sequences and Infinite Series

▶ **Example**

Express the following Riemann sums as definite integrals.

1. $\displaystyle\lim_{n\to\infty}\left(\dfrac{\sqrt{1}+\sqrt{2}+\sqrt{3}+\cdots+\sqrt{n}}{\sqrt{n^3}}\right)$

2. $\displaystyle\lim_{n\to\infty}\sum_{k=1}^{n}\left(\dfrac{3k}{n}\right)^{3}\cdot\dfrac{3}{n}$

---

1. $\displaystyle\lim_{n\to\infty}\left(\dfrac{\sqrt{1}+\sqrt{2}+\sqrt{3}+\cdots+\sqrt{n}}{\sqrt{n^3}}\right)=\lim_{n\to\infty}\sum_{k=1}^{n}\left(\sqrt{\dfrac{k}{n}}\right)\left(\dfrac{1}{n}\right)=\int_{0}^{1}\sqrt{x}\,dx$

2. $\displaystyle\lim_{n\to\infty}\sum_{k=1}^{n}\left(\dfrac{3k}{n}\right)^{3}\cdot\dfrac{3}{n}=\int_{0}^{3}x^{3}\,dx$

▶ **Example**

Express the following Riemann sums as definite integrals.

1. $\displaystyle\lim_{n\to\infty}\sum_{k=1}^{n}\left(1+\dfrac{2k}{n}\right)^{3}\cdot\dfrac{2}{n}$

2. $\displaystyle\lim_{n\to\infty}\sum_{k=1}^{\infty}\left(2+\dfrac{3k}{n}\right)^{2}\cdot\dfrac{3}{n}$

3. $\displaystyle\lim_{n\to\infty}\dfrac{1^{3}+2^{3}+3^{3}+\cdots+n^{3}}{n^{4}}$

4. $\displaystyle\lim_{n\to\infty}\dfrac{1}{n}\left\{\left(1+\dfrac{1}{n}\right)^{2}+\left(1+\dfrac{2}{n}\right)^{2}+\cdots+\left(1+\dfrac{n}{n}\right)^{2}\right\}$

---

1. $\displaystyle\lim_{n\to\infty}\sum_{k=1}^{n}\left(1+\dfrac{2k}{n}\right)^{3}\cdot\dfrac{2}{n}=\int_{1}^{3}x^{3}\,dx$

2. $\displaystyle\lim_{n\to\infty}\sum_{k=1}^{\infty}\left(2+\dfrac{3k}{n}\right)^{2}\cdot\dfrac{3}{n}=\int_{2}^{5}x^{2}\,dx$

3. $\displaystyle\lim_{n\to\infty}\dfrac{1^{3}+2^{3}+3^{3}+\cdots+n^{3}}{n^{4}}=\lim_{n\to\infty}\sum_{k=1}^{n}\left(\dfrac{k}{n}\right)^{3}\dfrac{1}{n}=\int_{0}^{1}x^{3}\,dx$

4. $\displaystyle\lim_{n\to\infty}\dfrac{1}{n}\left\{\left(1+\dfrac{1}{n}\right)^{2}+\left(1+\dfrac{2}{n}\right)^{2}+\cdots+\left(1+\dfrac{n}{n}\right)^{2}\right\}=\lim_{n\to\infty}\sum_{k=1}^{n}\left(1+\dfrac{k}{n}\right)^{2}\left(\dfrac{1}{n}\right)=\int_{1}^{2}x^{2}\,dx$

Chapter 11. Sequences and Infinite Series

▶ **Example**

Find the value of $\displaystyle\lim_{n\to\infty}\sum_{k=1}^{n}\left(2+\frac{3k}{n}\right)^2\left(\frac{3}{n}\right)$ using a definite integral.

$$\lim_{n\to\infty}\sum_{k=1}^{n}\left(2+\frac{3k}{n}\right)^2\left(\frac{3}{n}\right)=\int_{2}^{5}x^2\,dx=\left[\frac{x^3}{3}\right]_{2}^{5}=\frac{125-8}{3}=39$$

▶ **Example**

Find the value of $\displaystyle\lim_{n\to\infty}\sum_{k=1}^{n}\left(1+\frac{2k}{n}\right)^3\cdot\left(\frac{3}{n}\right)$ using a definite integral.

$$\lim_{n\to\infty}\sum_{k=1}^{n}\left(1+\frac{2k}{n}\right)^3\cdot\left(\frac{3}{n}\right)=\lim_{n\to\infty}\sum_{k=1}^{n}\frac{3}{2}\left(1+\frac{2k}{n}\right)^3\cdot\left(\frac{2}{n}\right)=\frac{3}{2}\int_{1}^{3}x^3\,dx=\frac{3}{2}\left[\frac{x^4}{4}\right]_{1}^{4}=\frac{3}{2}\left(\frac{4^4-1^4}{4}\right)=\frac{765}{8}$$

▶ **Example**

Find the value of $\displaystyle\lim_{n\to\infty}\sum_{k=1}^{2n}\frac{k^3}{n^4}$ using a definite integral.

$$\lim_{n\to\infty}\sum_{k=1}^{2n}\frac{k^3}{n^4}=\lim_{n\to\infty}16\sum_{k=1}^{2n}\frac{k^3}{(2n)^4}=\lim_{n\to\infty}16\sum_{k=1}^{2n}\frac{k^3}{(2n)^3}\left(\frac{1}{2n}\right)=\lim_{n\to\infty}16\sum_{k=1}^{2n}\left(\frac{k}{2n}\right)^3\left(\frac{1}{2n}\right)=16\int_{0}^{1}x^3\,dx=16\left[\frac{x^4}{4}\right]_{0}^{1}$$
$$=16\left(\frac{1}{4}\right)=4$$

Chapter 12. Maclaurin and Taylor Series

A. POWER SERIES

DEFINITION. **Power series** is a series with a variable x that is raised to some power n.

$$\sum_{n=0}^{\infty} a_n x^n = a_0 + a_1 x + a_2 x^2 + a_3 x^3 + \cdots \quad \text{is called a power series in "}x\text{".}$$

If f is represented by a **power series** centered at c (= constant), then

$$\sum_{k=0}^{\infty} a_n (x-c)^n = a_0 + a_1 (x-c) + a_2 (x-c)^2 + a_3 (x-c)^3 + a_4 (x-c)^4 + \cdots + a_n (x-c)^n + \cdots.$$

It is called a power series in "x", centered at "c".

Note 1. The power series create a function of x. $\rightarrow f(x) = \sum_{n=0}^{\infty} a_n (x-c)^n$

2. Domain of the function must be all values of x such that the power series

$$\sum_{n=0}^{\infty} a_n (x-c)^n \text{ converges.}$$

 Example

Find the domain of the function represented by the power series.

1. $f(x) = \sum_{n=0}^{\infty} x^n$
2. $f(x) = \sum_{n=0}^{\infty} (-1)^n x^{2n}$

---

1. $f(x) = \sum_{n=0}^{\infty} x^n = 1 + x + x^2 + \cdots \rightarrow$ Geometric series with a common ratio "x".

   Domain must be all x such that the series converges. $\rightarrow |x| < 1$ or $-1 < x < 1$

   The function created by the series is $f(x) = \dfrac{1}{1-x}$, where $-1 < x < 1$. (The first term is 1)

2. $f(x) = \sum_{n=0}^{\infty} (-1)^n x^{2n} = 1 - x^2 + x^4 - x^6 + \cdots \rightarrow$ Geometric series with a domain such the series converges.

   Common ratio $= -x^2$: $\quad -1 < -x^2 < 1 \rightarrow x^2 < 1 \rightarrow -1 < x < 1$

   The function created by the series is $f(x) = \dfrac{1}{1 - (-x^2)} = \dfrac{1}{1 + x^2}$, where $-1 < x < 1$.

Chapter 12. Maclaurin and Taylor Series

B. RADIUS AND INTERVAL OF CONVERGENCE

THEOREM. A function $f(x)$ can be represented by a power series as follows;

$$f(x) = \sum_{n=0}^{\infty} a_n (x-c)^n \text{, where the power series converges.}$$

Exactly one of the following statements is true.

1. The series converges only for $x = c$. $R = 0$
2. The series converges absolutely (and converges) for all real x. $R = \infty$
3. The series converges absolutely (and converges) for all x in some finite open interval $|x - c| < R$. R is the radius of convergence of the power series.
4. At either $x = c + R$ or $x = c - R$, the series may converges or diverges.

Note For a power series, if the radius of convergence is a finite number R, the convergence at the endpoints of the interval of convergence should be tested separately. (Convergent test)

 Example

Find the interval of convergence and radius of convergence of the series

1. $\sum_{n=1}^{\infty} \dfrac{(x-3)^n}{n^2}$
2. $\sum_{n=0}^{\infty} n! x^n$

---

1. Ratio test

$$\lim_{n \to \infty} \left| \dfrac{a_{n+1}}{a_n} \right| = \lim_{n \to \infty} \left| a_{n+1} \cdot \dfrac{1}{a_n} \right| = \lim_{n \to \infty} \left| \dfrac{(x-3)^{n+1}}{(n+1)^2} \cdot \dfrac{n^2}{(x-3)^n} \right| = \lim_{n \to \infty} \left| \left(\dfrac{n}{n+1}\right)^2 \dfrac{(x-3)}{1} \right| = \lim_{n \to \infty} \left(\dfrac{n}{n+1}\right)^2 |x-3| < 1$$

$|x-3| < 1 \to -1 < x - 3 < 1 \to 2 < x < 4$

End point test: At $x = 2 \to \sum_{n=1}^{\infty} \dfrac{(-1)^n}{n^2} \to$ Alternating series, convergent

At $x = 4 \to \sum_{n=1}^{\infty} \dfrac{1^n}{n^2} \to$ Convergent p-series $(p = 2)$

Therefore, the interval of convergence is $[2, 4]$ and the radius of convergence is $R = 1$.

2. Ratio test

$$\lim_{n \to \infty} \left| \dfrac{(n+1)! x^{n+1}}{n! x^n} \right| = \lim_{n \to \infty} |(n+1)x| = \lim_{n \to \infty} (n+1)|x| < 1 \text{ (convergent)}$$

We can see, if $x \neq 0$, then $\lim_{n \to \infty} (n+1)|x| = \infty$. Only for $x = 0$, $\lim_{n \to \infty} (n+1)|x| < 1$

Therefore, the power series converges only for $x = 0$ and radius of convergence is 0.

Chapter 12. Maclaurin and Taylor Series

▶ **Example**

Find the interval of convergence and radius of convergence of the series

1. $\sum_{n=0}^{\infty}(-1)^n \dfrac{x^{2n}}{(2n)!}$
2. $\sum_{n=0}^{\infty}\dfrac{x^n}{n}$

---

1. Ratio test

$$\lim_{n\to\infty}\left|\dfrac{(-1)^{n+1}x^{2(n+1)}}{[2(n+1)]!}\cdot\dfrac{(2n)!}{(-1)^n x^{2n}}\right|=\lim_{n\to\infty}\dfrac{(2n)!}{(2n+2)!}\left|\dfrac{x^2}{1}\right|=\lim_{n\to\infty}\dfrac{(2n)!}{(2n+2)(2n+1)(2n)!}|x^2|$$

$$\lim_{n\to\infty}\dfrac{1}{(2n+2)(2n+1)}|x^2|=\lim_{n\to\infty}\dfrac{x^2}{(2n+2)(2n+1)}=0 \text{ for all } x.$$

Therefore, interval of convergence is $(-\infty,\infty)$ and $R=\infty$.

2. Ratio test

$$\lim_{n\to\infty}\left|\dfrac{x^{n+1}}{n+1}\cdot\dfrac{n}{x}\right|=\lim_{n\to\infty}\dfrac{n}{n+1}|x|=|x|<1 \text{ or } -1<x<1 \rightarrow \text{ Convergent and } R=1.$$

Endpoint test: At $x=-1 \rightarrow \sum_{n=0}^{\infty}\dfrac{(-1)^n}{n} \rightarrow$ Alternating harmonic series, convergent

At $x=1 \rightarrow$ Harmonic series, divergent

Therefore, interval of convergence is $[-1, 1)$ and $R=1$.

▶ **Example**

Find the interval of convergence and radius of convergence of the series

1. $\sum_{n=1}^{\infty}\dfrac{(x-1)^n}{n^2 2^n}$

2. $\sum_{n=0}^{\infty}\dfrac{(-1)^n 4^n x^2}{\sqrt{n+1}}$

Chapter 12. Maclaurin and Taylor Series

1. Ratio test

    $$\lim_{n\to\infty}\left|\frac{(x-1)^{n+1}}{(n+1)^2 2^{n+1}} \cdot \frac{n^2 2^n}{(x-1)^n}\right| = \lim_{n\to\infty}\frac{1}{2}\left(\frac{n}{n+1}\right)^2 |x-1| = \frac{1}{2}|x-1| < 1 \;\to\; |x-1| < 2 \;\to\; R = 2$$

    $|x-1| < 2 \;\to\; -2 < x-1 < 2 \;\to\; -1 < x < 3$

    Endpoint test:

    $x = -1 \;\to\; \sum_{n=0}^{\infty}\frac{(-2)^n}{n^2 2^n} = \sum_{n=0}^{\infty}\left[\frac{(-2)^n}{2^n}\right]\frac{1}{n^2} = \sum_{n=0}^{\infty}(-1)^n \frac{1}{n^2} \;\to\;$ Alternating series: convergent.

    $x = 3 \;\to\; \sum_{n=0}^{\infty}\frac{(2)^n}{n^2 2^n} = \sum_{n=0}^{\infty}\frac{1}{n^2} = \sum_{n=0}^{\infty}\frac{1}{n^2} \;\to\;$ Convergent p-series ($p = 2$)

    Therefore, interval of convergence is $[-1, 3]$ and $R = 2$.

2. $\sum_{n=0}^{\infty}\left|\frac{(-1)^{n+1} 4^{n+1} x^{n+1}}{\sqrt{n+2}} \cdot \frac{\sqrt{n+1}}{(-1)^n 4^n x^n}\right| = \sum_{n=0}^{\infty}\frac{\sqrt{n+1}}{\sqrt{n+2}}\left|\frac{4x}{1}\right| = |4x| < 1 \;\to\; |x| < \frac{1}{4}$

    Endpoint: at $x = -\frac{1}{4}$, $\sum_{n=0}^{\infty}\frac{(-1)^n 4^n \left(-\frac{1}{4}\right)^n}{\sqrt{n+1}} = \sum_{n=0}^{\infty}\frac{(-1)^n 4^n \left(-\frac{1}{4}\right)^n}{\sqrt{n+1}} = \sum_{n=0}^{\infty}(-1)^n \frac{4^n \left(-\frac{1}{4}\right)^n}{\sqrt{n+1}} = \sum_{n=0}^{\infty}\frac{1}{\sqrt{n+1}}$

    By limit comparison test, $\sum_{n=0}^{\infty}\frac{1}{\sqrt{n}}$ is a p-series divergent series $\left(p = \frac{1}{2}\right)$

    $\lim_{n\to\infty}\frac{\frac{1}{\sqrt{n+1}}}{\frac{1}{\sqrt{n}}} = \lim_{n\to\infty}\frac{\sqrt{n}}{\sqrt{n+1}} = 1$ (Finite number) $\to$ The series diverges.

    At $x = \frac{1}{4}$, $\sum_{n=0}^{\infty}\frac{(-1)^n 4^n \left(\frac{1}{4}\right)^n}{\sqrt{n+1}} = \sum_{n=0}^{\infty}\frac{(-1)^n 4^n \left(\frac{1}{4}\right)^n}{\sqrt{n+1}} = \sum_{n=0}^{\infty}(-1)^n \frac{1^n}{\sqrt{n+1}} = \sum_{n=0}^{\infty}(-1)^n \frac{1}{\sqrt{n+1}}$

    By alternating series test, $\lim_{n\to\infty}\frac{1}{\sqrt{n+1}} = 0$ and terms are decreasing. The series converges.

    Therefore, interval of convergence is $\left(-\frac{1}{4}, \frac{1}{4}\right]$ and $R = \frac{1}{4}$.

Chapter 12. Maclaurin and Taylor Series

C. DERIVATIVES AND INTEGRALS OF POWER SERIES

DEFINITION.
If f is represented by a *power series* centered at c, then

$$f(x) = \sum_{k=0}^{\infty} a_n(x-c)^n = a_0 + a_1(x-c) + a_2(x-c)^2 + a_3(x-c)^3 + a_4(x-c)^4 + \cdots + a_n(x-c)^n + \cdots.$$

1. Derivative of $f(x)$ is

$$f'(x) = 0 + a_1 + 2a_2(x-c)^1 + 3a_3(x-c)^2 + 4a_4(x-c)^3 + \cdots \quad \text{Or}$$

$$f'(x) = \sum_{n=1}^{\infty} na_n(x-c)^{n-1}$$

2. Integral of $f(x)$ is

$$\int f(x)dx = a_0(x-c) + a_1\frac{(x-c)^2}{2} + a_2\frac{(x-c)^3}{3} + \cdots \quad \text{Or}$$

$$\int f(x)dx = \sum_{n=0}^{\infty} \frac{a_n(x-c)^{n+1}}{n+1} + C$$

▶ **Example**

Find a power series of $\dfrac{1}{1-x}$ on interval $(-1, 1)$.

Simply we can find a power series of $\dfrac{a}{1-x} = \sum_{n=1}^{\infty} ar^{n-1}$ or $\sum_{n=0}^{\infty} ar^n$

Geometric convergent series when $|x| < 1$.

The first term is 1 and common ratio is x. The power series of $\dfrac{1}{1-x}$ is

$$\frac{1}{1-x} = \sum_{n=0}^{\infty} x^n = 1 + x + x^2 + \cdots \text{ for } |x| < 1$$

Chapter 12. Maclaurin and Taylor Series

▶ **Example**

Find a power series of $\ln(1-x)$ using $\dfrac{1}{1-x} = \sum_{n=0}^{\infty} x^n$

Remember $[\ln(1-x)]' = \dfrac{-1}{(1-x)} \rightarrow \ln(1-x) = -\displaystyle\int \dfrac{1}{(1-x)} dx$

$\dfrac{1}{1-x} = \displaystyle\sum_{n=0}^{\infty} x^n$

$\ln(1-x) = -\displaystyle\int \left(\sum_{n=0}^{\infty} x^n\right) dx = -\sum_{n=0}^{\infty} \left(\int x^n dx\right) = -\sum_{n=0}^{\infty} \dfrac{x^{n+1}}{n+1} + C$, when $x = 0$, $\ln 1 = C \rightarrow C = 0$

Therefore, the power series of $\ln(1-x) = -\displaystyle\sum_{n=0}^{\infty} \dfrac{x^{n+1}}{n+1}$ on $(-1, 1)$.

Note $\ln(1-x) = -\displaystyle\sum_{n=0}^{\infty} \dfrac{x^{n+1}}{n+1} = -\left(\dfrac{x}{1} + \dfrac{x^2}{2} + \dfrac{x^3}{3} + \cdots\right) \rightarrow$ You can also use.

▶ **Example**

Find the value of $\displaystyle\sum_{n=1}^{\infty} \dfrac{(1/2)^n}{n}$ from $\ln(1-x) = -\displaystyle\sum_{n=0}^{\infty} \dfrac{x^{n+1}}{n+1}$.

When $x = \dfrac{1}{2}$, $\rightarrow$ [It is in the interval of convergence $(-1, 1)$]

We can see that $\displaystyle\sum_{n=1}^{\infty} \dfrac{\left(\dfrac{1}{2}\right)^n}{n} = \sum_{n=0}^{\infty} \dfrac{\left(\dfrac{1}{2}\right)^{n+1}}{n+1}$

Therefore, $\displaystyle\sum_{n=0}^{\infty} \dfrac{\left(\dfrac{1}{2}\right)^{n+1}}{n+1} = -\ln\left(1 - \dfrac{1}{2}\right) = -\ln\dfrac{1}{2} = \ln 2$

Note If $x = 3$, we cannot get the value of $\ln(1-3)$ from the series $\displaystyle\sum_{n=0}^{\infty} \dfrac{3^n}{n}$.

Because at $x = 3$, the series is not convergent.

Chapter 12. Maclaurin and Taylor Series

D. TAYLOR AND MACLAURIN SERIES

DEFINITION.

Taylor Series

If $f(x)$ has a power series centered at c,

$$f(x) = \sum_{n=0}^{\infty} a_n (x-c)^n = a_0 + a_1(x-c) + a_2(x-c)^2 + \cdots + a_n(x-c)^n + \cdots$$

We can see

$$a_0 = f(c), \quad a_1 = f'(c), \quad a_2 = \frac{f''(c)}{2!}, \quad a_3 = \frac{f'''(c)}{3!}, \cdots$$

Therefore, $a_n = \dfrac{f^n(c)}{n!}$

The power series can be represented by

$$\sum_{n=0}^{\infty} a_n (x-c)^n = \sum_{n=0}^{\infty} \frac{f^n(c)}{n!}(x-c)^n \quad \text{is called Taylor Series.}$$

Therefore, The Taylor series is

$$f'(x) = f(c) + f'(c)(x-c) + \frac{f''(c)}{2!}(x-c)^2 + \frac{f'''(c)}{3!}(x-c)^3 + \cdots + \frac{f^{(n)}(c)}{n!}(x-c)^n + \cdots$$

Maclaurin Series

If $c = 0$, this becomes Maclaurin series.

$$f'(x) = \sum_{n=0}^{\infty} \frac{f^n(0)}{n!} x^n = f(0) + f'(0)x + \frac{f''(0)}{2!}x^2 + \frac{f'''(0)}{3!}x^3 + \cdots + \frac{f^{(n)}(0)}{n!}x^n + \cdots$$

▶ **Example**

1. Find the nth Maclaurin series for $f(x) = e^x$.

2. What is the interval of convergence?

Chapter 12. Maclaurin and Taylor Series

Maclaurin series is "Taylor series at $c = 0$."
Since $f'(x) = f''(x) = f'''(x) = \cdots = f^{(n)}(x) = \cdots = e^x$, then $f(0) = 1$, $f'(0) = 1$, $f''(0) = 1, \cdots, f^{(n)}(0) = 1$.
Thus, the nth Maclaurin series is

$$f(x) = 1 + x + \frac{x^2}{2!} + \frac{x^3}{3!} + \cdots + \frac{x^n}{n!} + \cdots = \sum_{k=0}^{\infty} \frac{x^n}{n!}$$

Note the nth Maclaurin polynomials is $P_n = \sum_{n=0}^{n} \frac{x^n}{n!}$ (Partial Sum)

2. Ratio test

$$\lim_{n \to \infty} \left| \frac{x^{n+1}}{(n+1)!} \cdot \frac{n!}{x^n} \right| = \lim_{n \to \infty} \frac{1}{n+1} |x| = 0 \quad \to \quad \text{This means "for any value of } x \text{, the ratio is 0."}$$

Therefore, interval of convergence is $(-\infty, +\infty)$. → Means "We can get any $f(x)$ from the Taylor series."

▶ **Example**

1. Find the Taylor series for $f(x) = \ln x$ centered at $x = 1$. $(c = 1)$
2. Find the interval of convergence.

1.

$f(x) = \ln x$	$f'(x) = 1/x$	$f''(x) = -1/x^2$	$f'''(x) = 2/x^3$	$f^4(x) = -6/x^4$
$f(1) = 0$	$f'(1) = 1$	$f''(1) = -1$	$f'''(1) = 2$	$f^4(1) = -6$

From the table, we can see $f^n(1) = (-1)^{n-1}(n-1)!$
Therefore, the Taylor series is

$$\sum_{n=0}^{\infty} \frac{f^n(1)}{n!}(x-1)^n = \sum_{n=0}^{\infty} \frac{(-1)^{n-1}(n-1)!}{n!}(x-1)^n = \sum_{n=0}^{\infty} \frac{(-1)^{n-1}(n-1)!}{n(n-1)!}(x-1)^n = \sum_{n=0}^{\infty} \frac{(-1)^{n-1}}{n}(x-1)^n$$

2. Ratio test

$$\lim_{n \to \infty} \left| \frac{(-1)^n (x-1)^{n+1}}{n+1} \cdot \frac{n}{(-1)^{n-1}(x-1)^n} \right| = \lim_{n \to \infty} \frac{n}{n+1} |x-1| = |x-1| < 1 \quad \to \quad 0 < x < 2$$

Endpoint test:

$x = 0 \to \sum_{n=0}^{\infty} \frac{(-1)^{n-1}(-1)^n}{n} = \sum_{n=0}^{\infty} \frac{(-1)^{2n-1}}{n} = \sum_{n=0}^{\infty} \frac{(-1)}{n} = -\sum_{n=0}^{\infty} \frac{1}{n}$ Negative harmonic series (Divergent)

$x = 2 \to \sum_{n=0}^{\infty} \frac{(-1)^{n-1}(2-1)^n}{n} = \sum_{n=0}^{\infty} \frac{(-1)^{n-1}}{n}$ Alternating harmonic series (Convergent)

Therefore, interval of convergence is $(0, 2]$ and $R = 1$.
That means "the value of ln3 cannot be obtained from the Taylor series because 3 is not in the interval of convergence."

Chapter 12. Maclaurin and Taylor Series

▶ Example

1. Find the Maclaurin series for $f(x) = \sin x$ and interval of convergence of the series.

2. Use the result in (1) to find the Maclaurin series of $g(x) = \cos x$.

1.

$f(x) = \sin x$	$f'(x) = \cos x$	$f''(x) = -\sin x$	$f'''(x) = -\cos x$	$f^4(x) = \sin x$
$f(0) = 0$	$f'(0) = 1$	$f''(0) = 0$	$f'''(0) = -1$	$f^4(0) = 0$

Now, the Maclaurin series is

$$\sum_{n=0}^{\infty} \frac{f^n(x^n)}{n!} = 0 + \frac{x}{1!} - \frac{x^3}{3!} + \frac{x^5}{5!} - \frac{x^7}{7!} + \cdots = \frac{x}{1!} - \frac{x^3}{3!} + \frac{x^5}{5!} - \frac{x^7}{7!} + \cdots$$

Sigma Notation of the power series is

$$\frac{x}{1!} - \frac{x^3}{3!} + \frac{x^5}{5!} - \frac{x^7}{7!} + \cdots = \sum_{k=0}^{\infty} \frac{(-1)^k x^{2k+1}}{(2k+1)!}$$

Ratio test for interval of convergence

$$\lim_{k \to \infty} \left| \frac{(-1)^{k+1} x^{2(k+1)+1}}{(2(k+1)+1)!} \cdot \frac{(2k+1)!}{(-1)^k x^{2k+1}} \right| = \lim_{k \to \infty} \left| \frac{(-1)}{(2k+3)!} \cdot \frac{(2k+1)! \cdot x^{2k+3}}{1} \cdot \frac{1}{x^{2k+1}} \right| = \lim_{k \to \infty} \frac{1}{(2k+3)(2k+2)} x^2 = 0 < 1$$

Therefore, interval of convergence is $(-\infty, +\infty)$ and $R = \infty$.

2.

We know that $\dfrac{d(\sin x)}{dx} = \cos x$. Therefore, $\cos x = \dfrac{d}{dx}\left(x - \dfrac{x^3}{3!} + \dfrac{x^5}{5!} - \dfrac{x^7}{7!} + \cdots \right) = 1 - \dfrac{x^2}{2!} + \dfrac{x^4}{4!} - \dfrac{x^6}{6!} + \cdots$

$$\cos x = 1 - \frac{x^2}{2!} + \frac{x^4}{4!} - \frac{x^6}{6!} + \cdots = \sum_{n=0}^{\infty} \frac{(-1)^n x^{2n}}{(2n)!} \quad \text{on the interval } (-\infty, +\infty).$$

Chapter 12. Maclaurin and Taylor Series

E. SOME IMPORTANT MACLAURIN SERIES

Maclaurin Series	Interval of Convergence
$\dfrac{1}{1-x} = 1 + x + x^2 + x^3 + \cdots = \sum_{n=0}^{\infty} x^n$	$-1 < x < 1$
$e^x = 1 + x + \dfrac{x^2}{2!} + \dfrac{x^3}{3!} + \cdots = \sum_{n=0}^{\infty} \dfrac{x^n}{n!}$	$-\infty < x < +\infty$
$\ln(1+x) = x - \dfrac{x^2}{2} + \dfrac{x^3}{3} - \dfrac{x^4}{4} + \cdots = \sum_{n=1}^{\infty} (-1)^{n-1} \dfrac{x^n}{n}$	$-1 < x \leq 1$
$\sin x = x - \dfrac{x^3}{3!} + \dfrac{x^5}{5!} - \dfrac{x^7}{7!} + \cdots = \sum_{n=0}^{\infty} (-1)^n \dfrac{x^{2n+1}}{(2n+1)!}$	$-\infty < x < +\infty$
$\cos x = 1 - \dfrac{x^2}{2!} + \dfrac{x^4}{4!} - \dfrac{x^6}{6!} + \cdots = \sum_{n=0}^{\infty} (-1)^n \dfrac{x^{2n}}{(2n)!}$	$-\infty < x < +\infty$
$\arctan x = x - \dfrac{x^3}{3} + \dfrac{x^5}{5} - \dfrac{x^7}{7} + \cdots = \sum_{n=0}^{\infty} (-1)^n \dfrac{x^{2n+1}}{(2n+1)}$	$-1 \leq x \leq 1$

Chapter 12. Maclaurin and Taylor Series

▶ **Example**

1. The Maclaurin series $\dfrac{1}{1-x} = 1 + x + x^2 + x^3 + \cdots = \sum_{n=0}^{\infty} x^n$ on $(-1, 1)$. Find the Maclaurin series for $\dfrac{1}{1+x}$.

2. Using the result in (1), find the Taylor series of $\dfrac{1}{1+x}$ centered at $x = 2$.

---

1. We replace x by $-x$ in the series. $\dfrac{1}{1+x} = \dfrac{1}{1-(-x)} = 1 - x + x^2 - x^3 + \cdots = \sum_{n=0}^{\infty} (-1)^n x^n$.

   Interval of convergence: Replace x by $-x$ in the in interval. $-1 < -x < 1 \;\to\; -1 < x < 1 \;\to\; (-1, 1)$.

2. We need to create $(x-2)$ in the equation.

$$\dfrac{1}{1+x} = \dfrac{1}{1+[(x-2)+2]} = \dfrac{1}{3+(x-2)} = \dfrac{1}{3}\left[\dfrac{1}{1+\dfrac{(x-2)}{3}}\right]$$

$$\dfrac{1}{1+\dfrac{(x-2)}{3}} = 1 - \dfrac{(x-2)}{3} + \dfrac{(x-2)^2}{3^2} - \dfrac{(x-2)^3}{3^3} + \cdots = \sum_{n=0}^{\infty} \dfrac{(-1)^n (x-2)^n}{3^n}$$

Therefore, $\dfrac{1}{1+x} = \dfrac{1}{3}\sum_{n=0}^{\infty} \dfrac{(-1)^n (x-2)^n}{3^n} = \sum_{n=0}^{\infty} \dfrac{(-1)^n (x-2)^n}{3^{n+1}}$. Now check the interval of convergence.

Replace x by $\dfrac{x-2}{3} \;\to\; -1 < \dfrac{x-2}{3} < 1 \;\to\; -3 < x-2 < 3 \;\to\; -1 < x < 5 \;\to\; (-1, 5)$ and $R = 3$.

▶ **Example**

Find the Maclaurin series for $f(x) = x^3 \sin 2x$ and its interval of convergence.

$\sin x = x - \dfrac{x^3}{3!} + \dfrac{x^5}{5!} - \dfrac{x^7}{7!} + \cdots = \sum_{n=0}^{\infty} (-1)^n \dfrac{x^{2n+1}}{(2n+1)!}$ Replace x with $2x$.

$\sin 2x = (2x) - \dfrac{(2x)^3}{3!} + \dfrac{(2x)^5}{5!} - \cdots = \sum_{n=0}^{\infty} \dfrac{(-1)^n (2x)^{2n+1}}{(2n+1)!}$ So,

$x^2 \sin 2x = x^2 \sum_{n=0}^{\infty} \dfrac{(-1)^n (2x)^{2n+1}}{(2n+1)!} = x^2 \sum_{n=0}^{\infty} \dfrac{(-1)^n (2)^{n+1} (x)^{2n+1}}{(2n+1)!} = \sum_{n=0}^{\infty} \dfrac{(-1)^n 2^{n+1} x^{2n+3}}{(2n+1)!}$

The interval of convergence is $-\infty < (2x) < \infty \;\to\; -\infty < x < \infty \;\to\; (-\infty, \infty)$.

Chapter 12. Maclaurin and Taylor Series

▶ **Example**

Find the Maclaurin series for $f(x) = \sinh x$.

We can see that $\sinh x = \dfrac{e^x - e^{-x}}{2}$, $e^x = 1 + x + \dfrac{x^2}{2!} + \dfrac{x^3}{3!} + \cdots$, and $e^{-x} = 1 - x + \dfrac{x^2}{2!} - \dfrac{x^3}{3!} + \cdots$

Therefore, $\dfrac{1}{2}(e^x - e^{-x}) = \dfrac{1}{2}\left(2x + \dfrac{2x^3}{3!} + \dfrac{2x^5}{5!} + \dfrac{2x^7}{7!} + \cdots\right) = x + \dfrac{x^3}{3!} + \dfrac{x^5}{5!} + \dfrac{x^7}{7!} + \cdots = \sum_{n=0}^{\infty} \dfrac{x^{2n+1}}{(2n+1)!}$.

on the interval $(-\infty, \infty)$.

F. APPROXIMATION BY TAYLOR POLYNOMIAL

Taylor series of $f(x)$ centered at $x = c$ is

$$f(x) = f(c) + f'(c)(x-c) + \dfrac{f''(c)(x-c)^2}{2!} + \dfrac{f'''(c)(x-c)^3}{3!} + \cdots + \dfrac{f^n(x-c)^n}{n!} + R_n(x)$$

We defined nth degree Taylor polynomial $P_n(x)$ as follows.

$$P_n(x) = f(c) + f'(c)(x-c) + \dfrac{f''(c)(x-c)^2}{2!} + \dfrac{f'''(c)(x-c)^3}{3!} + \cdots + \dfrac{f^n(x-c)^n}{n!} \quad \text{and}$$

$$R_n(x) = \dfrac{f^{n+1}(c)(x-c)^{n+1}}{(n+1)!} + \dfrac{f^{n+2}(c)(x-c)^{n+2}}{(n+2)!} + \cdots .$$

Therefore, $R_n = \dfrac{f^{n+1}(z)(x-c)^{n+1}}{(n+1)!}$.

▶ **Example**

1. Use 4th degree Taylor polynomial for $f(x) = \ln x$ to approximate $\ln(1.1)$.

2. Find $|R_4|$.

Chapter 12. Maclaurin and Taylor Series

1. We can see $c = 1$ from $\ln(1+0.1)$. Taylor series centered at $x = 1$.
   Find derivatives.
   $f(x) = \ln x$, $f'(x) = \dfrac{1}{x}$, $f''(x) = \dfrac{-1}{x^2}$, $f'''(x) = \dfrac{2}{x^3}$, $f^4(x) = \dfrac{-6}{x^4}$, $f^5(x) = \dfrac{24}{x^5}$

   $f(1) = 0 \qquad f'(1) = 1 \qquad f''(1) = -1, \qquad f'''(1) = 2, \qquad f^4(1) = -6$

   Hence
   $$P_4 = f(1) + f'(1)(x-1) + \dfrac{f''(-1)(x-1)^2}{2!} + \dfrac{f'''(1)(x-1)^3}{3!} + \dfrac{f^4(1)(x-1)^4}{4!} \quad \text{So,}$$

   $$P_4 = 0 + (x-1) - \dfrac{(x-1)^2}{2!} + \dfrac{2(x-1)^3}{3!} - \dfrac{6(x-1)^4}{4!} = (x-1) - \dfrac{(x-1)^2}{2} + \dfrac{(x-1)^3}{3} - \dfrac{(x-1)^4}{4}$$

   Therefore, $\ln(1.1) \approx P_4 = (0.1) - \dfrac{(0.1)^2}{2} + \dfrac{(0.1)^3}{3} - \dfrac{(0.1)^4}{4} = 0.0953083333$

2. $R_4 = \dfrac{f^5(z)(x-1)^5}{5!}$

   We know that $f^5(z) = \dfrac{24}{z^5}$ and $1 \leq z \leq 1.1$ So, at $z = 1$, $f^5(1) = 24$. (Biggest number)

   $R_4 = \dfrac{f^5(z)(x-1)^5}{5!}$.

   $R_4(1.1) = \dfrac{f^5(z)(1.1-1)^5}{5}$ has a biggest error when $z = 1$. Therefore,

   $|R_4(1.1)| = \dfrac{24(0.1)^5}{5!} = \dfrac{0.1^5}{5} = 0.000002$

   Actual value of $\ln(1.1) = 0.953101798$

   We can see $|\ln 2 - P_4| = 0.000001846 \leq 0.000002$

▶ **Example**

1. Use 3rd-degree Maclaurin polynomial for $f(x) = e^x$ to approximate e.

2. Find $|R_3|$.

Chapter 12. Maclaurin and Taylor Series

1. $f(x) = e^x$, $c = 0$, and $z = [0, 1]$

   Since $e = 1 + x + \frac{x^2}{2!} + \frac{x^3}{3!} + \frac{x^4}{4!} + \cdots$, the 3rd Maclaurin polynomial is $p_3(x) = 1 + x + \frac{x^2}{2!} + \frac{x^3}{3!}$

   Therefore, $P_3(1) = 1 + 1 + \frac{1}{2!} + \frac{1}{3!} = \frac{8}{3}$

2. $f^4(z) = e^z$ on $[0, 1]$

   $R_3 = \frac{f^4(z) x^4}{4!}$ → has a biggest error at $z = 1$ that is $f^4(1) = e$

   Hence the error bound is

   $R_3 = \frac{e}{4!}(1)^4 = \frac{e}{4!} \approx 0.1132617429$

   Actual error is $|R_3(1)| = \left| e^1 - \frac{8}{3} \right| \approx 0.0516151618$

Chapter 12. Maclaurin and Taylor Series

G. A FUNCTION COMPLETELY REPRESENTED BY A TAYLOR SERIES

For Taylor series for a function f around $x = c$ is
$$f(x) = \sum_{n=0}^{\infty} \frac{f^n(c)(x-c)^n}{n!}$$
Recall $f(x) = P_n(x) + R_n(x)$ and $R_n(x) = f(x) - P_n(x)$
$$\lim_{n \to \infty} R_n(x) = \lim_{n \to \infty} \left[f(x) - P_n(x) \right]$$
If $\lim_{n \to \infty} R_n(x) = 0$, then $f(x)$ can be completely represented by a Taylor series at $x = c$.
Therefore, $f(x) = P_n$ as $n \to \infty$.

▶ **Example**

Show that the Maclaurin series for e^x converges to e^x for all x. $(-\infty, \infty)$

$R_n(x) = \dfrac{f^{n+1}(z) x^{n+1}}{(n+1)!}$ and $f^{n+1}(x) = e^x \;\to\; f^{n+1}(z) = e^z$

$R_n(x) = \dfrac{e^z x^{n+1}}{(n+1)!}$

If $x > 0$, $0 < z < x$, then $0 \le R_n(x) = \dfrac{e^z x^{n+1}}{(n+1)!} \le \dfrac{e^x x^{n+1}}{(n+1)!} \;\to\;$ because $f(x) = e^x$ increasing. $\left(e^z \le e^x\right)$

$0 \le \lim\limits_{n \to \infty} R_n(x) = \lim\limits_{n \to \infty} \dfrac{e^z x^{n+1}}{(n+1)!} \le \lim\limits_{n \to \infty} \dfrac{e^x x^{n+1}}{(n+1)!} = 0 \;\to\;$ Remember $e^x x^{n+1}$ is finite number

If $x < 0$, $x < z < 0$, then $\left| R_n(x) \right| = \left| \dfrac{e^z x^{n+1}}{(n+1)!} \right| \le \left| \dfrac{x^{n+1}}{(n+1)!} \right| \;\to\;$ because $e^z < e^0 = 1$.

$0 \le \lim\limits_{n \to \infty} \left| R_n(x) \right| = \lim\limits_{n \to \infty} \left| \dfrac{e^z x^{n+1}}{(n+1)!} \right| \le \lim\limits_{n \to \infty} \left| \dfrac{x^{n+1}}{(n+1)!} \right| = 0$

Therefore,

$\lim\limits_{n \to \infty} R_n(x) = \lim\limits_{n \to \infty} \dfrac{e^z x^{n+1}}{(n+1)!} = 0 \;\to\;$ means "The Maclaurin series create a function e^x for all x."

Chapter 12. Maclaurin and Taylor Series

H. LAGRANGE UPPER ERROR BOUND

Remainder (error) estimation provides a useful upper bound M.

$$|R_n(x)| = \left| \frac{f^{n+1}(c)(x-c)^{n+1}}{(n+1)!} \right| \leq \left| \frac{M(x-c)^{n+1}}{(n+1)!} \right|$$

M is an upper bound on the give interval.

▶ Example

Let $f(x) = \sin\left(5x + \frac{\pi}{4}\right)$ and let $p(x)$ be the third-degree Maclaurin polynomial for f.

Use the Lagrange error bound to show that $\left| f\left(\frac{1}{10}\right) - p\left(\frac{1}{10}\right) \right| < \frac{1}{100}$.

Lagrange error = $\left| R_3\left(\frac{1}{10}\right) \right| = \left| f\left(\frac{1}{10}\right) - P\left(\frac{1}{10}\right) \right|$

Find the derivatives: 1) $f'(x) = 5\cos\left(5x + \frac{\pi}{4}\right)$ 2) $f''(x) = -25\sin\left(5x + \frac{\pi}{4}\right)$

3) $f'''(x) = -125\cos\left(5x + \frac{\pi}{5}\right)$ 4) $f^{(4)}(x) = 625\sin\left(5x + \frac{\pi}{4}\right)$

$|R_3| = \left| \frac{f^4(z)(x)^4}{4!} \right|$

$|f^4(z)| = \left| 625\sin\left(5z + \frac{\pi}{4}\right) \right|$ → Roughly we can say $\left| \sin\left(5x + \frac{\pi}{4}\right) \right| \leq 1$ on the interval $\left[0, \frac{1}{10}\right]$

$|R_3| = \left| \frac{f^4(z)(x)^4}{4!} \right| \leq \frac{625x^4}{4!}$, where $M = 625$.

Therefore, $R_3\left(\frac{1}{10}\right) \leq \frac{625(1/10)^4}{4!} = 0.0026041667 < \frac{1}{100}$

The error is still less than $\frac{1}{100}$.

Chapter 12. Maclaurin and Taylor Series

More Practice

1. Let $p(x) = 8 - 3(x-4) + 4(x-4)^2 - 2(x-4)^3 + 3(x-4)^4$ be the fourth-degree Taylor polynomial for the function f around $x = 4$.

   a) Find $f'''(4)$.
   b) Write the third-degree polynomial for f' about $x = 4$.
   c) Use the third-degree polynomial to approximate $f'(4.2)$.
   d) Write the fourth-degree Taylor polynomial for $g(x) = \int_4^x f(t)\,dt$ around $x = 4$.

2. Let f be a function that has derivatives of all orders for all numbers and $f(1) = 2$, $f'(1) = -2$, $f''(1) = 3$, and $f'''(1) = 5$. Write the third-degree Taylor polynomial for f around $x = 1$ and use it to approximate $f(1.2)$.

3. Taylor series for $\sin x$ about 0 is given by
$$\sin x = x - \frac{x^3}{3!} + \frac{x^5}{5!} - \frac{x^7}{7!} + \cdots + \frac{(-1)^n x^{2n+1}}{(2n+1)!} + \cdots.$$

   a) Use the third-degree polynomial $P_3(x)$ to approximate $\sin(0.1)$.
   b) Use the Lagrange error bound to show that $|f(0.1) - P_3(0.1)| < \dfrac{5}{10^6}$.

4. Let f be the function given by $f(x) = \ln x$, and let $P(x)$ be the third-degree Taylor polynomial for f about $x = 1$.

   a) Find the $P(x)$.
   b) Use the third-degree polynomial to estimate $\ln(1.2)$.
   c) Use the Lagrange error bound to show that $|f(1.2) - P(1.2)| < 0.001$

Solution

1. Taylor series at $c = 4$ is $f(x) = f(4) + f'(4)(x-4) + \dfrac{f''(4)(x-4)^2}{2!} + \dfrac{f'''(4)(x-4)^3}{3!} + \cdots$

   a) Coefficient of $(x-4)^3$ must be equal. $-2 = \dfrac{f'''(4)}{3!} \rightarrow f'''(4) = (-2)(3!) = -12$

   b) $f(x) = 8 - 3(x-3) + 4(x-4)^2 - 2(x-4)^3 + 3(x-4)^4 - \cdots$
   $f'(x) = -3 + 8(x-4) - 6(x-4)^2 + 12(x-4)^3$

   c) $f'(4.2) = -3 + 8(0.2) - 6(0.2)^2 + 12(0.2)^3 \approx -1.544$

Chapter 12. Maclaurin and Taylor Series

d) $g(x) = \int_4^x f(t)dt = \int_4^x 8 - 3(t-4) + 4(t-4)^2 - 2(t-4)^3 + 3(t-4)^4 - \cdots dt$

$= \left[8t - \dfrac{3(t-4)^2}{2} + \dfrac{4(t-4)^3}{3} - \dfrac{1(t-4)^4}{2} + \cdots \right]_4^x = 8(x-4) - \dfrac{3(x-4)^2}{2} + \dfrac{4(x-4)^3}{3} - \dfrac{(x-4)^4}{2}$

Therefore, $G_4 = 8(x-4) - \dfrac{3(x-4)^2}{2} + \dfrac{4(x-4)^3}{3} - \dfrac{(x-4)^4}{2}$

2. $P_3 = f(1) + f'(1)(x-1) + \dfrac{f''(1)(x-1)^2}{2!} + \dfrac{f'''(1)(x-1)^3}{3!}$

Substitute derivatives in the polynomial.

$P_3 = 2 - 2(x-1) + \dfrac{3(x-1)^2}{2!} + \dfrac{5(x-1)^3}{3!}$ and

$P_3(1.2) = 2 - 2(1.2-1) + \dfrac{3(1.2-1)^2}{2} + \dfrac{5(1.2-1)^3}{6} = 2 - 2(0.2) + \dfrac{3(0.2)^2}{2} + \dfrac{5(0.2)^3}{6} \approx 1.667$

3. a) $P_3(x) = x - \dfrac{x^3}{3!} \rightarrow P_3(0.1) = 0.1 - \dfrac{0.1^3}{3!} = 0.0998333$

b) $f'(x) = \cos x$, $f''(x) = -\sin x$, $f'''(x) = -\cos x$, and $f^4(x) = \sin x$

$f^4(z) = \sin z$ on $[0, 0.1]$

The maximum value of $f^{(4)}(x) = \sin x$ on $[0, 0.1]$ is $\sin(0.1)$, since the graph of $y = f^4(x)$ is increasing on $[0, 0.1]$. Now

$|R_3(0.1)| \leq \left| \dfrac{\sin(0.1)}{4!}(0.1)^4 \right| \approx 4.159726 \times 10^{-7} < 5 \times 10^6$

Note Roughly we take 1 as the maximum of $\sin x$. $f^4(z) = \sin z \leq 1 \rightarrow M = 1$

$|R_3(0.1)| \leq \left| \dfrac{1}{4!}(0.1)^4 \right| = \dfrac{1}{240000} \approx 4.166667 \times 10^{-6} < 5 \times 10^{-6}$

The maximum error is still less than $\dfrac{5}{10^6}$.

4. From $\ln x = (x-1) - \dfrac{(x-1)^2}{2} + \dfrac{(x-1)^3}{3} - \dfrac{(x-1)^4}{4} + \cdots$ we obtain $P_3(x) = (x-1) - \dfrac{(x-1)^2}{2} + \dfrac{(x-1)^3}{3}$.

$P_3(1.2) = (1.2-1) - \dfrac{(1.2-1)^2}{2} + \dfrac{(1.2-1)^3}{3} = \dfrac{2}{10} - \dfrac{2}{100} + \dfrac{1}{375} = \dfrac{137}{750} \approx 0.182667$

Actual error is $\left| \ln(1.2) - \dfrac{137}{750} \right| \approx 0.00034511 < 0.001$

Now we apply the Lagrange error bound.

$|R_3(1.2)| \leq \left| \dfrac{f^4(z)}{4}(1.2-1)^4 \right|$ and $f'(x) = \dfrac{1}{x}$, $f''(x) = -\dfrac{1}{x^2}$, $f'''(x) = \dfrac{2}{x^3}$, $f^4(x) = -\dfrac{6}{x^4}$

The maxim value of $\left| f^{(4)}(x) \right| = \left| -\dfrac{6}{x^4} \right|$ on the interval $[1, 1.2]$ is $\left| f^{(4)}(1) \right| = \left| -\dfrac{6}{1} \right| = 6$.

Hence the Lagrange error bound is $|R_3(1.2)| \leq \dfrac{6}{4}(1.2-1)^4 = 0.0024 < 0.001$

Chapter 13. Polar Coordinates and Equation

A. POLAR COORDINATES AND POLAR EQUATION

1. The polar coordinates (r, θ) of a point are related to the rectangular coordinates (x, y) of the point as follows.
$$x = r\cos\theta, \quad y = r\sin\theta, \quad r^2 = x^2 + y^2, \quad \tan\theta = \frac{y}{x}$$

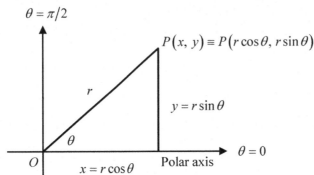

Note r can be negative. Pole is the origin. Polar axis is x-axis.

Example For $P(r, \theta)$, $P\left(-5, \dfrac{\pi}{3}\right) = P\left(5, \dfrac{\pi}{3} + \pi\right) \rightarrow P\left(-5, \dfrac{\pi}{3}\right) = P\left(5, \dfrac{4\pi}{3}\right)$

2. Graph of Polar equation
   $r = f(\theta)$ is called a polar equation.

   Symmetry for $r = f(\theta)$
   1) If $f(\theta) = f(-\theta)$, then r is symmetric about x-axis (polar axis)
   2) If $f(\pi - \theta) = f(\theta)$, then r is symmetric about y-axis.
   3) If $f(\pi + \theta) = -f(\theta)$, then r is symmetric about origin.

▶ **Example**

Sketch the graph of $r = 1 + \cos\theta$.

Chapter 13. Polar Coordinates and Equation

We can see $f(\theta) = f(-\theta)$, which is symmetric about polar axis.

θ	0	$\pi/4$	$\pi/2$	$3\pi/4$	π
r	$1+\cos 0 = 2$	$1+\cos \pi/4 \approx 1.7$	$1+\cos \pi/2 = 1$	$1+\cos 3\pi/4 \approx 0.3$	$1+\cos \pi = 0$

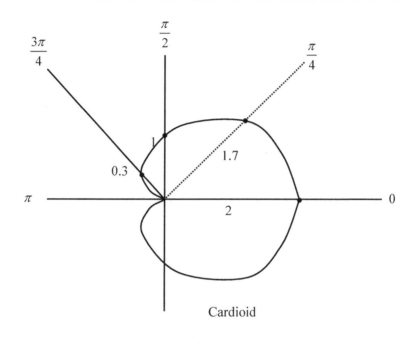

Cardioid

Chapter 13. Polar Coordinates and Equation

▶ **Example**

Sketch the graph of $r = 2\cos 2\theta$.

Check symmetry properties

1) $f(\theta) = f(-\theta)$, Symmetric about polar axis

2) $f(\pi - \theta) = f(\theta)$ → $2\cos[2(\pi-\theta)] = 2\cos(2\pi - 2\theta) = 2\cos 2\theta$, symmetric about $\theta = \dfrac{\pi}{2}$ (y-axis).

θ	0	$\pi/8$	$2\pi/8$	$3\pi/8$	π
r	$2\cos 0 = 2$	$2\cos 2(\pi/8) = \sqrt{2}$	$2\cos 2(\pi/4) = 0$	$2\cos 2(3\pi/8) = -\sqrt{2}$	$2\cos 2(\pi/2) = -2$

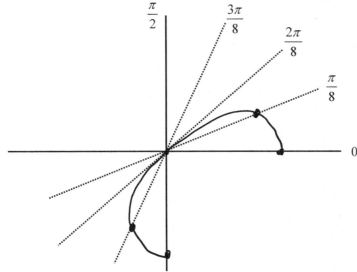

We got the graph from the table. But we know that it is symmetric about polar axis and $\theta = \dfrac{\pi}{2}$. Therefore, the graph of r must be as follows.

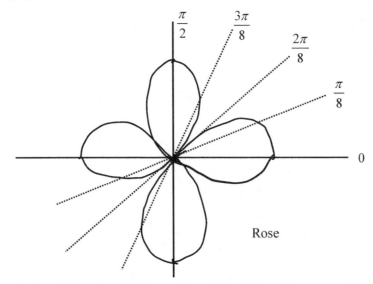

Rose

Chapter 13. Polar Coordinates and Equation

B. SLOPE OF TANGENT AND AREA IN POLAR COORDINATES

1. If f is a differentiable function of θ, then the slope of the line tangent to the graph of $r = f(\theta)$ at the point (r, θ) is

$$\frac{dy}{dx} = \frac{dy/d\theta}{dx/d\theta} = \frac{f'(\theta)\sin\theta + f(\theta)\cos\theta}{f'(\theta)\cos\theta - f(\theta)\sin\theta}$$

Proof: $y = r\sin\theta = f(\theta)\sin\theta$ and $x = r\cos\theta = f(\theta)\cos\theta$ (**Note** r is a function of θ.)

$$\frac{dy}{d\theta} = f'(\theta)\sin\theta + f(\theta)\cos\theta \quad \text{and} \quad \frac{dx}{d\theta} = f'(\theta)\cos\theta - f(\theta)\sin\theta$$

2. Area in Polar Coordinates
   If f is continuous and nonnegative on the interval $[\theta_1, \theta_2]$, then the area of the region enclosed by the graph of $r = f(\theta)$ on the interval $[\theta_1, \theta_2]$ is given by

$$A = \frac{1}{2}\int_{\theta_1}^{\theta_2} r^2 \, d\theta = \frac{1}{2}\int_{\theta_1}^{\theta_2} [f(\theta)]^2 \, d\theta$$

▶ **Example**

Find $\dfrac{dy}{dx}$ and slope of tangent at $\theta = \dfrac{\pi}{6}$ for $r = 1 + \cos\theta$.

We can see $\dfrac{dr}{d\theta} = -\sin\theta$ and substitute in the equation.

$$\frac{dy}{dx} = \frac{\dfrac{dr}{d\theta} \cdot \sin\theta + r\cos\theta}{\dfrac{dr}{d\theta} \cdot \cos\theta - r\sin\theta} = \frac{-\sin^2\theta + (1+\cos\theta)\cos\theta}{-\sin\theta\cos\theta - (1+\cos\theta)\sin\theta} = \frac{\cos^2\theta - \sin^2\theta + \cos\theta}{-2\sin\theta\cos\theta - \sin\theta} = -\left(\frac{\cos 2\theta + \cos\theta}{\sin 2\theta + \sin\theta}\right)$$

At $\theta = \dfrac{\pi}{6}$, $\dfrac{dy}{dx} = -\left(\dfrac{\cos\dfrac{\pi}{3} + \cos\dfrac{\pi}{6}}{\sin\dfrac{\pi}{3} + \sin\dfrac{\pi}{6}}\right) = -\left(\dfrac{\dfrac{1}{2} + \dfrac{\sqrt{3}}{2}}{\dfrac{\sqrt{3}}{2} + \dfrac{1}{2}}\right) = -1$

Chapter 13. Polar Coordinates and Equation

▶ **Example**

Find the slope of the tangent line to parametric equations $x = \cos\theta$, $y = \sin\theta$ at the point where $\theta = \dfrac{\pi}{6}$.

We can see $r = 1$. $\dfrac{dr}{d\theta} = 0$ and $\dfrac{dy}{dx} = \dfrac{\dfrac{dr}{d\theta} \cdot \sin\theta + r\cos\theta}{\dfrac{dr}{d\theta} \cdot \cos\theta - r\sin\theta} \rightarrow \dfrac{r\cos\theta}{-\sin\theta} = -\cot\theta$

Therefore, the slope at $\theta = \dfrac{\pi}{6}$ is $\left.\dfrac{dy}{dx}\right]_{\theta=\frac{\pi}{6}} = -\cot\theta\Big]_{\theta=\frac{\pi}{6}} = -\cot\dfrac{\pi}{6} = -\sqrt{3}$

C. CALCULUS OF PARAMETRIC EQUATION

For parametric equations $x = f(t)$, $y = g(t)$,

$$\dfrac{dy}{dx} = \dfrac{dy/dt}{dx/dt} \quad \left(\dfrac{dy}{dt} \neq 0\right)$$

On the curve of parametric equations

1. A horizon line occurs where $\dfrac{dy}{dt} = 0$ and $\dfrac{dx}{dt} \neq 0$.

$$m = 0 = \dfrac{dy}{dx} = \dfrac{dy/dt}{dx/dt} \rightarrow dy/dt = 0 \text{ and } dx/dt \neq 0$$

2. A vertical tangent line occurs where $\dfrac{dy}{dt} \neq 0$ and $\dfrac{dx}{dt} = 0$.

$$m = \text{undefined} = \dfrac{dy/dt}{dx/dt} \rightarrow dy/dt \neq 0 \text{ and } dx/dt = 0$$

Second derivative for parametric equations.

$$\dfrac{d^2y}{dx^2} = \dfrac{d}{dx}\left(\dfrac{dy}{dx}\right) = \dfrac{\dfrac{d\left(\dfrac{dy}{dx}\right)}{dt}}{\dfrac{dx}{dt}}$$

▶ **Example**

Find $\dfrac{dy}{dx}$ and $\dfrac{d^2y}{dx^2}$ at point $(1, 1)$ on the curves given by the parametric equations

$$x = t^2, \; y = t^3$$

Chapter 13. Polar Coordinates and Equation

We obtain $\dfrac{dy}{dx} = \dfrac{dy/dt}{dx/dt} = \dfrac{3t^2}{2t} = \dfrac{3}{4t}$ → $\dfrac{dy}{dt} = 3t^2$ and $\dfrac{dx}{dt} = 2t$

From the point $(1, 1)$ → $t^2 = 1$ and $t^3 = 1$ → $t = 1$

$$\left.\dfrac{dy}{dx}\right|_{t=1} = \left.\dfrac{3}{4t}\right|_{t=1} = \dfrac{3}{4}$$

$\dfrac{d^2y}{dx^2} = \dfrac{d\left(\dfrac{dy}{dx}\right)}{dx} = \dfrac{d(y')}{dx} = \dfrac{d(y')/dt}{dx/dt} = \dfrac{3}{4t}$ → $\dfrac{d(y')}{dt} = \dfrac{3}{2}$ and $\dfrac{dx}{dt} = 2t$

$\left.\dfrac{d^2y}{dx^2}\right|_{t=1} = \left.\dfrac{3}{4t}\right|_{t=1} = \dfrac{3}{4}$

▶ **Example**

Find the points on the curve of $r = 5$ at which there is a vertical tangent line.

$x = 5\cos\theta$, $y = 5\sin\theta$ and $\dfrac{dx}{d\theta} = -5\sin\theta$, $\dfrac{dy}{d\theta} = 5\cos\theta$

$\dfrac{dx}{d\theta} = -5\sin\theta = 0$ → $\sin\theta = 0$ → $\theta = 0, \pi$ on the interval $[0, 2\pi]$

At $\theta = 0$ and π, $\dfrac{dy}{d\theta} \neq 0$

Vertical tangent lines occur at $\theta = 0, \pi$.

▶ **Example**

If parametric equations $x = \sec t$ and $y = \tan t$ on $\left[-\dfrac{\pi}{2}, \dfrac{\pi}{2}\right]$, find the equation of tangent line at $t = \pi/4$.

At $t = \dfrac{\pi}{4}$ → $x = \sec\dfrac{\pi}{4} = \sqrt{2}$ and $\tan\dfrac{\pi}{4} = 1$ We got a point $(\sqrt{2}, 1)$

(Slope): $\dfrac{dy}{dx} = \dfrac{dy/dt}{dx/dt} = \dfrac{\sec^2 t}{\sec t \tan t} = \dfrac{\sec t}{\tan t} = \dfrac{1}{\sin t}$

$\dfrac{dy}{dx} = \dfrac{1}{\sin \pi/4} = \dfrac{1}{1/\sqrt{2}} = \sqrt{2}$

Hence the equation of tangent is $y - 1 = \sqrt{2}(x - \sqrt{2})$ → $y = \sqrt{2}x - 1$.

Dr. John Chung's AP Calculus AB/BC

Chapter 13. Polar Coordinates and Equation

▶ Example

For parametric equations $x = t^2$ and $y = t^3 - 3t$ $(t \geq 0)$
  1) Find the point where horizontal tangent occurs.
  2) Find the point where vertical tangent occurs.

1) $\dfrac{dy}{dt} = 3t^2 - 3 = 0 \rightarrow t^2 = 1 \rightarrow t = 1$

$\dfrac{dx}{dt} = 2t$, at $t = 1$, $\dfrac{dx}{dt} \neq 0$ Therefore, $x = 1$ and $y = -2$ $\rightarrow (1, -2)$

2) $\dfrac{dx}{dt} = 2t = 0 \rightarrow t = 0$ but $\dfrac{dy}{dt} = 3t^2 - 1 \neq 0$ Therefore, $x = 0$ and $y = 0$ $\rightarrow (0, 0)$

Chapter 13. Polar Coordinates and Equation

D. ARC LENGTH FOR POLAR CURVES

NOTE: Length of a curve $L = \int_a^b \sqrt{1 + \left[f'(x)\right]^2}\, dx$ for $a \leq x \leq b$

Parametrically $L = \int_a^b \sqrt{\left(\dfrac{dx}{dt}\right)^2 + \left(\dfrac{dy}{dt}\right)^2}\, dt$ for $a \leq t \leq b$

NOTE Arc Length on Polar Curve.
If r is continuous and differentiable with respect to θ on the interval $[\theta_1, \theta_2]$, then arc length L from θ_1 to θ_2 is

$$L = \int_{\theta_1}^{\theta_2} \sqrt{r^2 + \left[\dfrac{dr}{d\theta}\right]^2}\, d\theta$$

Proof

$L = \int_{\theta_1}^{\theta_2} \sqrt{\left(\dfrac{dx}{d\theta}\right)^2 + \left(\dfrac{dy}{d\theta}\right)^2}\, d\theta$ and $x = r\cos\theta,\ y = r\sin\theta$

We have $\dfrac{dx}{d\theta} = \dfrac{dr}{d\theta}\cdot\cos\theta - r\sin\theta$ and $\dfrac{dy}{d\theta} = \dfrac{dr}{d\theta}\cdot\sin\theta + r\cos\theta$

$\left(\dfrac{dx}{d\theta}\right)^2 + \left(\dfrac{dy}{d\theta}\right)^2$

$= \left(\dfrac{dr}{d\theta}\right)^2\cos^2\theta - 2r\sin\theta\cos\theta\cdot\dfrac{dr}{d\theta} + r^2\sin^2\theta + \left(\dfrac{dr}{d\theta}\right)^2\sin^2\theta + 2r\sin\theta\cos\theta\cdot\dfrac{dr}{d\theta} + r^2\cos^2\theta$

$= \left(\dfrac{dr}{d\theta}\right)^2\left(\cos^2\theta + \sin^2\theta\right) + r^2\left(\sin^2\theta + \cos^2\theta\right)$

$= \left(\dfrac{dr}{d\theta}\right)^2 + r^2$

$\left(\dfrac{dx}{d\theta}\right)^2 + \left(\dfrac{dy}{d\theta}\right)^2 = r^2 + \left(\dfrac{dr}{d\theta}\right)^2 \quad \rightarrow \quad$ Therefore,

$$L = \int_{\theta_1}^{\theta_2} \sqrt{\left(\dfrac{dx}{d\theta}\right)^2 + \left(\dfrac{dy}{d\theta}\right)^2}\, d\theta = \int_{\theta_1}^{\theta_2} \sqrt{r^2 + \left[\dfrac{dr}{d\theta}\right]^2}\, d\theta$$

Chapter 13. Polar Coordinates and Equation

▶ Example 1

1. The graph of the spiral $r(t) = e^t$ is shown below. Find the arc length of $r(t) = e^t$ between $t = 0$ and $t = \pi$.

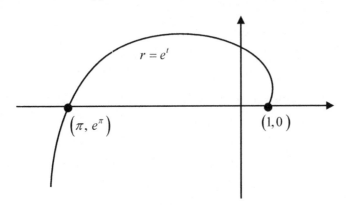

At $t = 0$, $r = e^0 = 1$ and at $t = \pi$, $r = e^\pi$

From $r^2 = e^{2t}$, $\dfrac{dr}{dt} = e^t$ → $L = \displaystyle\int_0^\pi \sqrt{r^2 + \left(\dfrac{dr}{dt}\right)^2}\, dt = \int_0^\pi \left(\sqrt{e^{2t} + e^{2t}}\right) dt = \int_0^\pi \left(\sqrt{2e^{2t}}\right) dt =$

$\displaystyle\int_0^\pi \sqrt{2}\,(e^t)\, dt = \left[\sqrt{2}\,(e^t)\right]_0^\pi = \sqrt{2}\,(e^\pi - 1)$

Or,

$x = r\cos t = e^t \cos t$ → $\dfrac{dx}{dt} = e^t \cos t - e^t \sin t = e^t(\cos t - \sin t)$

$y = r\sin t = e^t \sin t$ → $\dfrac{dy}{dt} = e^t \sin t + e^t \cos t = e^t(\sin t + \cos t)$

$\sqrt{\left(\dfrac{dx}{dt}\right)^2 + \left(\dfrac{dy}{dt}\right)^2} = \sqrt{e^{2t}(1 - 2\sin t \cos t) + e^{2t}(1 + 2\sin t \cos t)} = \sqrt{2e^{2t}} = e^t \sqrt{2}$

$L = \displaystyle\int_0^\pi e^t \sqrt{2}\, dt = \sqrt{2}\left[e^t\right]_0^\pi = \sqrt{2}\,(e^\pi - 1)$

Chapter 13. Polar Coordinates and Equation

E. AREA IN POLAR COORDINATES

If r is continuous on $[\theta_1, \theta_2]$, then the area of the region R enclosed by the polar curve $r = f(\theta)$ and the lines $\theta = \theta_1$ and $\theta = \theta_2$ is

$$\text{Area} = \int_{\theta_1}^{\theta_2} \frac{1}{2} r^2 \, d\theta$$

▶ Example

1. The polar curve r is given by $r(\theta) = 1 - \sin\theta$ for $0 \leq \theta \leq 2\pi$. Find the area in the second quadrant enclosed by the coordinate axes and the graph of r.

$$\text{Area} = \frac{1}{2}\int_{\pi/2}^{\pi} (1-\sin\theta)^2 \, d\theta = \frac{1}{2}\int_{\pi/2}^{\pi} (1 - 2\sin\theta + \sin^2\theta) \, d\theta = \frac{1}{2}\int_{\pi/2}^{\pi} \left(1 - 2\sin\theta + \frac{1-\cos 2\theta}{2}\right) d\theta$$

$$= \frac{1}{2}\int_{\pi/2}^{\pi} \left(\frac{3}{2} - 2\sin\theta - \frac{\cos 2\theta}{2}\right) d\theta = \frac{1}{2}\left[\frac{3}{2}x + 2\cos\theta - \frac{1}{4}\sin 2\theta\right]_{\frac{\pi}{2}}^{\pi}$$

$$= \frac{1}{2}\left[\left(\frac{3}{2}\pi + 2 - 0\right) - \left(\frac{3}{4}\pi + 0 - 0\right)\right] = \frac{1}{2}\left(\frac{3}{4}\pi + 2\right) = \frac{3}{8}\pi + 1$$

Remember. $\sin^2\theta = \dfrac{1-\cos 2\theta}{2}$ $\cos^2\theta = \dfrac{1+\cos 2\theta}{2}$

▶ Example

2. Let R be the region inside the graph of the polar curve $r = 2$ and outside the graph of the polar curve $r = 2(1 - \sin\theta)$. Find the area of R.

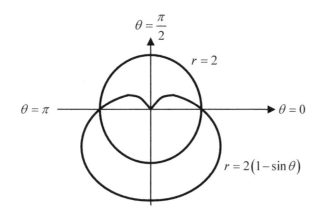

Dr. John Chung's AP Calculus AB/BC

Chapter 13. Polar Coordinates and Equation

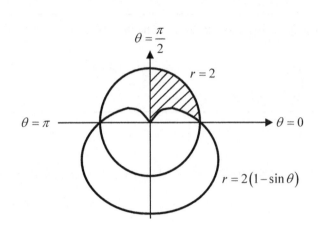

Area of the semicircle is $\dfrac{\pi(2)^2}{2} = 2\pi$

Area of the shaded region $= \pi - \displaystyle\int_0^\pi \dfrac{1}{2} \cdot 4(1-\sin\theta)^2\, d\theta = \pi - \int_0^\pi 2(1-\sin\theta)^2\, d\theta$

$\dfrac{A}{2} = \pi - 2\displaystyle\int_0^{\pi/2}(1-\sin\theta)^2\, d\theta = \pi - 2\int_0^{\pi/2}(1-2\sin\theta+\sin^2\theta)\, d\theta = \pi - 2\int_0^{\pi/2}\left(1-2\sin\theta + \dfrac{1-\cos 2\theta}{2}\right) d\theta$

$= \pi - 2\displaystyle\int_0^{\pi/2}\left(\dfrac{3}{2} - 2\sin\theta - \dfrac{\cos 2\theta}{d}\right) d\theta = \pi - 2\left[\dfrac{3}{2}\theta + 2\cos\theta - \dfrac{\sin 2\theta}{4}\right]_0^{\pi/2}$

$= \pi - 2\left[\left(\dfrac{3\pi}{4}+0-0\right)-(0+2-0)\right] = \pi - 2\left(\dfrac{3\pi}{4}-2\right) = 4 - \dfrac{\pi}{2}$

Now we obtain the area $A = 2\left(4 - \dfrac{\pi}{2}\right) = 8 - \pi$.

Calculus AB Practice Test

NO MATERIAL ON THIS PAGE

AP CALCULUS AB

TEST 1

AP CALCULUS AB

A CALCULATOR **CANNOT BE USED ON PART A OF SECTION 1**. A GRAPHING CALCULATOR FROM THE APPROVED LIST **IS REQUIRED FOR PART B OF SECTION 1 AND FOR PART A OF SECTION II** OF THE EXAMINATION.

SECTION 1
Time: 1 hour and 45 minutes
All questions are given equal weight.
Percent of total grade – 50

Part A: 55 minutes, 28 multiple-choice questions
A calculator is NOT allowed.
Part B: 50 minutes, 17 multiple-choice questions
A graphing calculator is required.

CALCULUS AB
SECTION 1, PART A
Time — 55 minutes
Number of questions — 28

A CALCULATOR MAY NOT BE USED ON THIS PART OF THE EXAMINATION.

Directions: Solve each of the following problems, using the available space for scratchwork. After examining the form of the choices, decide which is the best of the choices given and fill in the corresponding oval on the answer sheet. Do not spend too much time on any one problem.

In this test:

(1) Unless otherwise specified, the domain of a function f is assumed to be the set of all real numbers x for which $f(x)$ is a real number.

(2) The inverse of a trigonometric function f may be indicated using the inverse function notation f^{-1} or with the prefix "arc" (e.g, $\sin^{-1} x = \arcsin x$)

AP Calculus AB Test 1   Section 1

1. If $y = (2x^3 + 1)^3$, then $\dfrac{dy}{dx} =$

   (A) $(6x^3)^3$ (B) $3(2x^3+1)^2$ (C) $3(6x+1)^2$ (D) $18x^2(2x^3+1)^2$ (E) $3(6x^2)^2$

2. $\displaystyle\int_0^1 2^x\,dx =$

   (A) $\dfrac{1}{\ln 2}$ (B) 1 (C) $\ln 2$ (D) 2 (E) $2\ln 5$

3. If $f(x) = \dfrac{x^2+1}{x+2}$, then $f'(x) =$

   (A) x (B) $\dfrac{2x}{(x+2)^2}$ (C) $\dfrac{3x^2+4x+1}{(x+2)^2}$ (D) $\dfrac{x^2+4x-1}{(x+2)^2}$ (E) $\dfrac{2}{(x+2)^2}$

4. If $f(x) = \dfrac{2x+5}{x-2}$, then which of the following statements is not true?

   (A) $\lim\limits_{x \to -\infty} f(x) = 2$
   (B) $\lim\limits_{x \to 2} f(x) = \infty$
   (C) $\lim\limits_{x \to \infty} f(x) = 2$
   (D) $f(2)$ is undefined.
   (E) $x = 2$ is an vertical asymptote.

5. $\displaystyle\int_0^{\frac{\pi}{4}} \sin 2x\,dx =$

   (A) 1 (B) -1 (C) $-\dfrac{1}{2}$ (D) $\dfrac{1}{2}$ (E) $\dfrac{\sqrt{2}}{2}$

AP Calculus AB Test 1 Section 1

6. $\lim\limits_{x \to \infty} \dfrac{x^4 + 3x^2 - 1}{5x^4 - 3x^3 + 2x^2 - 1} =$

(A) −1   (B) 0   (C) 0.2   (D) 5   (E) ∞

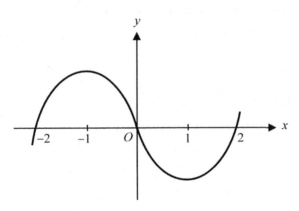

7. The graph of f' is shown above. Which of the following statements is true about f?

(A) f is decreasing for $-1 \leq x \leq 0$.
(B) f is increasing for $1 \leq x \leq 2$.
(C) f has a local maximum at $x = -1$.
(D) f has a local minimum at $x = -1$.
(E) f has a local maximum at $x = 0$.

---

8. $\displaystyle\int x^2 \sin(x^3)\, dx =$

(A) $\dfrac{1}{3}\cos(x^3) + C$

(B) $-\dfrac{1}{3}\cos(x^3) + C$

(C) $\dfrac{x^2}{3}\cos(x^3) + C$

(D) $-\dfrac{x^2}{3}\cos(x^3) + C$

(E) $-\dfrac{x^3}{3}\cos(3x^2) + C$

AP Calculus AB Test 1

9. If $f(x) = \ln\left(e^{3x} - e^{-3x} + x + 2\right)$, then $f'(0) =$

   (A) 3   (B) $\dfrac{7}{2}$ (C) 7   (D) $\dfrac{9}{2}$ (E) nonexistent

10. Using the substitution $u = 5x + 2$, $\displaystyle\int_0^2 \dfrac{10}{\sqrt{5x+2}}\,dx$ is equivalent to

    (A) $\displaystyle\int_0^2 \dfrac{10}{\sqrt{u}}\,du$ (B) $10\displaystyle\int_2^{12} \dfrac{1}{\sqrt{u}}\,du$ (C) $5\displaystyle\int_2^{12} \dfrac{1}{\sqrt{u}}\,du$ (D) $2\displaystyle\int_2^{12} \dfrac{1}{\sqrt{u}}\,du$ (E) $\dfrac{5}{2}\displaystyle\int_2^{12} \dfrac{1}{\sqrt{u}}\,du$

11. The rate of change of the population, P, of a certain town with respect to time, t, is directly proportional to the square root of the population. Which of the following differentiable equation shows the relation?

    (A) $\dfrac{P}{t} = \sqrt{P}$ (B) $P = k\sqrt{t}$ (C) $\dfrac{dP}{dt} = k\sqrt{P}$ (D) $\dfrac{dP}{dt} = k\sqrt{t}$ (E) $\dfrac{P}{\sqrt{P}} = kt$

12. If $f(x) > 0$, $f'(x) < 0$, and $f''(x) > 0$ for all real values x, which of the following could be the graph of f?

    (A)
    (B)
    (C)
    (D)
    (E)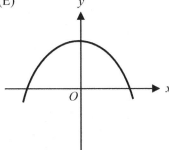

13. If $y = 3^x \sin 2x$, then $\dfrac{dy}{dx} =$

   (A) $3^x \cos 2x$
   (B) $2 \cdot 3^x \cos 2x$
   (C) $3^x (\ln 3 \cdot \sin 2x + 2\cos 2x)$
   (D) $3^x (3^x + 2\cos 2x)$
   (E) $3^x + 2\cos 2x$

14. Let f be the function given by $f(x) = x^3 - 3\ln x$. On which of the following interval is f decreasing?

   (A) $(-\infty, 0)$ (B) $(0, 2]$ (C) $(0, 1]$ (D) $(1, \infty)$ (E) $(0, \infty)$

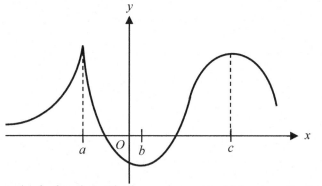

15. The graph of a function f is shown above. At which value of x does f have a local maximum?

   (A) a only   (B) b only   (C) c only   (D) a and c (E) b and c

16. If the line tangent to the graph of the function $f(x) = ax^2 + bx + 5$ at point $(2, 5)$ on the curve passes through the point $(-3, -5)$, then a and b are

   (A) $a = 2$ and $b = -1$
   (B) $a = 3$ and $b = -4$
   (C) $a = 1$ and $b = -2$
   (D) $a = -1$ and $b = 1$
   (E) $a = -2$ and $b = 4$

AP Calculus AB Test 1 Section 1

17. Let f be the function given by $f(x) = x^4 - 4x^3$. For which of the following intervals, the graph of f is concave down?

    (A) $(-\infty, 0)$ (B) $(0, 2)$ (C) $(0, 4)$ (D) $(2, 5)$ (E) $(5, \infty)$

---

18. A curve has slope $e^{2x} + 3x$ at each point (x, y) on the curve. If the curve passes through the point $(0, 3)$, Which of the following is the equation for this curve?

    (A) $y = \dfrac{e^{2x} + 3x + 5}{2}$

    (B) $y = \dfrac{e^{2x} + 3x^2 + 5x}{2}$

    (C) $y = \dfrac{e^{2x} + 3x^2 + 5}{2}$

    (D) $y = \dfrac{e^{2x} - 3x^2 + 3}{2}$

    (E) $y = \dfrac{e^{2x} - 3x^2 - 3}{2}$

$$f(x) = \begin{cases} x^2 + 1 & \text{if } x \leq 2 \\ 4x - 3 & \text{if } x > 2 \end{cases}$$

19. Function f is given above. Which of the following statements are true about f?

    I. f is continuous at $x = 2$.
    II. f is differentiable at $x = 2$.
    III. $\lim\limits_{x \to 2} f(x)$ exists.

    (A) I only   (B) II only   (C) III only   (D) I and III only   (E) I, II, and III

---

20. If $g'(x) = 4g(x)$ and $g(1) = 2$, then which of the following could be $g(x)$?

    (A) e^{4x}
    (B) $2e^{x-1}$
    (C) $2e^{4x-4}$
    (D) $4xe^{4x-4}$
    (E) $4e^{x-1} - 2$

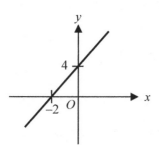

21. The graph of f'', the second derivative of f is shown above. If $f'(0) = 3$ and $f(0) = 6$, then $f(3) =$

    (A) −3 (B) 6 (C) 24 (D) 36 (E) 42

22. $\dfrac{d}{dx}\left(\displaystyle\int_0^{\sin x}(1-t^2)\,dt\right) =$

    (A) $-2t$ (B) $1-\sin^2 x$ (C) $\sin x^2$ (D) $\cos^2 x$ (E) $\cos^3 x$

23. Let f be the function defined by $y = \sqrt{x-1} + k$, where k is a constant. The equation of the line tangent to the graph of f is $y = \dfrac{x}{4}$. What is the value of k?

    (A) −2 (B) −0.75 (C) 1.25 (D) 3 (E) 5

24. Find the equation of the line tangent to the graph of $x^2 + 2y^2 - xy = 4$ at point $(1, -1)$.

    (A) $y + 1 = 3(x-1)$

    (B) $y + 1 = 3(x+1)$

    (C) $y + 1 = \dfrac{3}{5}(x-1)$

    (D) $y - 1 = \dfrac{3}{5}(x+1)$

    (E) $y + 1 = \dfrac{5}{3}(x-1)$

25. A particle moving along the x-axis so that at time $t \geq 0$, its position is given by $x(t) = t^3 - 6t^2 + 9t + 12$. During what time intervals is the particle moving to the left?

(A) $0 < t < 1$ (B) $1 < t < 3$ (C) $1 < t < 5$ (D) $3 < t < 5$ (E) $t > 5$

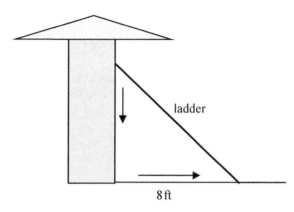

26. A ladder leans against a vertical wall of the house shown above. The base of the ladder is pulled away from the wall at a rate of 2 ft per second and the top of the ladder slips down the wall at a rate of 4 ft per second, when the base of the ladder is 8 feet from the wall. How long is the ladder?

(A) 8 ft (B) $4\sqrt{3}$ (C) $4\sqrt{5}$ (D) 12 (E) 15

27. Let f be the function defined by $f(x) = x^2 + x - 2$, where $x \geq 0$. If $g(x)$ is the inverse of f, what is the value of $g'(4)$?

(A) $\frac{1}{5}$ (B) 3 (C) 5 (D) $-\frac{1}{5}$ (E) -5

28. Let f be a twice-differentiable function with $f'(x) > 0$ and $f''(x) < 0$ for all real numbers x. If $f(2) = 10$ and $f(4) = 18$, which of the following is a possible value of $f(6)$?

(A) 30 (B) 28 (C) 26 (D) 24 (E) 16

END OF PART A OF SECTION 1

AP Calculus AB Test 1 Section 1

CALCULUS AB
SECTION 1, PART B
Time — 50minutes
Number of questions — 17

A GRAPHING CALCULATOR IS REQUIRED FOR SOME QUESTIONS ON THIS PART OF THE EXAMINATION.

Directions: Solve each of the following problems, using the available space for scratchwork. After examining the form of the choices, decide which is the best of the choices given and fill in the corresponding oval on the answer sheet. Do not spend too much time on any one problem.

In this test:

(1) The exact numerical value of the correct answer does not always appear among the choices given. When this happens, select from among the choices the number that best approximates the exact numerical value.

(2) Unless otherwise specified, the domain of a function f is assumed to be the set of all real numbers x for which $f(x)$ is a real number.

(3) The inverse of a trigonometric function f may be indicated using the inverse function notation f^{-1} or with the prefix "arc" (e.g, $\sin^{-1} x = \arcsin x$)

76. A particle moves along the x-axis so that its position $x(t) = x^2 + 2\cos(0.5t)$ at $t \geq 0$. What is the acceleration of the particle at time $t = 2$?

(A) 1.730   (B) 1.213   (C) 1.632   (D) 1.842   (E) 2.125

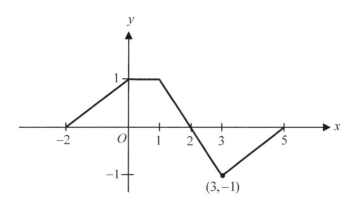

77. The graph of f is shown above for $-2 \leq x \leq 5$. What is the value of $\int_{-2}^{5}(f(x)+2)\,dx$?

(A) 1   (B) 4   (C) 6   (D) 15   (E) 19

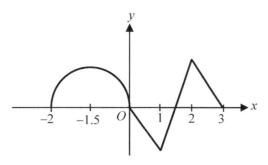

78. On the closed interval $[-2, 3]$, the graph of f', the derivative of the function f is shown above. For what value(s) does the function have a possible point of inflection?

(A) -1.5 only   (B) 0, 1.5,   (C) 1 and 2 only   (D) -1.5, 1, 2   (E) none

79. The function f is continuous on the closed interval $[-2, 5]$ and differentiable on the open interval $(-2, 5)$. If $f(-2) = -2$ and $f(5) = 12$, which of the following statements could be false?

   (A) On the interval $(-2, 5)$, there exists a value c, such that $f(c) = 2$.
   (B) On the interval $(-2, 5)$, there exists a value c, such that $f(c) = 4$.
   (C) On the interval $(-2, 5)$, there exists a value c, such that $\lim_{h \to 0} \frac{f(c+h) - f(c)}{h} = 2$
   (D) On the interval $[-2, 5]$, there exists a value c, such that $f(c) \leq f(x)$ for all x on $[-2, 5]$.
   (E) On the interval $(-2, 5)$, there exists a value c, such that $f''(c) > 0$.

80. If f be the function with derivative $f'(x) = e^{\sin x} - 2$. How many relative extrema does f have on the interval $0 < x < 10$?

   (A) None (B) One (C) Two (D) Three (E) Four

81. A particle travels along the x-axis so that its velocity is given by $v(t) = t^3 - 4t^2 + 5$ for $0 \leq t \leq 10$. Which of the following expressions gives the total distance traveled by the particle moving to the left?

   (A) $\int_0^{10} v'(t)\,dt$ (B) $\int_0^{1.382} v'(t)\,dt$ (C) $\int_{1.382}^{3.618} v(t)\,dt$ (D) $\int_0^{3.618} v(t)\,dt$ (E) $\int_{3.618}^{1.382} v(t)\,dt$

82. At time $t = 0$, a diver jumps from a diving board that is 40 feet above water. The position of the diver is given by $s(t) = -7t^2 + 14t + 40$ where s is measured in feet and t is measured in seconds. What is the diver's average speed from time $t = 0$ to $t = 3$?

   (A) 7 ft/sec (B) 9 ft/sec (C) $\frac{35}{3}$ ft/sec (D) 12 ft/sec (E) $\frac{46}{3}$ ft/sec

83. When a object is removed from a furnace and placed in a room with a constant temperature of $80°F$ at $t = 0$ minutes, its temperature is $1000°F$. The temperature of the object is changing at a rate of $-120te^{-0.5t}$ degrees Fahrenheit per minute. What is the temperature of the object at time $t = 10$ minutes to the nearest degree?

   (A) $167°F$ (B) $256°F$ (C) $365°F$ (D) $488°F$ (E) $539°F$

84. If both trapezoidal sum and right Riemann sum over-approximate $\int_a^b g(x)dx$, which of the following could be the graph of $y = g(x)$?

(A)

(B)

(C)

(D)

(E)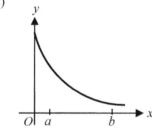

---

85. The base of a solid is the region in the first quadrant enclosed by the y-axis, the horizontal line $y = \dfrac{\pi}{2}$, and the graph of $y = \sin^{-1} x$. For this solid, each cross section perpendicular to the x-axis is a square. What is the volume of the solid?

(A) 0.995 (B) 1.142 (C) 1.547 (D) 2.345 (E) 2.546

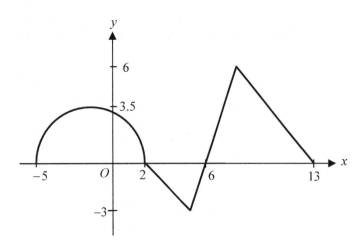

86. The graph of a function f is shown above. Which of the following could be the average value of f on the closed interval $[-5, 13]$?

    (A) 1.823   (B) 1.902   (C) 3.639   (D) 34.242   (E) 53.5

87. Let f' be the derivative of a function f. If f' is given by $f'(x) = \dfrac{\sqrt[3]{x^2}}{1+\sqrt{x}+x^3}$, what is the x-coordinate of the inflection point of the graph of f ?

    (A) 0   (B) 0.301   (C) 0.408   (D) 0.702   (E) 1.302

88. Let f be a differentiable function that is defined for all real numbers x with $f(1)=1$, $f'(1)=-2$, and $f''(1)=3$. If g is the function defined by $g(x)=(x^2+2)f(x)$, what is the value of $g''(1)$?

    (A) −3   (B) −1   (C) 1   (D) 3   (E) 5

AP Calculus AB Test 1   Section 1

89. Let f be the function that is defined for all real number x and that has the following properties.

(i) $f'(x) < 0$ (ii) $f''(x) > 0$ (iii) $f(x) > 0$

(A)

x	1	2	3	4	5
$f(x)$	7	8	10	13	17

(B)

x	1	2	3	4	5
$f(x)$	17	16	14	11	7

(C)

x	1	2	3	4	5
$f(x)$	13	8	4	1	−1

(D)

x	1	2	3	4	5
$f(x)$	3	5	8	12	17

(E)

x	1	2	3	4	5
$f(x)$	17	12	8	5	3

---

90. Let f be a function such that $f'(x) = xe^{1-x}$ for all real numbers x. If $f(1) = 5$, then $f(3) =$

(A) 6.459   (B) 6.785   (C) 7.125   (D) 8.292   (E) 12.581

---

91. Let f be a function defined by $f(x) = \ln(1 + 2^x)$. If $g(x) = \int_0^{\frac{x}{2}+5} f(t)\,dt$, then $g'(-2) =$

(A) $\ln 7$ (B) $2\ln 17$ (C) $4\ln 7$ (D) $\ln\sqrt{17}$ (E) $2\ln\dfrac{5}{4}$

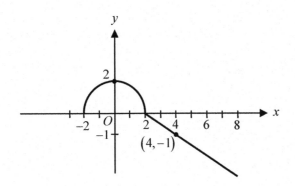

92. The graph of a function f consists of a semicircle and a line f as shown above. If g is the function given by $g(x) = \int_0^x f(t)\,dt$, what is an equation for the line tangent to the graph of g at $x = 4$?

(A) $y = -x + \pi + 3$
(B) $y = -x + \pi$
(C) $y = x - \pi + 1$
(D) $y = x + \pi - 1$
(E) $y = x + \pi - 3$

END OF SECTION 1

AP Calculus AB Test 1

Section II

CALCULUS AB

SECTION II
Time: 1 hour and 30 minutes
Percent of total grade – 50

Part A: 30 minutes, 2 problems
Part B: 60 minutes, 4 problems

Part A: A graphing calculator is required for some problems or parts of problems.
Part B: No calculator is allowed for these problems.

GENERAL INSTRUCTIONS FOR SECTION II PART A AND PART B

For each part of section II, you may wish to look over the problems before starting to work on them, since it is not expected that everyone will be able to complete all parts of all problems. All problems are given equal weight, but the parts of a particular problem are not necessary given equal weight.

- YOU SHOULD WRITE ALL WORK FOR EACH PART OF EACH PROBLEM IN THE SPACE PROVIDED FOR THAT PART IN THE TEST BOOKLET. Be sure to write clearly and legibly. If you make an error, you may save time by crossing it out rather than trying to erase it. Erased or cross-out work will not be graded.
- Show all work. Clearly label any functions, graphs, tables, or other object that you use. You will be graded on the correctness and completeness of your methods as well as your answers. Answers without supporting work may not receive credit.
- Justifications required that you give mathematical (noncalculator) reasons
- Your work must be expressed in standard mathematical notation rather than calculator syntax. For example, $\int_{1}^{5} x^2 \, dx$ may not be written as $\text{fnInt}(X^2, X, 1, 5)$
- Unless otherwise specified, answers (numeric or algebraic) need not be simplified.
- If you use decimal approximations in calculations, you will be graded on accuracy. Unless otherwise specified, your final answers should be accurate to three places after the decimal point.
- Unless otherwise specified, the domain of a function f is assumed to be the set of all real numbers x for which $f(x)$ is a real number.

CALCULUS AB
SECTION II, PART A
Time — 30 minutes
Number of questions — 2

A Graphing calculator is required for some problems or parts of problems.

Question 1

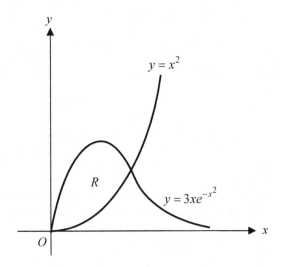

Let R be the region enclosed by the graphs of $y = 3xe^{-x^2}$ and $y = x^2$ as shown in the figure above.

a) Find the area of R.

b) Find the volume of the solid generated when R is revolved about the x-axis.

c) The region R is the base of a solid. For this solid, each cross section perpendicular to x-axis is a square. Find the volume of the solid.

AP Calculus AB Test 1 *Section II*

Question 2

A particle moves along the x-axis so that, at any time $t \geq 0$, its velocity is given by

$$v(t) = e^{(-t/5)} \cos t.$$

At time $t = 0$, the position of the particle is -20.

a) Find the acceleration of the particle at time $t = 3$.

b) At time $t = 3$, is the speed of the particle increasing or decreasing? Why?

c) Find all times t in the interval $0 < x < 6$ when the particle changes direction. Justify your answer.

d) Find the total distance traveled by the particle from $t = 0$ to $t = 3$.

e) Find $x(3)$, the position of the particle at time $t = 3$.

END OF PART A OF SECTION II

AP Calculus AB Test 1 — Section II

CALCULUS AB
SECTION II, PART B
Time — 60 minutes
Number of questions — 4

No calculator is required for these problems.

Question 3

t (day)	0	10	30	40	60	70
$R(t)$ (in thousand dollars per day)	10	15	40	50	55	80

A company introduces a new car and the profit in thousands of dollars over 70 days is given by a twice differentiable and strictly increasing function R of time t. A table of selected values of $R(t)$ for 70 days is given above.

a) Use the data from the table to find an approximation for $R'(50)$.

b) Approximate the value of $\int_0^{70} R(t)\,dt$ using a left Riemann sum with the five subintervals indicated by the data in the table.

c) Is the approximation in (b) greater than or less than the value of $\int_0^{70} R(t)\,dt$? Explain your reasoning.

d) Explain the meaning of $\int_0^{70} R(t)\,dt$.

e) Explain the meaning of $\dfrac{1}{70}\int_0^{70} R(t)\,dt$.

Question 4

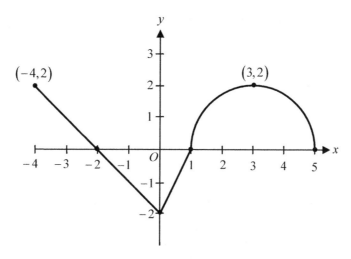

The graph of a function f defined on the closed interval $-4 \le x \le 5$ consists of a semicircle and two line segments as shown above. Let g be the function given by $g(x) = \int_0^x f(t)\,dt$.

a) Find the interval for which $g(x)$ is increasing.

b) Find $g(-2)$, $g(1)$, and $g(3)$.

c) Find all values of x on the open interval $(-4, 5)$ at which g have a relative maximum.

d) Find the x-coordinate of each point of inflection of the graph of g on the open interval $(-4, 5)$.

e) Write an equation for the line tangent to the graph of g at $x = 3$.

AP Calculus AB Test 1   Section II

Question 5

The formula for the volume of a sphere is $V = \frac{4}{3}\pi r^3$. The radius of r is increasing at the constant rate of 0.5 centimeters per minute.

a) Find $\frac{dV}{dt}$ in terms of r.

b) Find the rate of increase of the volume at the time when the radius of the sphere is 10 centimeters.

c) What is the rate of increase of the surface area of the sphere at the time when the volume of the sphere is 36π cubic centimeters? (The surface of a sphere is $A = 4\pi r^2$.)

d) If the volume and the surface area of the sphere are increasing at the same numerical rate, what is the radius?

Question 6

Let $P(t)$ represent the number of a certain type of bacterial in a population at time t hours, when $t \geq 0$. The rate of the population $P(t)$ is continuously increasing directly proportional to $1000 - P(t)$.

a) If $P(0) = 500$ and $P(5) = 900$, find $P(t)$ in terms of t.

b) Find the value of $P(10)$.

c) Find the value of $\lim_{t \to \infty} P(t)$.

END OF EXAM

AP Calculus AB Test 1 Answers

SECTION 1, PART A

1. **D**

   Apply Chain rule: $y = [u(x)]^n \Rightarrow \dfrac{dy}{dx} = n[u(x)]^{n-1} u'(x)$

   Therefore, $\dfrac{dy}{dx} = 3(2x^3+1)^2 (2x^3+1)' = 3(2x^3+1)^2 (6x^2) = 18x^2 (2x^3+1)^2$.

2. **A**

   Remember that $(2^x)' = 2^x \ln 2$. Therefore, $\displaystyle\int_0^1 2^x dx = \left[\dfrac{2^x}{\ln 2}\right]_0^1 = \dfrac{2}{\ln 2} - \dfrac{1}{\ln 2} = \dfrac{1}{\ln 2}$.

3. **D**

   Quotient Rule:

   $f(x) = \dfrac{x^2+1}{x+2}$, then $f'(x) = \dfrac{(x+2)(2x) - (x^2+1)(1)}{(x+2)^2} = \dfrac{2x^2+4x-x^2-1}{(x+2)^2} = \dfrac{x^2+4x-1}{(x+2)^2}$

4. **B**

   If $f(x) = \dfrac{2x+5}{x-2}$, then which of the following statements is not true?

   (A) $\displaystyle\lim_{x \to -\infty} f(x) = 2$ True

   (B) $\displaystyle\lim_{x \to 2} f(x) = \infty \Rightarrow \begin{cases} \displaystyle\lim_{x \to 2^+} f(x) = \infty \\ \displaystyle\lim_{x \to 2^-} f(x) = -\infty \end{cases}$ Limit doesn't exist (DNE) False

   (C) $\displaystyle\lim_{x \to \infty} f(x) = 2$ True

   (D) $f(2)$ is undefined. True

   (E) $x = 2$ is an vertical asymptote. True

   The graph of f is as follows.

   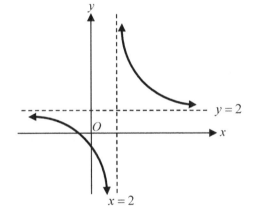

5. **D**

   $\displaystyle\int_0^{\pi/4} \sin 2x\, dx = \left[\dfrac{\cos 2x}{-2}\right]_0^{\pi/4} = -\dfrac{1}{2}\cos\dfrac{\pi}{2} - \left(-\dfrac{1}{2}\cos 0\right) = 0 - \left(-\dfrac{1}{2}\right) = \dfrac{1}{2}$

   If you don't know how to use the shortcut, then use u-substitution.

   Let $u = 2x$ and $du = 2dx$ and the upper and the lower bound be $x = 0 \to u = 0$ and $x = \dfrac{\pi}{4} \to u = \dfrac{\pi}{2}$.

   Therefore, $\displaystyle\int_0^{\pi/4} \sin 2x\, dx = \int_0^{\pi/2} \sin u \cdot \dfrac{du}{2} = \dfrac{1}{2}\int_0^{\pi/2} \sin u\, du = \dfrac{1}{2}[-\cos u]_0^{\pi/2} = \dfrac{1}{2}[-\cos \pi/2 - (-\cos 0)]$

   $= \dfrac{1}{2}(0+1) = \dfrac{1}{2}$.

Dr. John Chung's AP Calculus AB/BC

AP Calculus AB Test 1 Answers

6. C
   When $x \to \infty$, compare the leading coefficients.
   $$\lim_{x \to \infty} \frac{x^4 + 3x^2 - 1}{5x^4 - 3x^3 + 2x^2 - 1} = \lim_{x \to \infty} \frac{x^4}{5x^4} = \frac{1}{5} = 0.2, \text{ Or}$$

   $$\lim_{x \to \infty} \frac{x^4 + 3x^2 - 1}{5x^4 - 3x^3 + 2x^2 - 1} = \lim_{x \to \infty} \frac{(x^4 + 3x^2 - 1) \div x^4}{(5x^4 - 3x^3 + 2x^2 - 1) \div x^4} = \lim_{x \to \infty} \frac{\left(1 + \frac{3}{x^2} - \frac{1}{x^4}\right)}{\left(5 - \frac{3}{x} + \frac{2}{x^2} - \frac{1}{x^4}\right)} = \frac{1}{5} = 0.2$$

7. E
   Remember: This is the graph of f'.
   (A) f is decreasing for $-1 \le x \le 0$. → For $-1 \le x \le 0$, $f'(x) > 0$ $\Rightarrow$ $f(x)$ is increasing. False
   (B) f is increasing for $1 \le x \le 2$. → For $1 \le x \le 2$, $f'(x) < 0$ $\Rightarrow$ $f(x)$ is decreasing. False
   (C) f has a local maximum at $x = -1$. → At $x = -1$, $f(x)$ has a point of inflection. False
   (D) f has a local minimum at $x = -1$. False
   (E) f has a local maximum at $x = 0$. For $x < 0 \to f' > 0$, and for $x > 0 \to f' < 0$. True

8. B
   Use u-substitution: $u = x^3 \to \begin{cases} du = 3x^2 dx \\ x^2 dx = \dfrac{du}{3} \end{cases}$

   $$\int x^2 \sin(x^3) dx = \frac{1}{3} \int \sin u \, du = -\frac{1}{3} \cos u + C \to (\text{Replace } u \text{ with } x^3) \to -\frac{1}{3} \cos(x^3) + C$$

9. B
   $$f(x) = \ln(e^{3x} - e^{-3x} + x + 2) \Rightarrow f'(x) = \frac{3e^{3x} + 3e^{3x} + 1}{e^{3x} - e^{-3x} + x + 2} \Rightarrow f'(0) = \frac{7}{2}$$

10. D
    $u = 5x + 2 \Rightarrow \begin{cases} du = 5dx \\ dx = \dfrac{du}{5} \end{cases}$ upper bound: $u = 12$ when $x = 2$

    lower bound: $u = 2$ when $x = 0$

    $\displaystyle\int_0^2 \frac{10}{\sqrt{5x+2}} dx$ is equivalent to $\displaystyle\int_2^{12} \frac{10}{\sqrt{u}} \cdot \frac{du}{5} = 2\int_2^{12} \frac{1}{\sqrt{u}} du$.

11. C
    The rate of change of the population P is $\dfrac{dP}{dt}$. Therefore, $\dfrac{dP}{dt} = k\sqrt{P}$.

12. D
    $f(x) > 0 \Rightarrow$ The grpah of $y = f(x)$ exists in the first and second quadrants.
    $f'(x) < 0 \Rightarrow$ The graph decreasing over the x-axis.
    $f''(x) > 0 \Rightarrow$ The graph of $y = f(x)$ is concave up over the x-axis.
    In addition to those, we can see that $\lim_{x \to \infty} f(x) = 0$.

13. C
    Product rule: $y = 3^x \sin 2x$, then $\dfrac{dy}{dx} = (3^x)'(\sin 2x) + (3^x)(\sin 2x)' = 3^x \ln x \cdot \sin 2x + 3^x (2\cos 2x)$
    $= 3^x (\ln 3 \cdot \sin 2x + 2 \cos 2x)$

14. C

234

AP Calculus AB Test 1 Answers

We need to find critical points from $f'(x) = 0$, where $x > 0$.

$$f'(x) = 3x^2 - \frac{3}{x} = 3\left(x^2 - \frac{1}{x}\right) = 3\left(\frac{x^3 - 1}{x}\right) = 3\left(\frac{(x-1)\left|x^2 + x + 1\right|}{x}\right) = 0$$

We can see that f have critical points at $x = 0$ or $x = 1$. f has an asymptote at $x = 1$.

x	0	(0,1)	1	(1, ∞)
f'	asymptote	−	0	+

f is increasing on $(0, 1]$.

15. **D**

The graph of f have local maximums at $x = a$ and $x = b$, because f' changes in sign from + to −.
At $x = b$, f has a local minimum.

16. **C**

The slope between the two point is $\frac{5-(-5)}{2-(-3)} = \frac{10}{5} = 2$. $f'(x) = 2ax + b$ and $f'(2) = 4a + b$

Therefore $4a + b = 2$, because the slopes are equal at $x = 2$. Since point $(2,5)$ lies on the curve, we have
$f(2) = 5 \rightarrow 4a + 2b + 5 = 5 \rightarrow 4a + 2b = 0$. Solve the equations $4a + b = 2$ and $4a + 2b = 0$.
Therefore, $b = -2$ and $a = 1$.

17. **B**

If $f''(x) < 0$, then the graph of f is concave down.
$f(x) = x^4 - 4x^3 \rightarrow f'(x) = 4x^3 - 12x^2 \rightarrow f''(x) = 12x^2 - 24x$.
Since the graph is concave down, $f''(x) = 12x(x-2) < 0 \rightarrow 0 < x < 2$ or $(0, 2)$.

18. **C**

Because $f'(x) = e^{2x} + 3x$, $f(x) = \int (e^{2x} + 3x)\,dx = \frac{e^{2x}}{2} + \frac{3x^2}{2} + C$. Since the graph of f passes through point
$(0, 3)$, $f(0) = \frac{1}{2} + 0 + C = 3$ and $C = \frac{5}{2}$.

Therefore, $f(x) = \frac{e^{2x}}{2} + \frac{3x^2}{2} + \frac{5}{2} = \frac{e^{2x} + 3x^2 + 5}{2}$.

19. **E**

$f(x) = \begin{cases} x^2 + 1 & \text{if } x \leq 2 \rightarrow f'(x) = 2x \\ 4x - 3 & \text{if } x < 2 \rightarrow f'(x) = 4 \end{cases}$

I. f is continuous at $x = 2$.

$f(x) = \begin{cases} x^2 + 1 & \text{if } x \leq 2 \rightarrow f(2) = 5 \\ 4x - 3 & \text{if } x > 2 \rightarrow \lim_{x \to 2^+}(4x - 3) = 5 \end{cases}$

f is continuous $x = 2$, because $\lim_{x \to 2} f(x) = f(2)$.

II. f is differentiable at $x = 2$.

$f(x) = \begin{cases} x^2 + 1 & \text{if } x \leq 2 \rightarrow f'(x) = 2x \rightarrow f'(2) = 4 \\ 4x - 3 & \text{if } x > 2 \rightarrow f'(x) = 4 \rightarrow f'(2) = 4 \end{cases}$

III. $\lim_{x \to 2} f(x) = 5$ exists, because f is continuous at $x = 2$.

AP Calculus AB Test 1 Answers

20. C

Solve the differential equation $\dfrac{g'}{g} = 4$ → $\displaystyle\int \dfrac{g'}{g}dx = \int 4\,dx$ → $\ln|g| = 4x + C$ → $|g| = \pm e^{4x+C}$

→ $|g| = \pm e^{C}\left(e^{4x}\right)$ → $g = Ae^{4x}$

Now we got a solution form $g(x) = Ae^{4x}$.

Using the initial condition $g(1) = 2$, find the value of A. $Ae^{4} = 2$ or $A = \dfrac{2}{e^{4}} = 2e^{-4}$ Substituting in A, thus $g = \left(2e^{-4}\right)e^{4x} = 2e^{4x-4}$.

21. E

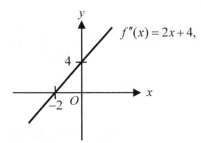

$f''(x) = 2x + 4$,

Since $f''(x) = 2x + 4$, then $f'(x) = \displaystyle\int (2x+4)\,dx$ → $f'(x) = x^2 + 4x + C$ and $f'(0) = C = 3$.

Therefore, $f'(x) = x^2 + 4x + 3$ and

$f(3) = f(0) + \displaystyle\int_0^3 f'(x)\,dx = 6 + \int_0^3 (x^2 + 4x + 3)\,dx = 6 + \left[\dfrac{x^3}{3} + 2x^2 + 3x\right]_0^3$

$= 6 + \left[\dfrac{27}{3} + 2(3)^2 + 3(3)\right] = 6 + 36 = 42$.

Or, $f(x) = \displaystyle\int (x^2 + 4x + 3)\,dx = \dfrac{x^3}{3} + 2x^2 + 3x + C$ and $f(0) = 6$ → $C = 6$

22. E

The Second Fundamental Theorem of Calculus is $\dfrac{d}{dx}\displaystyle\int_a^{u(x)} f(t)\,dt = f(u)u'$.

Therefore, $\dfrac{d}{dx}\left(\displaystyle\int_0^{\sin x} (1 - t^2)\,dt\right) = \left(1 - (\sin x)^2\right)(\sin x)' = \left(1 - (\sin x)^2\right)\cos x = \cos^3 x$.

23. B

Slope of the line tangent to the curve is $f'(x) = \dfrac{1}{2\sqrt{x-1}} = \dfrac{1}{4}$ because the slope of the tangent is $\dfrac{1}{4}$.

From the equation we got $2\sqrt{x-1} = 4$ → $\sqrt{x-1} = 2$ → $x - 1 = 4$ → $x = 5$.

When $x = 5$, $y = \left.\dfrac{x}{4}\right|_{x=5} = 1.25$. Therefore, the point of tangency is (5, 1.25).

This point of tangency also lies on the graph of $y = \sqrt{x-1} + k$, substitute (5, 1.25) into the equation. Hence $1.25 = \sqrt{5-1} + k$ → $1.25 = 2 + k$ or $k = 1.25 - 2 = -0.75$.

AP Calculus AB Test 1 Answers

24. C

    Differentiate implicitly to get $\dfrac{dy}{dx}(=y')$ as follows.

    $x^2+2y^2-xy=4 \;\;\to\;\; 2x+4yy'-(x'y+xy')=0 \;\;\to\;\; y'(4y-x)=y-2x$

    Therefore, $y'=\dfrac{y-2x}{4y-x}$ and $y'=\dfrac{y-2x}{4y-x}\bigg|_{\substack{x=1\\y=-1}}=\dfrac{-1-2(1)}{4(-1)-1}=\dfrac{3}{5}$.

    Hence the equation of the line tangent at $(1,-1)$ is $y-(-1)=\dfrac{3}{5}(x-1)$ or $y+1=\dfrac{3}{5}(x-1)$.

25. B

    When the particle moving to the left, $v(t)=x'(t)<0$.

    $x'(t)=3t^2-12t+9<0 \;\;\to\;\; 3(t^2-4t+3)<0 \;\;\to\;\; 3(t-3)(t-1)<0$

    ```
       f' > 0      f' < 0      f' > 0
    ───────○───────────○───────────
           t = 1       t = 3
    ```

    Therefore, the particle is moving to the left for $1<t<3$.

26. C

    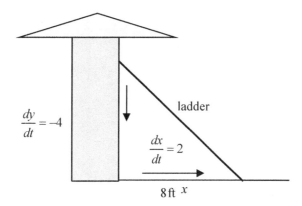

    Related rate: The primary equation is $x^2+y^2=k^2$ where k is the length of the ladder.
    Differentiate the primary equation with respect to time t.

    $2x\dfrac{dx}{dt}+2y\dfrac{dy}{dt}=0$

    When $x=8$, $\dfrac{dx}{dt}=2$, $\dfrac{dy}{dt}=-4$. Substitute the givens into the equation.

    $2(8)(2)+2y(-4)=0 \;\;\to\;\; 8y=32 \;\;\to\;\; y=4$

    Therefore, when $x=8$, $y=4$.

    Hence the length of the ladder $k=\sqrt{x^2+y^2}=\sqrt{64+16}=\sqrt{80}=4\sqrt{5}$

27. A

    If $g(4)=k$, then $f(k)=4$ because $g(x)$ is the inverse of f.

    Let's find the value of k from the equation $k^2+k-2=4 \;\;\to\;\; k^2+k-6=0 \;\;\to\;\; (k+3)(k-2)=0$.

    $k=-3, 2$ Since $x\geq 0$, $k=2$.

    $f'(x)=2x+1$ and $f'(2)=5$

    Therefore, $g'(4)=\dfrac{1}{f'(2)}=\dfrac{1}{5}$.

AP Calculus AB Test 1 Answers

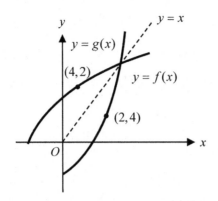

28. D

    f is the function with $f'(x) > 0$ and $f''(x) < 0$ for all real numbers x and $f(2) = 10$ and $f(4) = 18$.
    The graph of f is increasing and concave down for all numbers x as shown below.

    The equation for line ℓ: slope $m = \dfrac{18-10}{4-2} = 4 \;\to\; y - 10 = 4(x-2) \;\to\; f(x) = 4x + 2$

    $\therefore\; f(6) = 26$

    Therefore, the value of k must be greater than 18 but less than 26.

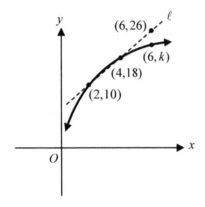

SECTION 1, PART B

76. A

    $v'(t) = x'(t) = 2x - 2\sin(0.5t) \cdot (0.5) = 2x - \sin(0.5t)$

    $a(t) = v'(t) = 2 - 0.5\cos(0.5t)$

    Therefore, $a(2) = 2 - 0.5\cos(1) = 1.7298488 \approx 1.730$.

77. D

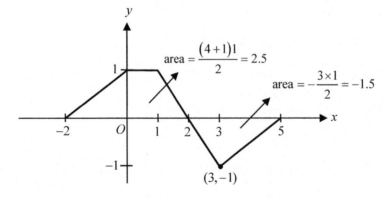

AP Calculus AB Test 1 Answers

$$\text{Area} = \int_{-2}^{5}(f(x)+2)\,dx = \int_{-2}^{5}f(x)\,dx + \int_{-2}^{5}2\,dx = (2.5-1.5) + [2x]_{-2}^{5} = 1 + (10+4) = 15$$

78. D

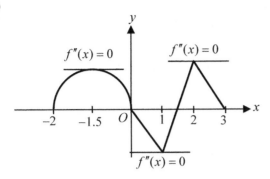

x		-1.5		1		2	
f''	$+$	Point of inflection	$-$	Point of inflection	$+$	Point of inflection	$-$

There are three points of inflection.

79. E

The function f is continuous on the closed interval $[-2, 5]$ and differentiable on the interval $(-2, 5)$.

The graph of f : $\begin{cases} 1)\text{ continuous} \\ 2)\text{ differentiable} \to \text{No cusp and no sharp corner} \end{cases}$

(A) On the interval $(-2, 5)$, there exists c such that $f(c) = 2$. True (Intermediate Theorem)

(B) On the interval $(-2, 5)$, there exists c such that $f(c) = 4$. True (Intermediate Theorem)

(C) On the interval $(-2, 5)$, there exists c such that $\lim_{h \to 0} \dfrac{f(c+h)-f(c)}{h} = 2 \;\to\; f'(c) = 2$

True (The Mean value Theorem)

The slope of line ℓ is $\dfrac{12-(-2)}{5-(-2)} = 2$.

Therefore, there exists at least one c such that $\dfrac{f(b)-f(a)}{b-a} = f'(c)$.

(D) On the interval $[-2, 5]$, there exists c such that $f(c) \le f(x)$ for all x on $[-2, 5]$.

At $c = -2$, $f(c) \le f(x)$ on the closed interval $[-2, 5]$. True

$f(c) \le f(x)$ imply that the graph of f has an absolute minimum.

(E) On the interval $(-2, 5)$, there exists c such that $f''(c) > 0$. False

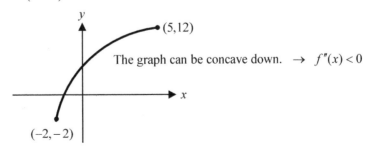

The graph can be concave down. $\to f''(x) < 0$

AP Calculus AB Test 1 Answers

80. E

    Use your graphing calculator:

    The graph of $f'(x) = e^{\sin x} - 2$ is shown below.

    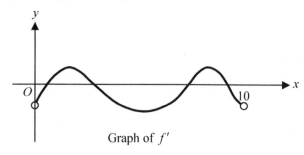

    Graph of f'

    The graph of f has four x-intercepts.

81. E

    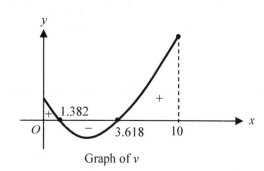

    Graph of v

    For $1.382 \leq t \leq 3.618$ as shown above, the particle travels to the left.

    Therefore, $D = \int_{1.382}^{3.618} |v(t)| \, dt = \int_{1.382}^{3.618} -v(t) \, dt = \int_{3.618}^{1.382} v(t) \, dt$

82. C

    Average speed $= \dfrac{\text{Total traveled distance}}{\text{Total Time}}$

    Using a graphic utility, the graph of $s(t) = -7t^2 + 14t + 40$ is as follows.

    Total distance $= (47 - 40) + (47 - 19) = 35$ and $\Delta t = 3 - 0 = 3$ seconds

    Therefore, the average speed is $\dfrac{35}{3}$ feet/second.

83. E

    Fundamental Theorem of Calculus: $F(b) = F(a) + \int_a^b f(x) \, dx$ for $F'(x) = f(x)$.

    $T = 1000 + \int_0^{10} \left(-120 t e^{-0.5t} \right) dt \approx 539.40529$

240

AP Calculus AB Test 1 Answers

84. C

The graph of g must be increasing (right Riemann sum over approximation) and concave up (trapezoidal sum over approximation)

$\int_a^b g(x)\,dx$

Right Riemann sum

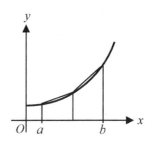
Trapezoidal sum

85. B

The base of a solid is the region in the first quadrant enclosed by the y-axis, the horizontal line $y=\dfrac{\pi}{2}$, and the graph of $y=\sin^{-1}x$. The graph of the region is shown below. Use your graphing calculator.

$\dfrac{\pi}{2}=\sin^{-1}x \;\Rightarrow\; \sin\dfrac{\pi}{2}=x \;\Rightarrow\; x=1$

Therefore, the volume of the solid is

$V=\int_0^1\left(\dfrac{\pi}{2}-\sin^{-1}x\right)^2 dx \approx 1.1415927 \approx 1.142$

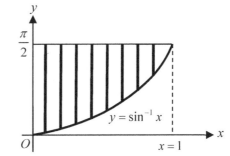

86. B

Average value of f over the interval from a to b is $\dfrac{1}{b-a}\int_a^b f(x)\,dx$.

Compare:

Average rate of change of function f between two points $(a, f(a))$ and $(b, f(b))$ is $\dfrac{f(b)-f(a)}{b-a}$.

It is the slope of the secant line connecting the two points.

Therefore, average value is $\dfrac{1}{13-(-5)}\int_{-5}^{13} f(x)\,dx = \dfrac{1}{18}\left(\dfrac{\pi\cdot(3.5^2)}{2}-\dfrac{4\cdot 3}{2}+\dfrac{7\cdot 6}{2}\right)=1.902$

87. D

Use your graphing calculator:
If $f''(a)=0$, then $(a, f(a))$ is a point of inflection.

AP Calculus AB Test 1 Answers

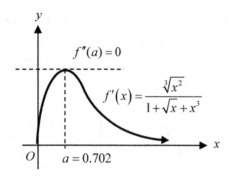

The x-coordinate of the inflection point of the graph of f in the figure shown above is 0.702.

88. D

$g(x) = (x^2 + 2)f(x) \Rightarrow g'(x) = 2x \cdot f(x) + (x^2 + 2)f'(x)$

$g''(x) = 2 \cdot f(x) + 2x \cdot f'(x) + 2x \cdot f'(x) + (x^2 + 2)f''(x)$

$g''(1) = 2f(1) + 2(1)f'(1) + 2(1)f'(1) + (1+2)f''(1) = 2 - 4 - 4 + 9 = 3$

89. E

For all real numbers x, the graph of f is (i) $f'(x) < 0$ decreasing
(ii) $f''(x) > 0$ concave up
(iii) $f(x) > 0$.

Example:

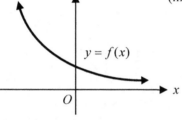

90. A

Use your graphing calculator.

$f(3) = f(1) + \int_1^3 xe^{1-x}\,dx \approx 6.459$

91. D

$g(x) = \int_0^{\frac{x}{2}+5} f(t)\,dt = \int_0^{\frac{x}{2}+5} \ln(1+2^t)\,dt$

$g'(x) = \ln\left(1 + 2^{\frac{x}{2}+5}\right)\left(\frac{x}{2}+5\right)' \Rightarrow \ln\left(1+2^{\frac{x}{2}+5}\right) \cdot \frac{1}{2} \Rightarrow g'(-2) = \frac{1}{2}\ln(1+2^4) = \frac{1}{2}\ln 17 = \ln\sqrt{17}$

AP Calculus AB Test 1 Answers

92. A

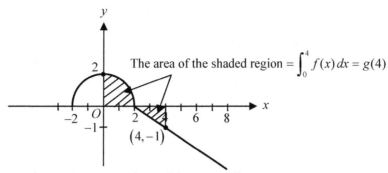

$g'(x) = f(x) \Rightarrow g'(4) = f(4) = -1$: Slope of the tangent line at $x = 4$

The equation for the line tangent to the graph of g at $x = 4$ is $y - y_1 = m(x - 4)$.

$y_1 = g(4) = \int_0^4 f(x)\,dx = \dfrac{4\pi}{4} - \dfrac{2 \cdot 1}{2} = \pi - 1$

Therefore, $y - (\pi - 1) = -1 \cdot (x - 4) \Rightarrow y = -x + \pi + 3$

SECTION II, PART A

Question 1

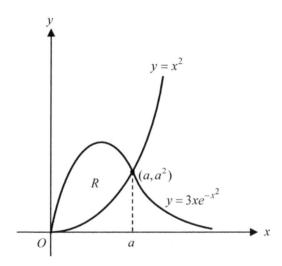

a) The point of intersection using your calculator. $a = 1.033$

$\text{Area} = \int_0^{1.033} \left(3xe^{-x^2} - x^2\right)dx \approx 0.617$

b) $\text{Volume} = \pi \int_0^{1.033} \left(\left(3xe^{-x^2}\right)^2 - \left(x^2\right)^2\right)dx = 0.845\pi$

c) $\text{Volume} = \int_0^{1.033} \left(3xe^{-x^2} - x^2\right)^2 dx = 0.448 \text{ or } 0.449$

AP Calculus AB Test 1 Answers

Question 2

a) $a(t) = v'(t) = -\dfrac{1}{5}e^{-t/5}\cos t - e^{-t/5}\sin t$

$a(3) = -\dfrac{1}{5}e^{-3/5}\cos 3 - e^{-3/5}\sin 3 = 0.031$

b) $v(3) = e^{-3/5}\cos 3 = -0.543 < 0$ and $a(3) = 0.031 > 0$. The particle's speed is decreasing, because the velocity and acceleration of the particle have opposite sign.

c) The graph of v:

At time $t = 1.571$ and $t = 4.712$, the particle changes direction.
At $t = 1.571$: The particle moves from right to left.
At $t = 4.712$: The particle moves from left to right.

d) Find the total distance traveled by the particle from $t = 0$ to $t = 3$.

$d = \left|\displaystyle\int_0^6 e^{-t/5}\cos t\, dt\right| = \displaystyle\int_0^{1.571} e^{-t/5}\cos t\, dt - \displaystyle\int_{1.571}^3 e^{-t/5}\cos t\, dt = 1.418$

e) Find $x(3)$, the position of the particle at time $t = 3$.

$x(3) = x(0) + \displaystyle\int_0^3 v(t)\, dt = -20 + \displaystyle\int_0^3 e^{-t/5}\cos t\, dt = -19.629$

Question 3

a) $R'(50) \approx \dfrac{R(60) - R(40)}{60 - 40} = \dfrac{55000 - 50000}{20} = \$250 / day$

b)

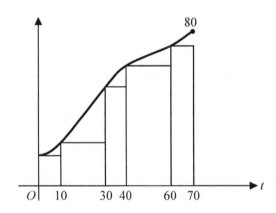

$S = (10\times 10) + (20\times 15) + (10\times 40) + (20\times 50) + (10\times 55) = 2350$ Thousand dollars

c) The approximation is less than $\displaystyle\int_0^{70} R(t)\, dt$, because the graph of R is strictly increasing.

d) It is total amount of profit during the 70 days.

e) Average profit in thousand dollars per day

AP Calculus AB Test 1 Answers

Question 4

a) $g'(x) = f(x) > 0$ → $g(x)$ is increasing on the intervals $[-4, -2]$ and $[1, 5]$.

b) $g(-2) = \int_0^{-2} f(t)\,dt = -\int_{-2}^0 f(t)\,dt = -\left(-\frac{2 \times 2}{2}\right) = 2$, $g(1) = \int_0^1 f(t)\,dt = -\frac{1 \times 2}{2} = -1$

$g(3) = \int_0^3 f(t)\,dt = \pi - 1$

c) $\begin{cases} x < -2, & g'(x) > 0 \\ x > -2, & g'(x) < 0 \end{cases}$ → g has a relative maximum at $x = -2$.

d) $x = 0, 3$

At $x = 0$, $g'(x)$ changes from decreasing to increasing.

At $x = 3$, $g'(x)$ changes from increasing to decreasing.

e) $g'(3) = 2$ and $g(3) = \pi - 1$

Therefore, $y - (\pi - 1) = 2(x - 3)$.

Question 5

a) $\dfrac{dr}{dt} = 0.5$

$\dfrac{dV}{dt} = 4\pi r^2 \dfrac{dr}{dt} = 4\pi r^2 \cdot (0.5) = 2\pi r^2$

b) $\dfrac{dV}{dt} = 2\pi r^2 \Big|_{r=10} = 200\pi$ cm^3/minute

c) Since surface area is $A = 4\pi r^2$, $\dfrac{dA}{dt} = 8\pi r \dfrac{dr}{dt}$

$V = \dfrac{4\pi r^3}{3} = 36\pi \Rightarrow r = 3$

Therefore, $\dfrac{dA}{dt}\Big|_{r=3} = 8\pi(3)(0.5) = 12\pi$ cm^2/minute.

d) If the volume and the surface area of the sphere are increasing at the same numerical rate, what is the radius?

$\dfrac{dV}{dt} = \dfrac{dA}{dt} \Rightarrow 4\pi r^2 \dfrac{dr}{dt} = 8\pi r \dfrac{dr}{dt} \Rightarrow 4r^2 = 8r \Rightarrow r(r-2) = 0$

Therefore, $r = 2$ $(r > 0)$

Question 6

a) $\dfrac{dP}{dt} = k(1000 - P)$

$\dfrac{dP}{1000 - P} = k\,dt$ → $-\ln(1000 - P) = kt + C_1$ → $\ln(1000 - P) = -kt + C_2$

$1000 - P = Ce^{-kt}$ → $P = 1000 - Ce^{-kt}$

Initial condition: $P(0) = 500$ → $1000 - Ce^0 = 500$ → $1000 - C = 500$ → $C = 500$

Therefore, $P(t) = 1000 - 500e^{-kt}$.

AP Calculus AB Test 1 Answers

Find the constant k. $P(5) = 1000 - 500e^{-5k} = 900 \Rightarrow 500e^{-5k} = 100 \Rightarrow e^{-5k} = \frac{1}{5}$

$$-5k = \ln\frac{1}{5} \Rightarrow -5k = -\ln 5 \Rightarrow k = \frac{\ln 5}{5}$$

Therefore, $P(t) = 1000 - 500e^{\frac{\ln 5}{5}t}$.

b) $P(10) = 1000 - 500e^{\frac{-\ln 5}{5}(10)} = 1000 - 500e^{-2\ln 5} = 1000 - 500e^{\ln\frac{1}{25}} = 1000 - 500 \cdot \frac{1}{25} = 1000 - 20 = 980$.

c) $\lim_{t \to \infty} \left(1000 - 500e^{-\frac{\ln 5}{5}t}\right) = 1000$

The graph of P is shown as follows.

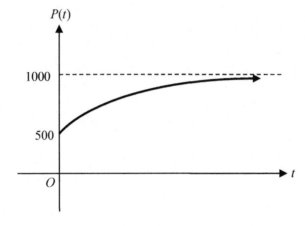

AP CALCULUS AB

TEST 2

AP Calculus AB Test 2 Section 1

AP CALCULUS AB

A CALCULATOR **CANNOT BE USED ON PART A OF SECTION 1**. A GRAPHIC CALCULATOR FROM THE APPROVED LIST **IS REQUIRED FOR PART B OF SECTION 1 AND FOR PART A OF SECTION II** OF THE EXAMINATION.

SECTION 1
Time: 1 hour and 45 minutes
All questions are given equal weight.
Percent of total grade – 50

Part A: 55 minutes, 28 multiple-choice questions
A calculator is NOT allowed.
Part B: 50 minutes, 17 multiple-choice questions
A graphing calculator is required.

CALCULUS AB
SECTION 1, PART A
Time — 55 minutes
Number of questions — 28

A CALCULATOR MAY NOT BE USED ON THIS PART OF THE EXAMINATION.

Directions: Solve each of the following problems, using the available space for scratchwork. After examining the form of the choices, decide which the best of the choices given and fill in the corresponding oval on the answer sheet. No credit will be given for anything written in the test book. Do not spend too much time on any one problem.

In this test:

(1) Unless otherwise specified, the domain of a function f is assumed to be the set of all real numbers x for which $f(x)$ is a real number.

(2) The inverse of a trigonometric function f may be indicated using the inverse function notation f^{-1} or with the prefix "arc" (e.g, $\sin^{-1} x = \arcsin x$)

AP Calculus AB Test 2

Section 1

1. $\lim_{x \to 1} \dfrac{\sqrt{x+8} - 3}{x - 1} =$

   (A) 0   (B) $\dfrac{1}{6}$ (C) $\dfrac{1}{3}$ (D) 5   (E) nonexistent

2. $\displaystyle\int \dfrac{4x}{x^2 + 1} dx =$

   (A) $\dfrac{2}{3} x^3 + C$ (B) $\ln(x^2 + 1) + C$ (C) $2\ln(x^2 + 1) + C$ (D) $2x^{-3} + C$ (E) $\dfrac{2}{3} x^{-3} + C$

3. If $f(x) = \cos^2(x^3 + 1)$, then $f'(x) =$

   (A) $2\sin(x^3 + 1)$
   (B) $3x^2 \sin(x^3 + 1)$
   (C) $-2\cos(x^3 + 1) \cdot \sin(x^3 + 1)$
   (D) $6x^2 \cos(x^3 + 1) \cdot \sin(x^3 + 1)$
   (E) $-6x^2 \cos(x^3 + 1) \cdot \sin(x^3 + 1)$

4. $\displaystyle\int_0^1 \dfrac{x}{\sqrt{x^2 + 1}} dx =$

   (A) 0   (B) $\sqrt{2} - 1$ (C) 1   (D) $\sqrt{2} + 1$ (E) 4

5. $\lim_{x \to 1} \dfrac{x - 1}{|x - 1|} =$

   (A) -1 (B) 0   (C) 1   (D) 2   (E) nonexistent

AP Calculus AB Test 2 — Section 1

$$f(x) = \begin{cases} \dfrac{x^2-4}{x-2} & \text{if } x \neq 2 \\ k-2x & \text{if } x = 2 \end{cases}$$

6. Let f be the function defined above. Find the value of k so that the function is continuous for all real numbers x.

   (A) 2 (B) 4 (C) 6 (D) 8 (E) nonexistent

7. If $x^2 + xy + y^2 = 12$, then which of the following could be the value of $\dfrac{dy}{dx}$ at $x = 2$?

   (A) –2 (B) –1 (C) 1 (D) 2 (E) 4

8. For what value of k will $kx + \dfrac{4}{x}$ have a relative minimum at $x = 1$?

   (A) –2 (B) –1 (C) 0 (D) 2 (E) 4

9. If $f(x) = e^{-\frac{x}{2}}$, then $f'(\ln 4) =$

   (A) $-\dfrac{1}{8}$ (B) $-\dfrac{1}{4}$ (C) $-\dfrac{1}{2}$ (D) $\dfrac{1}{2}$ (E) $\dfrac{1}{4}$

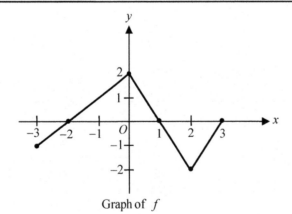

Graph of f

10. Let f be the piecewise linear function in the figure above. If $g(x) = \int_0^x f(t)\,dt$, which of the following values is greatest?

    (A) $g(-3)$ (B) $g(-2)$ (C) $g(0)$ (D) $g(1)$ (E) $g(3)$

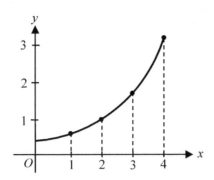

11. The graph of the function is shown above for $0 \leq x \leq 4$. Which of the following has the greatest value?

    (A) Left Riemann sum approximation with using 4 equal subintervals
    (B) Midpoint Riemann sum approximation using 4 equal subintervals
    (C) Right Riemann sum approximation using 4 equal subintervals
    (D) Trapezoidal sum approximation using 4 equal subintervals
    (E) $\int_0^4 f(x)dx$

---

12. If $f(x) = 4^{1/x}$, then $f'(2) =$

    (A) $-\ln 2$ (B) -1 (C) $\ln 2$ (D) 2 (E) $2\ln 2$

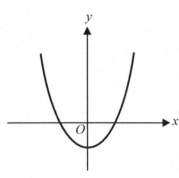

Graph of f'

13. The graph of f', the derivative of f, is shown above. Which of the following could be the graph of f?

(A)

(B)

(C)

(D)

(E)

x	0	2	4	6
$f''(x)$	−2	0	3	−1

14. The function f is continuous and differentiable for all real x and selected values of its second derivative f'' are shown in the table above. Which of the following statements must be true?

    (A) f is increasing on the interval $(2, 3)$.
    (B) f is increasing on the interval $(0, 2)$.
    (C) f has a local maximum at $x = 4$.
    (D) The graph of f has a point of inflection at $x = 2$.
    (E) The graph of f has at least two points of inflection in the interval $(0, 6)$.

---

15. $\int \tan x \, dx =$

    (A) $\sec^2 x + C$ (B) $\csc^2 x + C$ (C) $\ln(\sin x) + C$ (D) $\ln|\cos x| + C$ (E) $\ln|\sec x| + C$

---

16. If the line $3x - y + k = 0$, where k is a constant, is tangent to the graph of $y = x^2 - 3x + 5$, what is the value of k?

    (A) −6   (B) −4   (C) 3   (D) 6   (E) 9

---

17. If $x \cos^2 y = 1$, evaluate the derivative at $\left(2, \dfrac{\pi}{3}\right)$.

    (A) $\dfrac{\sqrt{3}}{12}$ (B) $\dfrac{\sqrt{3}}{2}$ (C) 2   (D) $2\sqrt{3}$ (E) 4

18. The position of a particle moving in the xy-plane is given by the parametric equations $x = \dfrac{t^3}{3}$ and $y = \dfrac{t^3 - 4}{t}$. What is the equation of the line tangent to the graph $y = f(x)$ at $t = 2$?

(A) $y = 2\left(x - \dfrac{8}{3}\right)$

(B) $y + 2 = \dfrac{4}{5}\left(x - \dfrac{8}{3}\right)$

(C) $y - 2 = \dfrac{5}{4}\left(x - \dfrac{8}{3}\right)$

(D) $y - 2 = \dfrac{4}{5}\left(x - \dfrac{8}{3}\right)$

(E) $y - 2 = \dfrac{4}{5}\left(x - \dfrac{8}{3}\right)$

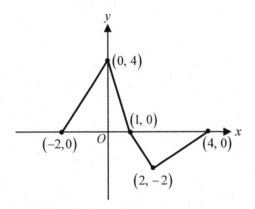

19. The graph of the function f shown above. Let g be the function defined by $g(x) = \displaystyle\int_{-2}^{x} f(t)\,dt$. Find the equation of the line tangent to the graph of g at $x = 2$.

(A) $y = -2x + 3$
(B) $y = -2x + 5$
(C) $y = -2x + 9$
(D) $y = 2x + 3$
(E) $y = 2x + 5$

AP Calculus AB Test 2　　　　　　　　　　　　　　　　Section 1

20. Find all asymptotes of the graph of $y = \dfrac{1+e^x}{1-e^x}$ in the xy-plane.

(A) $x = 0$ only
(B) $x = 0$ and $y = -1$ only
(C) $y = -1$ only
(D) $y = -1$ and $y = 1$ only
(E) $x = 0$, $y = -1$, and $y = 1$

---

21. Let f be the function defined by $f(x) = \dfrac{x^6}{30} - \dfrac{x^4}{12} + x + 3$. What are the x-coordinates of the points of inflection of the graph of f ?

(A) 0 only　(B) -1 and 1 only　(C) 0 and 1 only　(D) 0, 1, and 3　(E) -1, 0, and 1

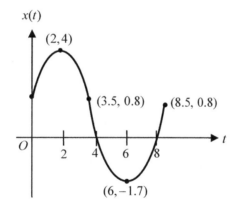

22. A particle moves along a straight line. The graph of the particle's position $x(t)$ is shown above for $0 \le t \le 8.5$ and the graph has a point of inflection at $(3.5, 0.8)$. For what value of t is the velocity of the particle decreasing?

(A) $2 < t < 4$
(B) $2 < t < 6$
(C) $2 < t < 3.5$ only
(D) $0 < t < 3.5$
(E) $4 < t < 8$

AP Calculus AB Test 2

Section 1

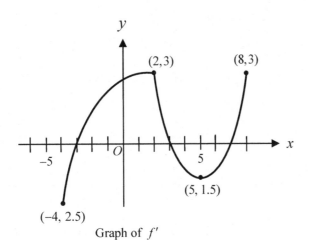

Graph of f'

23. The graph of f', the derivative of f is shown above for $-4 \leq x \leq 8$. For what values of x, the graph of f is concave upward?

(A) $-3 < x < 3$ only
(B) $2 < x < 8$
(C) $-4 < x < 2$ only
(D) $-4 < x < 2$ and $5 < x < 8$
(E) $2 < x < 5$

24. Which of the following is the solution to the differential equation $\dfrac{dy}{dx} = \dfrac{2x+1}{y}$ with the initial condition $y(2) = -4$?

(A) $y = 4x + 2x$
(B) $y = -x^2 + 8x - 8$
(C) $y = \sqrt{2x^2 + 8x - 8}$
(D) $y = -\sqrt{2x^2 + 2x + 4}$
(E) $y = -\sqrt{4x^2 + 2x - 8}$

25. Let f be the function given by $y = \sqrt{x} - 1$. What is the value of the approximation of $f(3.8)$ using the line tangent to the graph of f at $x = 4$?

(A) 0.8 (B) 0.95 (C) 0.97 (D) 1.02 (E) 1.13

AP Calculus AB Test 2   Section 1

26. If $\int_0^2 e^{x^2} dx = k$ where k is constant, then which of the following is the value of $\int_{-2}^2 \left(e^{x^2} + 2\right) dx$?

(A) $k+2$ (B) $k+4$ (C) $2k+2$ (D) $2k+4$ (E) $2k+8$

---

27. $\int_0^1 \frac{2x+1}{x^2+1} dx =$

(A) $\ln 2 + \frac{\pi}{4}$ (B) $\ln 2$ (C) $\sqrt{2}-1$ (D) $1 + \ln 2$ (E) $\sqrt{2} + \frac{\pi}{4}$

---

28. The function g is the inverse of f and f is given by $f(x) = -\sqrt{x^3 + 2x^2 + 6}$. What is the value of $g'(-3)$?

(A) $-\frac{7}{6}$ (B) $-\frac{6}{7}$ (C) $-\frac{1}{6}$ (D) $\frac{1}{6}$ (E) $\frac{1}{3}$

END OF PART A OF SECTION 1

AP Calculus AB Test 2   *Section 1*

CALCULUS AB
SECTION 1, PART B
Time — 50 minutes
Number of questions — 17

A GRAPHING CALCULATOR IS REQUIRED FOR SOME QUESTIONS ON THIS PART OF THE EXAMINATION.

Directions: Solve each of the following problems, using the available space for scratchwork. After examining the form of the choices, decide which is the best of the choices given and fill in the corresponding oval on the answer sheet. No credit will be given for anything written in the test book. Do not spend too much time on any one problem.

In this test:

(1) The exact numerical value of the correct answer does not always appear among the choices given. When this happens, select from among the choices the number that best approximates the exact numerical value.

(2) Unless otherwise specified, the domain of a function f is assumed to be the set of all real numbers x for which $f(x)$ is a real number.

(3) The inverse of a trigonometric function f may be indicated using the inverse function notation f^{-1} or with the prefix "arc" (e.g, $\sin^{-1} x = \arcsin x$)

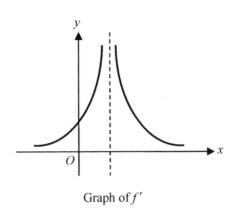

Graph of f'

76. The graph of f', the derivative of f, is shown above. Which of the following could be the graph of f?

(A)

(B)

(C)

(D)

(E)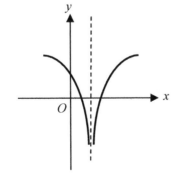

AP Calculus AB Test 2 — Section 1

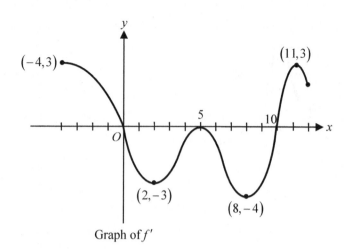

Graph of f'

77. The graph of f', the derivative of f, is shown above for $-4 \leq x \leq 12$. On what intervals is the graph of f concave upward?

   (A) $-4 < x < 0$ only
   (B) $0 < x < 5$ and $5 < x < 10$
   (C) $-4 < x < 0$ and $2 < x < 8$
   (D) $2 < x < 5$ only
   (E) $2 < x < 5$ and $8 < x < 11$

---

78. The derivative of the function is given by $f'(x) = e^{2\sin x} - 1$ for $0 \leq x \leq 8$. On what intervals of f is the graph of f decreasing?

   (A) $0 \leq x \leq 3.142$
   (B) $3.142 \leq x \leq 5$ only
   (C) $3.142 \leq x \leq 6.283$
   (D) $5 \leq x \leq 8$
   (E) $0 \leq x \leq 3.142$ and $6.283 \leq x \leq 8$

AP Calculus AB Test 2

Section 1

79. Let f be the function defined by

$$f(x) = \begin{cases} x^2 & \text{for } x < 2 \\ 4x - 4 & \text{for } x \geq 2 \end{cases}.$$

Which of the following are true about f?

I. $\lim_{x \to 2} f(x)$ exists.

II. $f(2)$ exists.

III. $f(x)$ is continuous at $x = 2$.

(A) I only   (B) I and II only   (C) II only   (D) II and III only   (E) I, II, and III

---

80. If $\int_{-2}^{4} f(x)\,dx = 10$ and $\int_{6}^{4} f(x)\,dx = 2$, what is the value of $\int_{0}^{8} (f(x-2)+1)\,dx$?

(A) −8   (B) −2   (C) 0   (D) 8   (E) 16

---

81. If $F'(x) = x^2 \cos x$ and $F(0) = -5$, then $F(4) =$

(A) −20.824   (B) −10.156   (C) −8.842   (D) 4.412   (E) 8.842

---

82. A particle moves along a straight line with velocity given by $v(t) = te^{-t}$ at time $t \geq 0$. Which of the following is true about the speed of the particle at time $t = 3$?

(A) The speed of the particle is constant.
(B) The speed of the particle is decreasing.
(C) The speed of the particle is increasing.
(D) The speed of the particle is zero.
(E) None of these.

---

83. What is the area enclosed by the graphs of $y = 3x^3 - x^2 - 10x$ and $y = -x^2 + 2x$?

(A) 18.432   (B) 20.852   (C) 24   (D) 28   (E) 35.883

AP Calculus AB Test 2   Section 1

x	f	f'	g	g'
0	2	1	−10	1
2	4	3	−7	3
4	8	9	−4	6
6	18	27	−1	9

84. The table above gives values of functions f and g, and their derivatives at selected values of x. The functions f and g are differentiable on the closed interval $[0, 6]$. If $K = f(2x) \cdot g(2x)$, then $K'(2) =$

    (A) −12   (B) 12   (C) 24   (D) 36   (E) 46

---

85. The volume of a cone is increasing at the rate of 30π cubic unit per second. When the radius of the cone is 5 units, its volume is 20π cubic units. If the radius of the base is increasing at the rate of 2 units, what is the rate of change of its height h, at the instant when the radius of the cone is 5 units?

    (A) 1.7 units per second
    (B) 2.2 units per second
    (C) 2.5 units per second
    (D) 3.5 units per second
    (E) 4.2 units per second

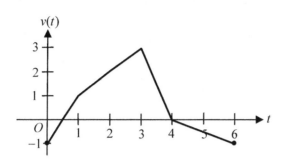

86. The graph of the velocity $v(t)$ of a particle moving along the x-axis is shown above. At time $t=0$, the particle is at the origin. Which of the following could be the graph of the position, $x(t)$, of the particle for $0 \le t \le 6$?

(A)

(B)

(C)

(D)

(E)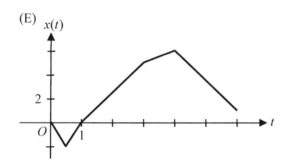

AP Calculus AB Test 2

Section 1

87. Let R be the region enclosed by the graphs of $y = \dfrac{10}{x^2+1}$ and $y = 0$. Find the area of R.

    (A) 15.708   (B) 22.143   (C) 23.987   (D) 24.152   (E) 31.416

---

88. What is the average value of $y = \dfrac{3\cos x}{x^2+1}$ for $0 \leq x \leq 1$?

    (A) 1.125   (B) 1.673   (C) 2.049   (D) 3.481   (E) 4.925

---

89. When the equation is defined by $y + \cos y + 1 = x$ for $0 \leq y \leq 2\pi$, find an equation for the vertical tangent to the curve.

    (A) $x = -\pi$

    (B) $x = -\dfrac{\pi}{2}$

    (C) $x = \dfrac{\pi}{2} - 1$

    (D) $x = \dfrac{\pi}{2} + 1$

    (E) $x = \pi + 2$

---

90. The function f is defined by $f(x) = -|x-2| + 5$ for $-3 \leq x \leq 7$. Which of the following statements must be true?

    (A) For $-3 \leq x \leq 7$, $f(x)$ is increasing.
    (B) For $-3 \leq x \leq 7$, $f(x)$ is decreasing.
    (C) For $-3 \leq x \leq 7$, $f'(x)$ exists.
    (D) For $-3 \leq x \leq 7$, relative maximum exists.
    (E) For $-3 \leq x \leq 7$, $\lim\limits_{x \to 2} f'(x)$ exists.

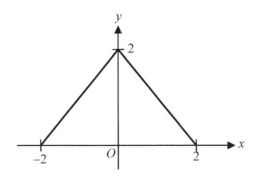

91. The graph of f', the derivative of a function f is shown above. The domain of f is the set of all x such that $-2 < x < 2$. Which of the following could be the graph of f?

(A)

(B)

(C)

(D)

(E)

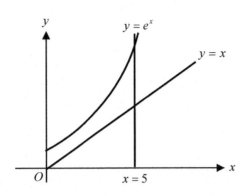

92. Let R be the region enclosed by the graphs of $y = e^x$, $y = x$, and the lines $x = 0$ and $x = 5$ as shown above. Which of the following expressions gives the volume of the solid generated when R is evolved about the y axis?

(A) $\int_0^5 (e^x - x) dx$

(B) $\pi \int_0^5 (5 - e^x)^2 dx$

(C) $2\pi \int_0^5 x(e^x - x) dx$

(D) $2\pi \int_0^5 (e^x - x) dx$

(E) $\pi \int_0^5 (e^{2x} - x^2) dx$

END OF SECTION 1

AP Calculus AB Test 2   **Section II**

CALCULUS AB

SECTION II
Time: 1 hour and 30 minutes
Percent of total grade – 50

Part A: 30 minutes, 2 problems
Part B: 60 minutes, 4 problems

Part A: A graphing calculator is required for some problems or parts of problems.
Part B: No calculator is allowed for these problems.

GENERAL INSTRUCTIONS FOR SECTION II PART A AND PART B

For each part of section II, you may wish to look over the problems before starting to work on them, since it is not expected that everyone will be able to complete all parts of all problems. All problems are given equal weight, but the parts of a particular problem are not necessary given equal weight.

- YOU SHOULD WRITE ALL WORK FOR EACH PART OF EACH PROBLEM IN THE SPACE PROVIDED FOR THAT PART IN THE TEST BOOKLET. Be sure to write clearly and legibly. If you make an error, you may save time by crossing it out rather than trying to erase it. Erased or cross-out work will not be graded.
- Show all work. Clearly label any functions, graphs, tables, or other object that you use. You will be graded on the correctness and completeness of your methods as well as your answers. Answers without supporting work may not receive credit.
- Justifications required that you give mathematical (noncalculator) reasons
- Your work must be expressed in standard mathematical notation rather than calculator syntax.
  For example, $\int_{1}^{5} x^2 \, dx$ may not be written as $\text{fnInt}(X^2, X, 1, 5)$
- Unless otherwise specified, answers (numeric or algebraic) need not be simplified.
- If you use decimal approximations in calculations, you will be graded on accuracy. Unless otherwise specified, your final answers should be accurate to three places after the decimal point.
- Unless otherwise specified, the domain of a function f is assumed to be the set of all real numbers x for which $f(x)$ is a real number.

AP Calculus AB Test 2 — Section II

CALCULUS AB
SECTION II, PART A
Time — 30 minutes
Number of questions — 2

A Graphing calculator is required for some problems or parts of problems.

Question 1

At time $t = 0$, there were 200 people in an auditorium for a musical comedy. Let R, in people per minute, be the function of the rate at which people enter the auditorium for $0 \le t \le 10$, and R is modeled by the function $R(t) = 100e^{-0.5t}\left(t^2 + t + 1\right)$.

(a) How many people enter in the auditorium at time $t = 10$?

(b) How many people are in the auditorium at time $t = 10$?

(c) At what time t will the rate at which people enter be a maximum?

(d) Find $\int_0^{10} R(t)\,dt$ and explain the meaning of $\int_0^{10} R(t)\,dt$.

(e) Find the average rate at which people enter the auditorium over the interval $0 \le t \le 10$.

AP Calculus AB Test 2

Section II

Question 2

An internet shopping mall sells a new product computer (JCK-model) for $500 per one. The cost of the computer varies every day from the first day. Let C be the function given by $C(t) = 300 - 10\sqrt{t}$ and N, the number of computer sold, be the function given by $N(t) = 3t^2 - 10t + 5$.

(a) How many computers did the shopping mall sell over the first 10 days?

(b) Find the shopping mall's profit over the first 10 days.

(c) Explain the meaning of $\int_5^{10} C(t)\,dt$ in the context of this problem.

(d) Write an expression, involving an integral that represents the shopping mall's profit over the first k days.

END OF PART A OF SECTION II

AP Calculus AB Test 2 — Section II

CALCULUS AB
SECTION II, PART B
Time — 60 minutes
Number of questions — 4

No calculator is required for these problems.

Question 3

A particle moves in the xy-plane for $0 \le t \le 10$, in minutes. Its velocity, in miles per minute, is modeled by a semicircle and the piecewise-linear functions whose graph shown above.

(a) Find the acceleration of the particle at time $t = 5.5$ minutes.

(b) Find the values of $\int_0^{10} v(t)\,dt$ and explain the meaning of $\int_0^{10} v(t)\,dt$.

(c) Find the value of $\int_0^{10} |v(t)|\,dt$ and explain the meaning of $\int_0^{10} |v(t)|\,dt$.

(d) At what time the particle changes its direction?

AP Calculus AB Test 2

Section II

Question 4

The function f is defined by $f(x) = \sqrt{x^3 + 9}$ for $x \geq 0$.

(a) Find $f'(x)$.

(b) Write an equation for the line tangent to the curve at $x = 3$. Use this line to approximate $f(3.2)$.

(c) Let g be the function defined by $g(x) = \begin{cases} \sqrt{x^3 + 9} & \text{for } 0 \leq x \leq 3 \\ ax + b & \text{for } x > 3 \end{cases}$.

If g is differentiable at $x = 3$, what are the values of a and b?

(d) Find the value of $\int_0^3 x^2 \sqrt{x^3 + 9}\, dx$.

Question 5

The number of a certain type of bacteria increases at a rate proportional to the difference 500 and the number present. There are 100 bacteria at time $t = 0$. If $P(t)$ is the number of bacteria at time t minutes, then

$$\frac{dP}{dt} = \frac{1}{10}(500 - P).$$

(a) If $y = P(t)$ is the particular solution to the differential equation at time $t \geq 0$, then find $y = P(t)$.

(b) Find the value of $\lim_{t \to \infty} P(t)$.

(c) Is the graph of $y = P(t)$ is concave down? Explain your reasoning.

(d) Find $\dfrac{d^2 P}{dt^2}$ in terms of P. Is the value of $\dfrac{d^2 P}{dt^2}$ always negative? Explain your reasoning.

AP Calculus AB Test 2 — Section II

Question 6

A particle moves along the x-axis. The acceleration of the particle at time t is given by $a(t) = 4\sin\left(\dfrac{t}{4}\right)$. At time $t = 0$, the velocity of the particle is $v(0) = -16$ and its position is $x(0) = 5$.

(a) Write an equation for the velocity $v(t)$ of the particle.

(b) For $0 \leq t \leq 8\pi$, find the values of t for which the particle is at rest.

(c) For $0 \leq t \leq 8\pi$, when is the particle moving to the right?

(d) For $0 \leq t \leq 8\pi$, when is the speed of the particle increasing?

(e) Find the total distance traveled by the particle from $t = 0$ to 4π.

(f) Find the position of the particle at time $t = 2\pi$.

END OF EXAM

AP Calculus AB Test 2 Answers

SECTION 1, PART A

1. B

$$\lim_{x\to 1}\frac{\sqrt{x+8}-3}{x-1} = \lim_{x\to 1}\frac{(\sqrt{x+8}-3)(\sqrt{x+8}+3)}{(x-1)(\sqrt{x+8}+3)} = \lim_{x\to 1}\frac{x-1}{(x-1)(\sqrt{x+8}+3)} = \lim_{x\to 1}\frac{1}{(\sqrt{x+8}+3)} = \frac{1}{6}$$

2. C

$$\int \frac{4x}{x^2+1}dx = 2\int \frac{2x}{x^2+1}dx = 2\int \frac{(x^2)'}{x^2+1}dx = 2\ln(x^2+1) + C$$

Or, use u-substitution. $\begin{cases} u = x^2+1 \\ du = 2xdx \end{cases}$

3. E

$$f(x) = \cos^2(x^3+1) \;\;\to\;\; f'(x) = 2\cos(x^3+1)\cdot(-\sin(x^3+1))(3x^2) = -6x^2\cos(x^3+1)\sin(x^3+1)$$

4. B

u-substitution: $\begin{cases} u = x^2+1 \\ du = 2xdx \;\;\Rightarrow\;\; xdx = \dfrac{du}{2} \end{cases}$

$$\int_0^1 \frac{x}{\sqrt{x^2+1}}dx = \frac{1}{2}\int_1^2 \frac{1}{\sqrt{u}}du = \frac{1}{2}\int_1^2 u^{-1/2}du = \frac{1}{2}\left[2u^{1/2}\right]_1^2 = \frac{1}{2}(2\sqrt{2}-2) = \sqrt{2}-1$$

5. E

$$\lim_{x\to 1}\frac{x-1}{|x-1|} = \begin{cases} \lim_{x\to 1^+}\dfrac{x-1}{x-1} = 1 \\ \lim_{x\to 1^-}\dfrac{x-1}{-(x-1)} = -1 \end{cases}$$

Since $\lim_{x\to 1^+}\dfrac{x-1}{x-1} \neq \lim_{x\to 1^-}\dfrac{x-1}{-(x-1)}$, limit does not exist (DNE).

6. D

$$f(x) = \begin{cases} \dfrac{x^2-4}{x-2} & \text{if } x \neq 2 \\ k-2x & \text{if } x = 2 \end{cases}$$

$f(2) = k-4$ and $\lim_{x\to 2}\dfrac{x^2-4}{x-2} = \lim_{x\to 2}\dfrac{(x+2)(x-2)}{x-2} = \lim_{x\to 2}\dfrac{(x+2)}{1} = 4$

Continuity: $\lim_{x\to a}f(x) = f(a)$

$k - 4 = 4 \;\;\Rightarrow\;\; k = 8$

7. B

$x^2 + xy + y^2 = 12 \;\;\Rightarrow\;\; 2x + y + xy' + 2yy' = 0 \;\;\Rightarrow\;\; (x+2y)y' = -2x - y$

$y' = \dfrac{-2x-y}{x+2y}$

If $x = 2$, then $2^2 + 2y + y^2 - 12 = 0 \;\;\to\;\; y^2 + 2y - 8 = 0 \;\;\to\;\; (y+4)(y-2) = 0$.

$y = -4$ or $y = 2$

AP Calculus AB Test 2 Answers

Therefore, $y' = \dfrac{-2x-y}{x+2y}\bigg|_{(2,-4)} = \dfrac{-4+4}{2-8} = 0$ or $y' = \dfrac{-2x-y}{x+2y}\bigg|_{(2,2)} = \dfrac{-4-2}{2+4} = -1$

8. E

For what value of k will $kx + \dfrac{4}{x}$ have a relative minimum at $x=1$?

$\left(kx+\dfrac{4}{x}\right)'_{x=-1} = 0 \;\rightarrow\; k - \dfrac{4}{x^2} = 0 \;\rightarrow\; \dfrac{kx^2-4}{x^2} = 0$

At $x=1$, $\dfrac{k-4}{1} = 0 \;\Rightarrow\; k=4$.

Graph of $y = 4x + \dfrac{4}{x}$:

9. B

$f(x) = e^{-\frac{x}{2}} \;\Rightarrow\; f'(x) = -\dfrac{1}{2}e^{-\frac{x}{2}} \;\Rightarrow\; f'(\ln 4) = -\dfrac{1}{2}e^{-\frac{\ln 4}{2}} = -\dfrac{1}{2}e^{\ln\left(\frac{1}{2}\right)} = -\dfrac{1}{2}\left(\dfrac{1}{2}\right) = -\dfrac{1}{4}$

10. D

(A) $g(-3) = \int_0^{-3} f(t)\,dt = -\int_{-3}^0 f(t)\,dt = -(-0.3+3) = -2.5$

(B) $g(-2) = \int_0^{-2} f(t)\,dt = -\int_{-2}^0 f(t)\,dt = -3$ (C) $g(0) = \int_0^0 f(t)\,dt = 0$

(D) $g(1) = \int_0^1 f(t)\,dt = 1$ (E) $g(3) = \int_0^3 f(t)\,dt = 1-2 = -1$

---

11. C

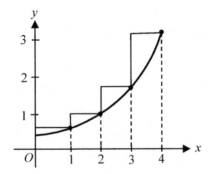

12. A

Since $f(x) = 4^{1/x} \;\Rightarrow\; f'(x) = 4^{1/x} \cdot \ln 4 \cdot \left(-\dfrac{1}{x^2}\right)$, then

$f'(2) = 4^{1/2} \cdot \ln 4 \cdot \left(-\dfrac{1}{4}\right) = -\dfrac{1}{2}\ln 4 = -\dfrac{1}{2}\ln 2^2 = -\ln 2$

AP Calculus AB Test 2 Answers

13. D

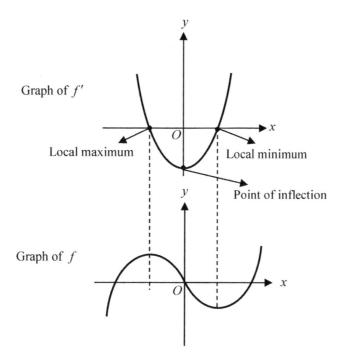

14. E
    - (A) f is increasing on the interval $(2, 4)$. Not always true.
    - (B) f is increasing on the interval $(0, 2)$. Not always true.
    - (C) f has a local maximum at $x = 4$. Not always true.
    - (D) The graph of f has a point of inflection at $x = 2$. Not always true.
    - (E) The graph of f has at least two points of inflection in the interval $(0, 6)$. True
      Intermediate Theorem

15. E
    $$\int \tan x \, dx = \int \frac{\sin x}{\cos x} dx = -\int \frac{(-\sin x)}{\cos x} dx = -\ln|\cos x| + C = \ln \frac{1}{|\cos x|} + C = \ln|\sec x| + C$$

16. B
    $3x - y + k = 0 \;\rightarrow\; y = 3x + k \;\rightarrow\; \text{slope} = 3$
    $y = x^2 - 3x + 5 \;\rightarrow\; f'(x) = \dfrac{dy}{dx} = 2x - 3$
    Point of tangency: $(a, f(a))$
    $f'(a) = 2a - 3 = 3 \;\Rightarrow\; a = 3$ and $f(3) = 9 - 9 + 5 = 5$
    Therefore, point of tangency is $(3, 5)$.
    Point $(3, 5)$ is on the tangent line. $3(3) - 5 + k = 0 \;\Rightarrow\; k = -4$

17. A
    $x\cos^2 y = 1 \;\rightarrow\; \cos^2 y + x(2\cos y)\cdot(-\sin y)\cdot y' = 0$
    At $\left(2, \dfrac{\pi}{3}\right)$, $\cos^2\left(\dfrac{\pi}{3}\right) + 2(2)\cos\left(\dfrac{\pi}{3}\right)\left(-\sin\left(\dfrac{\pi}{3}\right)\right)y' = 0 \;\Rightarrow\; \left(\dfrac{1}{2}\right)^2 + 4\left(\dfrac{1}{2}\right)\cdot\left(-\dfrac{\sqrt{3}}{2}\right)y' = 0$
    $\Rightarrow\; \dfrac{1}{4} - \sqrt{3}y' = 0 \;\Rightarrow\; y' = \dfrac{1}{4\sqrt{3}} = \dfrac{\sqrt{3}}{12}$

Dr. John Chung's AP Calculus AB/BC

AP Calculus AB Test 2 Answers

18. C

$$\frac{dy}{dx} = \frac{dy/dt}{dx/dt}, \quad \frac{dx}{dt} = t^2, \text{ and } \frac{dy}{dt} = \frac{3t^3 - (t^3 - 4)}{t^2} = \frac{2t^3 + 4}{t^2}$$

$$\frac{dy}{dx} = \frac{dy/dt}{dx/dt} = \frac{2t^3 + 4}{t^4} \Rightarrow \left.\frac{dy}{dx}\right|_{t=2} = \frac{20}{16} = \frac{5}{4}$$

At time $t = 2$, $x = \frac{8}{3}$ and $y = \frac{8-4}{2} = 2$

Therefore, the equation of the line tangent to the graph $y = f(x)$ at $t = 2$ is

$$y - 2 = \frac{5}{4}\left(x - \frac{8}{3}\right).$$

19. C

$g'(x) = f(x) \rightarrow g'(2) = f(2) = -2 \rightarrow$ slope $= -2$

$g(2) = \int_{-2}^{2} f(t)\,dt = \frac{3 \times 4}{2} - \frac{1 \times 2}{2} = 5 \rightarrow$ point $(2, 5)$

Therefore, the equation of the line tangent to the graph is $y - 5 = -2(x - 2) \Rightarrow y = -2x + 9$.

20. E

$y = \frac{1+e^x}{1-e^x}$

Vertical asymptote: $1 - e^x = 0 \Rightarrow e^x = 1 \Rightarrow x = 0$

Horizontal asymptote: $\lim\limits_{x \to \infty} \frac{1+e^x}{1-e^x} = -1$ and

$$\lim\limits_{x \to -\infty} \frac{1+e^x}{1-e^x} = \frac{1+e^{-\infty}}{1-e^{-\infty}} = 1$$

Graph of $y = \frac{1+e^x}{1-e^x}$ →

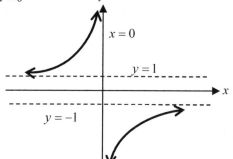

21. B

$f(x) = \frac{x^6}{30} - \frac{x^4}{12} + x + 3 \rightarrow f'(x) = \frac{x^5}{5} - \frac{x^3}{3} + 1 \rightarrow f''(x) = x^4 - x^2 = 0.$

$\rightarrow x^2(x-1)(x+1) = 0 \rightarrow$ point of inflection: $x = -1, x = 1$

Remember: at $x = 0$, no point of inflection (There is no change of concavity).

AP Calculus AB Test 2 Answers

22. D

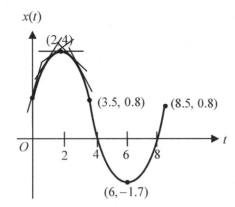

The slope of the line tangent to the curve is decreasing on the interval $(0, 3.5)$.

23. D

$f''(x) > 0$ for concave up

Therefore, $f''(x) > 0$ over the intervals $(-4, 2)$ and $(5, 8)$.

24. D

$ydy = (2x+1)dx \Rightarrow \dfrac{y^2}{2} = x^2 + x + C_1 \rightarrow y^2 = 2x^2 + 2x + C \rightarrow y = \pm\sqrt{2x^2 + 2x + C}$

Initial condition $y(2) = -4 \rightarrow -4 = -\sqrt{8+4+C} \rightarrow C = 4$

Therefore, $y = -\sqrt{2x^2 + 2x + 4}$

25. B

$y = \sqrt{x} - 1 \Rightarrow y' = \dfrac{1}{2\sqrt{x}} \Rightarrow f'(4) = \dfrac{1}{2\sqrt{4}} = \dfrac{1}{4}$.

$f(4) = \sqrt{4} - 1 = 1$

Therefore, the equation of the line tangent to the curve is $y - 1 = \dfrac{1}{4}(x-4) \rightarrow y = \dfrac{1}{4}x$

The value of the approximation of $f(3.8)$ is $\dfrac{1}{4}(3.8) = 0.95$.

26. E

$\displaystyle\int_{-2}^{2}\left(e^{x^2} + 2\right)dx = 2\int_{0}^{2}\left(e^{x^2} + 2\right)dx = 2\int_{0}^{2}e^{x^2}dx + 2[2x]_{0}^{2} = 2k + 8$

Remember $f(x) = e^{x^2} + 2$ is an even function.

27. A

$\displaystyle\int_{0}^{1}\dfrac{2x+1}{x^2+1}dx = \int_{0}^{1}\dfrac{2x}{x^2+1}dx + \int_{0}^{1}\dfrac{1}{x^2+1}dx = \left[\ln(x^2+1)\right]_{0}^{1} + \left[\tan^{-1}\right]_{0}^{1} = \ln 2 + \tan^{-1}1 - \tan^{-1}0 = \ln 2 + \dfrac{\pi}{4}$

28. B

$-3 = -\sqrt{x^3 + 2x^2 + 6} \Rightarrow 9 = x^3 + 2x^2 + 6 \rightarrow x^3 + 2x^2 - 3 = 0 \rightarrow (x-1)(x^2 + 3x + 3) = 0$

$x = 1$ and $f'(x) = -\dfrac{3x^2 + 4x}{2\sqrt{x^3 + 2x^2 + 6}}$, $f'(1) = -\dfrac{7}{6}$

AP Calculus AB Test 2 Answers

Since $g'(-3) = \dfrac{1}{f'(1)}$, $g'(-3) = -\dfrac{6}{7}$

SECTION 1, PART B

76. C

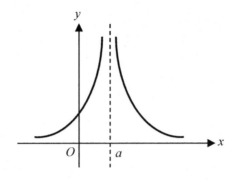

Graph of f'

For $x < a$, $f' > 0$ and $f'' > 0$ → The graph of f is strictly increasing and concave up.

For $x > a$, $f' > 0$ and $f'' < 0$ → The graph of f is strictly increasing and concave down.

77. E

$f''(x) > 0$: Slope of tangent line must be positive.

78. C

$f'(x) = e^{2\sin x} - 1 < 0$. Use a graphic calculator.

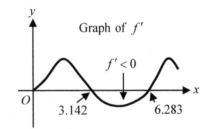

79. E

  I. $\lim\limits_{x \to 2} f(x)$ exists. $\lim\limits_{x \to 2^+} 4x - 4 = 4$ and $\lim\limits_{x \to 2^-} x^2 = 4$ → $\lim\limits_{x \to 2} = 4$ True

  II. $f(2)$ exists. $f(2) = 4(2) - 4 = 4$ True

  III. $f(x)$ is continuous at $x = 2$. $f(2) = \lim\limits_{x \to 2} f(x)$ True

80. E

$\int_0^8 f(x-2) + 1\, dx = \int_{-2}^6 (f(x) + 1)\, dx$ → $\int_{-2}^6 f(x)\, dx + \int_{-2}^6 1\, dx = \int_{-2}^4 f(x)\, dx + \int_4^6 f(x)\, dx + \int_{-2}^6 1\, dx$

$= 10 - 2 + 8 = 16$

Remember: $\int_4^6 f(x)\, dx = -\int_6^4 f(x)\, dx$

81. A

$F(4) = F(0) + \int_0^4 \left(x^2 \cos x\right) dx = -5 + \int_0^4 \left(x^2 \cos x\right) dx \approx -20.824$

82. B

$a(t) = e^{-t} - te^{-t}$

278

AP Calculus AB Test 2 Answers

$v(3) = 3e^{-3} > 0$ and $a(3) = e^{-3} - 4e^{-3} = -3e^{-3} < 0$

Since the velocity and the acceleration of the particle have opposite sign, the particle's speed is decreasing.

83. C

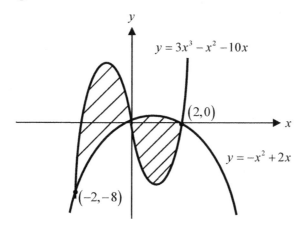

$$\text{Area} = \int_{-2}^{0}\left[(3x^3 - x^2 + 10x) - (-x^2 + 2x)\right]dx + \int_{0}^{2}\left[(-x^2 + 2x) - (3x^3 - x^2 + 10x)\right]dx = 24$$

84. C

$K'(x) = f'(2x) \cdot 2 \cdot g(2x) + f(2x) \cdot g'(2x) \cdot 2$

$K'(2) = f'(4) \cdot 2 \cdot g(4) + f(4) \cdot g'(4) \cdot 2 = (9)(2)(-4) + (8)(6)(2) = 24$

85. A

The volume of a cone is $V = \dfrac{\pi r^2 h}{3}$.

$\dfrac{dV}{dt} = \dfrac{\pi}{3}\left(2rh\dfrac{dr}{dt} + r^2\dfrac{dh}{dt}\right)$ and $\dfrac{dr}{dt} = 2$

When $r = 5$ and $V = 30\pi$, what is the value of h?

$20\pi = \dfrac{\pi \cdot 25 \cdot h}{3} \Rightarrow h = 2.4$

$30\pi = \dfrac{\pi}{3}\left((2)(5)(2.4)(2) + 25 \cdot \dfrac{dh}{dt}\right) \Rightarrow 30\pi = \dfrac{\pi}{3}\left(48 + 25\dfrac{dh}{dt}\right) \Rightarrow 90 = 48 + 25\dfrac{dh}{dt}$

Therefore, $\dfrac{dh}{dt} = 1.68$

86. C

$x\left(\dfrac{1}{2}\right) = \int_{0}^{1/2} f(t)\,dt = \dfrac{1}{4}$, $x(1) = \int_{0}^{1} f(t)\,dt = 0$

87. E

$\text{Area} = \lim_{b\to\infty} 10\int_{-b}^{b}\dfrac{1}{x^2+1}dx = \lim_{b\to\infty} 20\int_{0}^{b}\dfrac{1}{x^2+1}dx = \lim_{b\to\infty} 20\left[\tan^{-1}b - \tan^{-1}0\right] = 20\left(\dfrac{\pi}{2} - 0\right) = 10\pi$

≈ 31.416

88. C

The average value of $y = \dfrac{3\cos x}{x^2+1}$ for $0 \leq x \leq 1$

AP Calculus AB Test 2 Answers

$$= \frac{1}{1-0} \int_0^1 \frac{3\cos x}{x^2+1} dx \approx 2.049$$

89. D

For the vertical tangent to the curve, $\frac{dy}{dx}$ must be undefined.

$$y' - \sin y \cdot y' = 1 \quad \rightarrow \quad y' = \frac{1}{1-\sin y} \quad \rightarrow \quad 1 - \sin y = 0 \quad \rightarrow \quad \sin y = 1 \quad \rightarrow \quad y = \frac{\pi}{2}$$

At $y = \frac{\pi}{2}$, $\frac{\pi}{2} + \cos\left(\frac{\pi}{2}\right) + 1 = x \quad \Rightarrow \quad x = \frac{\pi}{2} + 1$

90. D

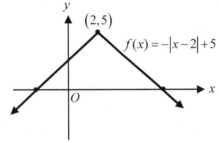

(A) For $-3 \le x \le 7$, $f(x)$ is increasing. False (Increasing on $[-3, 2]$ and decreasing on $[2, 7]$)
(B) For $-3 \le x \le 7$, $f(x)$ is decreasing. False
(C) For $-3 \le x \le 7$, $f'(x)$ exists. False ($f'(2)$ does not exist)
(D) For $-3 \le x \le 7$, relative maximum exists. True (At $x = 2$, relative maximum)
(E) For $-3 \le x \le 7$, $\lim_{x \to 2} f'(x)$ exists. False

91. D

For $-2 < x < 0$, the graph of f is increasing and concave up. $f''(x) = 1 > 0$

For $0 < x < 2$, the graph of f is increasing and concave down. $f''(x) < 0$

At $x = 0$, the graph of f has a point of inflection.

92. C

Cylindrical Shell: $V = 2\pi \int_a^b xy \, dx$ and $y = e^x - x$

$$V = 2\pi \int_0^5 x(e^x - x) \, dx$$

END OF SECTION 1

AP Calculus AB Test 2 Answers

SECTION II, PART A

Question 1

(a) $R(10) = 100e^{-5}(100+10+1) \approx 75$ people

(b) Number of people in the auditorium $= 200 + \int_0^{10} 100e^{-0.5t}(t^2+t+1)dt = 2183.0381 \approx 2183$

(c) Graph of $R(t)$ is as follows.

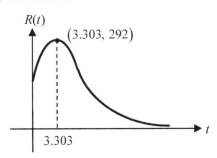

At $t = 3.303$ minute, the graph of R has a local maximum. $R(3.303) = 292$.

(d) In part (b), $\int_0^{10} R(t)dt = 2183 - 200 = 1983$ people.

$\int_0^{10} R(t)dt$ is the number of people who enter the auditorium over the first 10 minutes.

(e) Let $F'(t) = R(t)$.

Average rate at which people enter the auditorium over the interval $0 \leq t \leq 10$ is

Average rate $= \dfrac{F(10)-F(0)}{10-0} = \dfrac{\int_0^{10} R(t)dt}{10} = \dfrac{1983}{10} = 198.3$ or 198 people per minute.

Question 2

(a) $\int_0^{10} N(t)dt = \int_0^{10}(3t^2 - 10t + 5)dt = 550$

(b) Profit
$= \int_0^{10} 500 \cdot N(t)dt - \int_0^{10}(N(t) \cdot C(t))dt = \int_0^{10} 500(3t^2 - 10t + 5)dt - \int_0^{10}(300 - 10\sqrt{t})(3t^2 - 10t + 5)dt$
≈ 122510.22

(c) $\int_5^{10} C(t)dt$ is total amount of change in cost from the 5th day to the 10th day.

(d) Write an expression, involving an integral that represents the shopping mall's profit over the first k days.

$\int_0^k \left[500 \cdot (3t^2 - 10t + 5) - (300 - 10\sqrt{t})(3t^2 - 10t + 5)\right] dt$
$= \int_0^k \left[(200 + 10\sqrt{t})(3t^2 - 10t + 5)\right] dt$

AP Calculus AB Test 2 Answers

Question 3

(a) $a(5.5) = v'(5.5) = 2 \text{ miles/minute}^2$

(b) $\int_0^{10} v(t)\,dt = \frac{\pi}{2} - \left(\frac{2 \times 1}{2}\right) + \left(\frac{1 \times 1}{2}\right) + \left(\frac{(1+3) \cdot 1}{2}\right) + \left(\frac{(3+2) \cdot 1}{2}\right) + (2 \times 2) + \left(\frac{1 \times 2}{2}\right) = 9 + \frac{\pi}{2}$

Semicircle-triangle-triangle-trapezoid-trapezoid-square-triangle

$\int_0^{10} v(t)\,dt$ is the change in position from $t = 0$ to $t = 10$.

(c) Find the value of $\int_0^{10} |v(t)|\,dt$ and explain the meaning of $\int_0^{10} |v(t)|\,dt$.

$\int_0^{10} |v(t)|\,dt$ is the total distance traveled by the particle from $t = 0$ to $t = 10$.

$\int_0^{10} |v(t)|\,dt = \frac{\pi}{2} + \left(\frac{2 \times 1}{2}\right) + \left(\frac{1 \times 1}{2}\right) + \left(\frac{(1+3) \cdot 1}{2}\right) + \left(\frac{(3+2) \cdot 1}{2}\right) + (2 \times 2) + \left(\frac{1 \times 2}{2}\right)$

$= 11 + \frac{\pi}{2}$ miles

(d) At what time the particle changes its direction?

$v(t) = 0 \;\rightarrow\; t = 2, 4$

Question 4

(a) $f(x) = \sqrt{x^3 + 9} \;\Rightarrow\; f'(x) = \dfrac{3x^2}{2\sqrt{x^3 + 9}}$

(b) $f'(3) = \dfrac{9}{4}$

When $x = 3$, the value of y is $f(3) = \sqrt{27 + 9} = 6$.

The equation for the line tangent to the curve at $x = 3$ is $y - 6 = \dfrac{9}{4}(x - 3) \;\Rightarrow\; y = \dfrac{9x - 3}{4}$

Therefore, $f(3.2) \approx \dfrac{9(3.2) - 3}{4} = 6.45$

(c) Let g be the function defined by $g(x) = \begin{cases} \sqrt{x^3 + 9} & \text{for } 0 \le x \le 3 \\ ax + b & \text{for } x > 3 \end{cases}$.

If g is differentiable at $x = 3$, what are the values of a and b?

$g(x)$ should be continuous at $x = 3$ and $g'(3)$ should exist.

$g(3) = 6$ and $\lim\limits_{x \to 3} ax + b = 3a + b \;\Rightarrow\; 3a + b = 6$

$g'(3) = \dfrac{9}{4}$ for $0 \le x \le 3$

$g'(x) = a$ for $x > 3 \;\rightarrow\; a = \dfrac{9}{4}$

Therefore, $b = 6 - 3a = 6 - 3\left(\dfrac{9}{4}\right) = -\dfrac{3}{4}$.

$a = \dfrac{9}{4}$ and $b = -\dfrac{3}{4}$

AP Calculus AB Test 2 Answers

(d) Find the value of $\int_0^3 x^2 \sqrt{x^3+9}\, dx$.

$\begin{cases} u = x^3+9 \\ du = 3x^2 dx \end{cases} \Rightarrow \dfrac{1}{3}\int_9^{36} \sqrt{u}\, du = \dfrac{1}{3}\left[\dfrac{2}{3} u^{\frac{3}{2}}\right]_9^{36} = \dfrac{2}{9}\left[36^{\frac{3}{2}} - 9^{\frac{3}{2}}\right] = \dfrac{2}{9}(189) = 42$

Question 5

(a) If $y = P(t)$ is the particular solution to the differential equation at time $t \geq 0$, then find $y = P(t)$.

$\dfrac{dP}{500-P} = \dfrac{1}{10} dt \Rightarrow \int \dfrac{dP}{500-P} = \int \dfrac{1}{10} dt \Rightarrow -\ln|500-P| = \dfrac{1}{10}t + C_1$

The solution is $500 - P = Ae^{-\frac{1}{10}t} \to P = 500 - Ae^{-\frac{1}{10}t}$

$P = 500 - Ae^{-\frac{t}{10}}$. With initial condition $(0, 100)$. $100 = 500 - Ae^0 \to A = 400$

Therefore, $P(t) = 500 - 400e^{-\frac{t}{10}}$.

(b) $\lim\limits_{t \to \infty} \left(500 - 400e^{-\frac{t}{10}}\right) = 500$

(c) $P(t) = 500 - 400e^{-\frac{t}{10}} \to P'(t) = 40e^{-\frac{t}{10}} \to P''(t) = -4e^{-\frac{t}{10}}$

Since $P''(t) < 0$ over all t, the graph of $y = P(t)$ is concave down.

(d) Find $\dfrac{d^2P}{dt^2}$ in terms of P. Is the value of $\dfrac{d^2P}{dt^2}$ always positive? Explain your reasoning.

$\dfrac{dP}{dt} = \dfrac{1}{10}(500 - P) \to \dfrac{d^2P}{dt^2} = -\dfrac{1}{10} \cdot \dfrac{dP}{dt} = -\dfrac{1}{10}\left(\dfrac{500-P}{10}\right) = -\dfrac{500-P}{100}$

Since $100 \leq P < 500$, $\dfrac{d^2P}{dt^2} = \dfrac{P-500}{100} < 0$. The value of $\dfrac{d^2P}{dt^2}$ is always negative.

Question 6

(a) $v(t) = v(0) + \int_0^t 4\sin\left(\dfrac{t}{4}\right) = -16 + \left[-16\cos\left(\dfrac{t}{4}\right)\right]_0^t = -16 + \left[-16\cos\left(\dfrac{t}{4}\right) - \left(-16\cos\left(\dfrac{0}{4}\right)\right)\right]$

$= -16\cos\left(\dfrac{t}{4}\right)$

(b) $v(t) = -16\cos\left(\dfrac{t}{4}\right) = 0 \Rightarrow \cos\left(\dfrac{t}{4}\right) = 0$

$\dfrac{t}{4} = \dfrac{\pi}{2}, \dfrac{3\pi}{2}, \dfrac{5\pi}{2} \cdots \Rightarrow t = 2\pi, 6\pi$

(c) For $2\pi < t < 6\pi$, $v(t) > 0$.
The particle moves to the right.

(d) For $2\pi < t < 4\pi$, they are both positive.
For $6\pi < t < 8\pi$, they are both negative.
Therefore, the speed of the particle is increasing for the time $2\pi < t < 4\pi$ and $6\pi < t < 8\pi$.

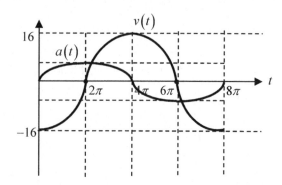

(e) Find the total distance traveled by the particle from $t = 0$ to 4π.

$$d = \int_0^{4\pi} |v(t)|\, dt = 2\int_0^{2\pi} 16\cos\left(\frac{t}{4}\right) dt = 32\left[4\sin\frac{t}{4}\right]_0^{2\pi} = 128\sin\frac{2\pi}{4} = 128$$

(f) Find the position of the particle at time $t = 2\pi$.

$$x(2\pi) = x(0) + \int_0^{2\pi} -16\cos\left(\frac{t}{4}\right) dt = 5 + \left[-64\sin\left(\frac{t}{4}\right)\right]_0^{2\pi} = 5 + \left[-64\sin\left(\frac{2\pi}{4}\right)\right] = -59$$

AP CALCULUS AB

TEST 3

AP CALCULUS AB

A CALCULATOR **CANNOT BE USED ON PART A OF SECTION 1**. A GRAPHIC CALCULATOR FROM THE APPROVED LIST **IS REQUIRED FOR PART B OF SECTION 1 AND FOR PART A OF SECTION II** OF THE EXAMINATION.

SECTION 1
Time: 1 hour and 45 minutes
All questions are given equal weight.
Percent of total grade – 50

Part A: 55 minutes, 28 multiple-choice questions
A calculator is NOT allowed.
Part B: 50 minutes, 17 multiple-choice questions
A graphing calculator is required.

CALCULUS AB
SECTION 1, PART A
Time— 55minutes
Number of questions— 28

A CALCULATOR MAY NOT BE USED ON THIS PART OF THE EXAMINATION.

Directions: Solve each of the following problems, using the available space for scratchwork. After examining the form of the choices, decide which is the best of the choices given and fill in the corresponding oval on the answer sheet. No credit will be given for anything written in the test book. Do not spend too much time on any one problem.

In this test:

(1) Unless otherwise specified, the domain of a function f is assumed to be the set of all real numbers x for which $f(x)$ is a real number.

(2) The inverse of a trigonometric function f may be indicated using the inverse function notation f^{-1} or with the prefix "arc" (e.g, $\sin^{-1} x = \arcsin x$)

AP Calculus AB Test 3 — Section 1

1. If $f(x) = \begin{cases} \dfrac{e^x - 1}{x} & \text{for } x \neq 0 \\ \dfrac{k}{2} & \text{for } x = 0 \end{cases}$ is continuous at $x = 0$, then $k =$

   (A) -1 (B) 0   (C) 1   (D) 2   (D) 3   (E) 4

---

2. $\displaystyle\int_0^4 \dfrac{3x+1}{\sqrt{x}}\,dx =$

   (A) 4   (B) 8   (C) 12   (D) 16   (E) 20

---

3. If $\sqrt{x} + \sqrt{y} = 4$, then what is the value of $\dfrac{dy}{dx}$ at $x = 1$?

   (A) -5 (B) -3 (C) 1   (D) 3   (E) 5

---

4. $\displaystyle\lim_{h \to 0} \dfrac{\sin^2\left(\dfrac{\pi}{4} + h\right) - \sin^2\left(\dfrac{\pi}{4}\right)}{h} =$

   (A) 0   (B) 1   (C) 1.5   (D) $\dfrac{\pi}{2}$ (E) π

---

5. For what value of x will $f(x) = \dfrac{x^2}{x+1}$ have a relative maximum?

   (A) -4 (B) -2 (C) 0   (D) 1   (E) 2

---

6. $\displaystyle\lim_{x \to 0} \dfrac{x}{\sin x + \tan x} =$

   (A) $-\infty$ (B) -2 (C) $\dfrac{1}{2}$ (D) 2   (E) ∞

AP Calculus AB Test 3 — Section 1

7. If the parametric equations are $x(t) = e^t$ and $y(t) = t$, which of the following could be the graph of $y = f(x)$?

(A)

(B)

(C)

(E)

(E)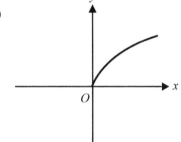

---

8. If the function f defined by $f(x) = e^{2x-1}$ and $g(x)$ is the inverse function of f, then $g'(e) = $

(A) $\dfrac{1}{2e}$ (B) 1 (C) $\dfrac{e}{2}$ (D) $2e$ (E) $\dfrac{5}{2}$

AP Calculus AB Test 3 Section 1

9. What is the equation of the line tangent to the graph of $y = \arctan x$ at $x = 1$?

(A) $y - 1 = 2(x - 1)$

(B) $y - \dfrac{\pi}{2} = 2(x - 1)$

(C) $y - \dfrac{\pi}{4} = \dfrac{1}{2}(x - 1)$

(D) $y - 1 = \dfrac{1}{2}\left(x - \dfrac{\pi}{4}\right)$

(E) $y - \dfrac{\pi}{4} = 2(x - 1)$

---

10. Let the function f be defined by $f(x) = \dfrac{1}{4}x^4 - 2x^2 + 2$. Which of the following is true about f at $x = 1$?

(A) f is increasing.
(B) f has a relative maximum.
(C) f has a relative minimum.
(D) f is discontinuous.
(E) f is concave downward.

---

11. $\dfrac{d}{dx}\left(e^x + 3^x\right) =$

(A) $e^x + 3^x$ (B) $(3e)^x$ (C) $\dfrac{e^x + 3^x}{\ln 3}$ (D) $e^x + x3^{x-1}$ (E) $e^x + 3^x \ln 3$

---

12. Let R be the region enclosed by the graphs of $y = \sin(\pi x)$, y-axis, and the lines $y = 2$ and $x = 1$. Find the area of R.

(A) $\dfrac{1}{\pi}$ (B) $\dfrac{2}{\pi}$ (C) $\dfrac{2\pi - 2}{\pi}$ (D) π (E) 2π

---

13. If a parametric equations are $x(t) = \sqrt{2t - 1}$ and $y(t) = 2t$, what is $\dfrac{dy}{dx}$ in terms of x?

(A) $2x$ (B) $\dfrac{x+1}{2}$ (C) $x + 2$ (D) $x^2 - 1$ (E) $\dfrac{x^2 - 1}{2}$

AP Calculus AB Test 3

Section 1

14. Find the equation of the normal to the graph of $y = \ln x$ at $x = e$.

   (A) $y = ex - e$
   (B) $y = -e^x + e^2 + 1$
   (C) $y = -ex + e^2 + 1$
   (D) $y = ex - e^2 - 1$
   (E) $y = ex + e^2 + 1$

15. $\int_0^4 \sqrt{x^2 - 4x + 4}\, dx =$

   (A) 0   (B) 2   (C) 4   (D) 6   (E) 8

16. The function f defined by $f(x) = 3\cos x - 4\sin x$. What is the amplitude of f?

   (A) -1 (B) 1   (C) 3.5   (D) 5   (E) 7

17. $\int_3^4 \frac{1}{(x-1)(x-2)}\, dx =$

   (A) $-\frac{2}{3}$ (B) $\ln\frac{4}{3}$ (C) $\ln\frac{3}{2}$ (D) $\ln 6$ (E) $\frac{3}{2}$

AP Calculus AB Test 3

Section 1

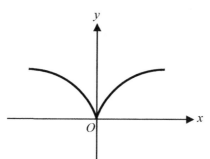

18. The graph of f', the derivative of f is shown in the figure above. Which of the following is true about the graph of f at $x = 0$?

   (A) f is not differentiable at $x = 0$.
   (B) A point of inflection exists at $x = 0$.
   (C) f is discontinuous at $x = 0$.
   (D) f has a relative minimum at $x = 0$.
   (E) $\lim_{x \to 0} f(x)$ does not exist.

19. The position of a particle on the x-axis at time t is given by $x(t) = 8t - t^2 + 2$. What is the total distance travelled by the particle for $0 \le t \le 5$?

   (A) 15   (B) 16   (C) 17   (D) 18   (E) 20

20. Which of the following functions are differentiable for all real numbers x?

   I. $y = x^{\frac{2}{3}}$
   II. $y = x^{\frac{1}{3}}$
   III. $y = |x - 1|$

   (A) None
   (B) I only
   (C) I and II only
   (D) I and III only
   (E) I, II, and III

AP Calculus AB Test 3 Section 1

21. If $f(x) = \dfrac{x^6\sqrt{2x-1}}{(x+1)^2}$, then $f'(1) =$

    (A) $\dfrac{1}{2}$ (B) $\dfrac{2}{3}$ (C) $\dfrac{3}{2}$ (D) $\dfrac{5}{2}$ (E) $\dfrac{7}{2}$

---

22. If $\lim_{x \to a} f(x) = f(a)$, then which of the following must be true?

    I. f is continuous at $x = a$.
    II. $f'(a)$ exists.
    III. $\lim_{x \to a^+} f(x) = \lim_{x \to a^-} f(x)$

    (A) I only (B) II only (C) I and II only (D) I and III only (E) I, II, and III

---

23. A particle moves along the x-axis and it's acceleration at time t is given by $a(t) = \dfrac{1}{t^2}$. When $t = 1$, the position of the particle is 5 and velocity is 2. Find the position of the particle at time $t = e$.

    (A) 10 (B) $3e$ (C) $3e + 1$ (D) $3e + 5$ (E) 20

---

24. Let R be the region enclosed by the graphs of $x^2 - 9y^2 = 9$ and the line $x = 9$. What is the volume of the solid by revolving R about the x-axis?

    (A) 8π (B) 12π (C) 16π (D) 20π (E) 24π

---

25. If $f(x) = x^2 e^{-x}$, then the graph of f is increasing for all x such that

    (A) $x < 0$ (B) $0 < x < 2$ (C) $-2 < x < 2$ (D) $0 < x < 4$ (E) $x > 2$

26. Let f be the function defined by $f(x) = 2 - \ln(x^2)$. What is the equation for the line tangent to the graph of f at $x = 1$?

(A) $y = -2x$
(B) $y = -2x + 4$
(C) $y = 2x$
(D) $y = 2x + 1$
(E) $y = 2x + 4$

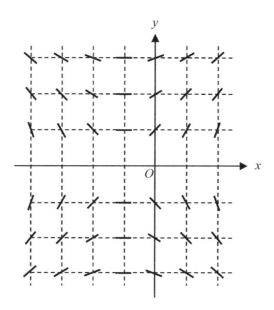

27. A slope field is shown above. Which of the following could be the differential equation for the slope field?

(A) $\dfrac{dy}{dx} = xy - y$

(B) $\dfrac{dy}{dx} = \dfrac{x+1}{y-1}$

(C) $\dfrac{dy}{dx} = xy + y$

(D) $\dfrac{dy}{dx} = \dfrac{x+1}{y}$

(E) $\dfrac{dy}{dx} = \dfrac{x+1}{x}$

x	f	f'	g	g'
−8	4	2	−2	6
−6	6	4	2	9
−2	8	5	6	18
2	12	6	12	24

28. Let f and g be the functions that are differentiable and have values shown in the table above. If $A(x) = f(g(3x))$, then $A'(-2) =$

    (A) 27   (B) 54   (C) 108   (D) 144   (E) 162

END OF PART A OF SECTION 1

AP Calculus AB Test 3   *Section 1*

CALCULUS AB
SECTION 1, PART B
Time— 50minutes
Number of questions— 17

A GRAPHING CALCULATOR IS REQUIRED FOR SOME QUESTIONS ON THIS PART OF THE EXAMINATION.

Directions: Solve each of the following problems, using the available space for scratchwork. After examining the form of the choices, decide which is the best of the choices given and fill in the corresponding oval on the answer sheet. No credit will be given for anything written in the test book. Do not spend too much time on any one problem.

In this test:

(1) The exact numerical value of the correct answer does not always appear among the choices given. When this happens, select from among the choices the number that best approximates the exact numerical value.

(2) Unless otherwise specified, the domain of a function f is assumed to be the set of all real numbers x for which $f(x)$ is a real number.

(3) The inverse of a trigonometric function f may be indicated using the inverse function notation f^{-1} or with the prefix "arc" (e.g, $\sin^{-1} x = \arcsin x$)

AP Calculus AB Test 3

Section 1

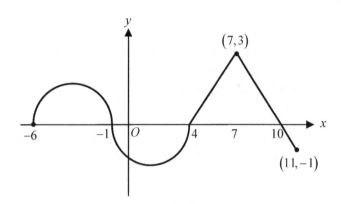

76. The graph of the function f on the closed interval $-6 \leq x \leq 11$ consists of two semicircles and two line segments as shown above. Which of the following is not true?

    (A) $f'(-3.5) = 0$
    (B) $f'(-1)$ exists
    (C) $f'(4)$ does not exist.
    (D) f has a point of inflection at $x = -1$.
    (E) f has a relative maximum at $x = 7$.

---

77. The function f is defined by $f(x) = x^3 - 3x^2 + 5 + \ln x$. For what value of x does the graph of f change its concavity?

    (A) 0   (B) 1.130   (C) 1.412   (D) 2.275   (E) 3.125

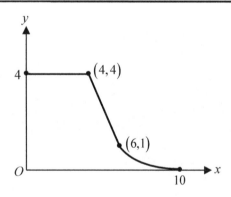

78. The graph of f is shown above. If $\int_6^{10} f(x)dx = 1.8$ and $g'(x) = f(x)$, then $g(10) - g(4) =$

    (A) 5   (B) 5.3   (C) 6.8   (D) 14.8   (E) 22.8

AP Calculus AB Test 3

Section 1

79. Let f be the function given by $f(x) = e^{\sqrt{x}}$. If the slope of the line tangent to the graph of f at point $(a, f(a))$ is 5, what is the value of a?

(A) 8.027   (B) 8.234   (C) 8.887   (D) 9.252   (E) 12.796

---

80. Find the point on the graph of $f(x) = x^2 + 1$ that is nearest to the point $(3, 0)$.

(A) $(-1, 2)$ (B) $(0.735, 1.540)$ (C) $(0.735, 2.739)$ (D) $(2, 5)$ (E) $(3, 10)$

---

81. Let f be the function given by $f(x) = \dfrac{4x^2 - 16}{x^2 - 9}$. Find the intervals on which f is increasing.

(A) $(-\infty, 0)$
(B) $(-3, 3)$
(C) $(-\infty, -3]$ or $(-3, 0]$
(D) $(-\infty, -3)$ or $(-3, 0)$
(E) $(-\infty, 0)$ or $(3, \infty)$

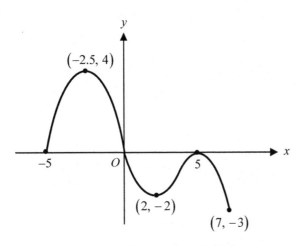

82. The graph of f', the derivative of a function f is shown above. If $f(-5)=0$, $\int_{-5}^{0} f'(x)\,dx = 12.8$, $\int_{0}^{5} f'(x)\,dx = -6.8$, and $\int_{5}^{7} f'(x)\,dx = -2.3$, which of the following could be the graph of f?

(A)

(B)

(C)

(D)

(E)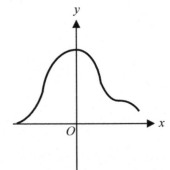

AP Calculus AB Test 3

Section 1

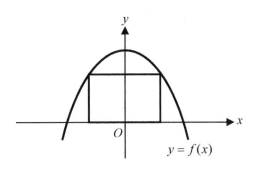

83. Let $A(x)$ be the area of the rectangle inscribed under the curve $f(x) = 4 - x^2$ and the x-axis as shown above. What is the greatest value of $A(x)$?

   (A) 6.158   (B) 6.786   (C) 8.125   (D) 10.562   (E) 12.775

---

84. If the function g is given by $g(x) = \dfrac{d}{dx}\displaystyle\int_0^x \left(e^t - t^3\right) dt$, at which of the following values of x does g have a relative maximum?

   (A) −0.459   (B) 0.910   (C) 1.857   (D) 2.547   (E) 4.536

---

85. What is the equation of the line tangent to the graph of $f(x) = \sqrt{1 - \sin x}$ at $x = 0$?

   (A) $y = -2x + 1$

   (B) $y = -\dfrac{1}{2}x + 1$

   (C) $y = \dfrac{1}{2}x - 1$

   (D) $y = 2x - 1$

   (E) $y = 2x + 1$

---

86. Let R be the region enclosed by the graphs of $y = \ln x$, $x = 4$, and the x-axis. What is the volume of the solid generated by revolving region R about x-axis?

   (A) 8.158   (B) 9.125   (C) 10.765   (D) 11.291   (E) 15.998

AP Calculus AB Test 3   Section 1

87. If $\int_{-2}^{4} f(x)\,dx = 15$ and $\int_{4}^{6} f(x)\,dx = -3$, what is the value of $\int_{0}^{8} \frac{f(x-2)}{3}\,dx$?

   (A) −8   (B) −2   (C) 4   (D) 8   (E) 12

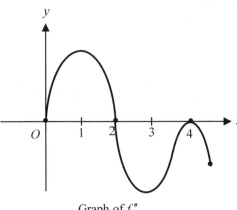

Graph of f''

88. The graph of f'', a second derivative of a function is shown above for $0 \le x \le 5$. If $f'(0) = 0$ and $\int_{0}^{2} f''(x)\,dx + \int_{2}^{4} f''(x)\,dx = 0$, at which of the following values of x does f have a relative maximum?

   (A) 1 only   (B) 2 only   (C) 3 only   (D) 4 only   (E) none

---

89. Let f be a function such that $f''(x) = 2x - 1$. If the graph of f is tangent to the line $3x + y = 2$ at point $(0, 2)$, find $f(x)$.

   (A) $f(x) = x^3 - x^2 - x + 2$

   (B) $f(x) = \dfrac{x^3}{3} - \dfrac{x^2}{2} - 3x + 2$

   (C) $f(x) = x^3 + 2x + 2$

   (D) $f(x) = -x^3 - 2x^2 - 2x + 2$

   (E) $f(x) = -\dfrac{x^3}{3} - \dfrac{x^2}{2} + 3x + 2$

AP Calculus AB Test 3

Section 1

90. Find the arc length of the graph of $x = e^{-y}$ over the interval $0 \leq y \leq 2$.

(A) 1.835 (B) 1.962 (C) 2.221 (D) 2.718 (E) 3.372

---

91. At any time $t \geq 0$, the velocity of a particle traveling in a straight line is $v(t) = e^{2x} \cos x$. If the initial position is $x(0) = 3$, what is the position of the particle at time $t = \dfrac{\pi}{2}$?

(A) 7.228 (B) 4.147 (C) 1.283 (D) −1.325 (E) −4.147

AP Calculus AB Test 3

Section 1

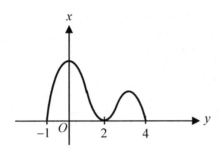

92. The graph of f', the derivative of a function f is shown above. Which of the following could be the graph of f?

(A)

(B)

(C)

(D)

(E)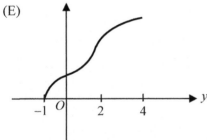

END OF SECTION 1

AP Calculus AB Test 3

Section II

CALCULUS AB

SECTION II
Time: 1 hour and 30 minutes
Percent of total grade – 50

Part A: 30 minutes, 2 problems
Part B: 60 minutes, 4 problems

Part A: A graphing calculator is required for some problems or parts of problems.
Part B: No calculator is allowed for these problems.

GENERAL INSTRUCTIONS FOR SECTION II PART A AND PART B

For each part of section II, you may wish to look over the problems before starting to work on them, since it is not expected that everyone will be able to complete all parts of all problems. All problems are given equal weight, but the parts of a particular problem are not necessary given equal weight.

- YOU SHOULD WRITE ALL WORK FOR EACH PART OF EACH PROBLEM IN THE SPACE PROVIDED FOR THAT PART IN THE BOOKLET. Be sure to write clearly and legibly. If you make an error, you may save time by crossing it out rather than trying to erase it. Erased or cross-out work will not be graded.
- Show all work. Clearly label any functions, graphs, tables, or other object that you use. You will be graded on the correctness and completeness of your methods as well as your answers. Answers without supporting work may not receive credit.
- Justifications required that you give mathematical (noncalculator) reasons
- Your work must be expressed in standard mathematical notation rather than calculator syntax. For example, $\int_{1}^{5} x^2 \, dx$ may not be written as $\text{fnInt}(X^2, X, 1, 5)$
- Unless otherwise specified, answers (numeric or algebraic) need not be simplified.
- If you use decimal approximations in calculations, you will be graded on accuracy. Unless otherwise specified, your final answers should be accurate to three places after the decimal point.
- Unless otherwise specified, the domain of a function f is assumed to be the set of all real numbers x for which $f(x)$ is a real number.

AP Calculus AB Test 3

Section II

**CALCULUS AB
SECTION II, PART A
Time— 30 minutes
Number of questions— 2**

A Graphing calculator is required for some problems or parts of problems.

Question 1

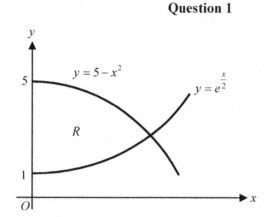

Let R be the region enclosed by the y-axis and the graphs of $y = e^{\frac{x}{2}}$ and $y = 5 - x^2$ as shown in the figure above.

(a) Find the area of R.

(b) The base of a solid is region R. Each cross section of the solid to the x-axis is an isosceles right triangle whose hypotenuse lies in the xy-plane. Write an expression involving one or more integrals that gives the volume of the solid. Do not evaluate.

(c) The vertical line $x = k$ divides region R into two equal areas. Write, but do not solve, an equation involving one or more integrals in terms of k.

(d) Find the volume of the solid generated when R is revolved about the y-axis.

AP Calculus AB Test 3

Section II

Question 2

t(hours)	0	3	5	9	15
$G(t)$(gallons)	220	260	320	350	440

The amount of water flows into a swimming pool through a pipe at time t is strictly increasing. At time $t = 0$, the amount of water in the swimming pool is 220 gallons. The amount of water, $G(t)$, is twice-differentiable function is measured in gallons and t is measured in hours. The table above shows the amount of water at selected time t over 15 hour period.

(a) Use the data from the table to find an approximation for $G'(7)$. Show the computations that lead to your answer. Indicate units of measure.

(b) Approximate the average amount of the water over the time interval $0 \le t \le 15$ hours by using a right Riemann sum with the four subintervals in the table. Does this approximation overestimate or underestimate the average amount of the water over 15 hours? Explain your reasoning.

(c) Evaluate $\int_0^{15} G'(t)\,dt$. Using correct units, interpret the meaning of $\int_0^{15} G'(t)\,dt$, and the meaning of $G(0) + \int_0^{15} G'(t)\,dt$.

(d) For $0 \le t \le 15$, the function G is given by $G(t) = 220 + te^{\frac{t}{5}}$, as a model for the amount of the water in the swimming pool at time t. Find the average value of the amount of the water in the pool over the time interval $0 \le t \le 15$.

(e) For $t \ge 15$, the amount of the water in the pool is decreasing with the rate $G'(t) = \dfrac{50\ln t}{t}$. Based on the model, find $G(20)$?

END OF PART A OF SECTION II

CALCULUS AB
SECTION II, PART B
Time— 60 minutes
Number of questions— 4

No calculator is required for these problems.

Question 3

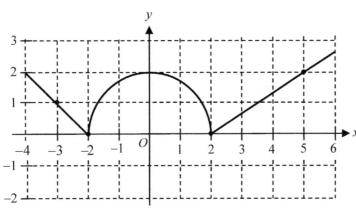

Graph of f

Let g be the function given by $g(x) = 3 + \int_{2}^{x} f(t)\,dt$. The figure above shows the graph of f, the derivative of a function g defined on $[-4, 6]$. The graph consists of two line segments and a semicircle centered at origin.

(a) Find the values of $g(0)$ and $g(5)$.

(b) Find the values of $g'(-2)$, $g''(-2)$. If the value does not exist, explain your reasoning.

(c) Write an equation for the line tangent to the graph of g at $x = 5$.

(d) For $-4 < x < 6$, find all values of x for which the graph of g has a point of inflection.

(e) Find the average value of g' over the closed interval $[-3, 5]$.

AP Calculus AB Test 3 — Section II

Question 4

A lake is stocked with 1000 fish, and their population strictly decreases according to the curve $P(t)$ whose differential equation $\dfrac{dP}{dt} = \dfrac{1}{5}(200 - P)$, where t is measured in months from the beginning.

(a) Find the particular solution to the differential equation $\dfrac{dP}{dt} = \dfrac{1}{5}(200 - P)$ with initial condition $P(0) = 1000$.

(b) Find the value of $\lim\limits_{t \to \infty} P(t)$.

(c) Write an equation of the line tangent to the graph of P at $t = 0$.

(d) Use the line tangent from part (a) to approximate the fish population at $t = 1$.

(e) Use $\dfrac{d^2 P}{dt^2}$ in terms of P to determine whether your answer in part (b) is an underestimate or an overestimate of the fish population at time $t = 1$. Explain your reasoning.

AP Calculus AB Test 3 — Section II

Question 5

The differential equation is given by $\dfrac{dy}{dx} = \dfrac{x-2}{y}$.

(a) Sketch a slope field for the given differential equation $\dfrac{dy}{dx} = \dfrac{x-2}{y}$ on the axis at the given points.

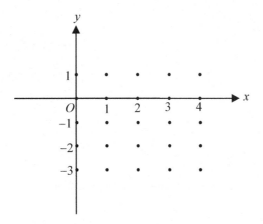

(b) Find the particular solution $y = f(x)$ to the differential equation $\dfrac{dy}{dx} = \dfrac{x-2}{y}$ with $f(0) = -2$.

Sketch the graph of $y = f(x)$ in the xy-plane provided above.

(c) Find the values of $\lim\limits_{x \to 2} f(x)$ and $f'(2)$.

(d) Does the graph of $y = f(x)$ have a relative maximum? Explain your reasoning.

Question 6

Two particles move along the x-axis. For time $t \geq 0$, the position of particle A is given by $A(t) = -4\sin\left(\dfrac{\pi}{4}t\right)$, and the position of particle B is given by $B(t) = t^3 - 9t^2 + 24t + 2$.

(a) Find the velocity for each particle.

(b) For $0 \leq t \leq 8$, find all time t during which the two particles move in same directions.

(c) For $0 \leq t \leq 8$, find all time t during which particle A is speeding up. Explain your reasoning.

(d) Find the distance traveled by particle A from $t = 0$ to $t = 8$.

(e) Find the average velocity of the particle B over the interval $0 \leq t \leq 4$.

END OF EXAM

AP Calculus AB Test 3 Answers

SECTION 1, PART A

1. **D**

   $$\lim_{x\to 0}\frac{e^x-1}{x}=\lim_{x\to 0}\frac{(e^x-1)'}{x'}=\lim_{x\to 0}\frac{e^x}{1}=\frac{e^0}{1}=1$$

   Therefore, $\frac{k}{2}=1$ or $k=2$.

2. **E**

   $$\int_0^4\frac{3x+1}{\sqrt{x}}dx=\int_0^4\left(3\sqrt{x}+\frac{1}{\sqrt{x}}\right)dx=\left[2x^{\frac{3}{2}}+2x^{\frac{1}{2}}\right]_0^4=2(4)^{\frac{3}{2}}+2(4)^{\frac{1}{2}}=16+4=20$$

3. **B**

   $\sqrt{x}+\sqrt{y}=4 \Rightarrow \frac{1}{2\sqrt{x}}+\frac{1}{2\sqrt{y}}\frac{dy}{dx}=0 \Rightarrow \frac{dy}{dx}=-\frac{\sqrt{y}}{\sqrt{x}}$

   If $x=1$, then $\sqrt{1}+\sqrt{y}=4 \Rightarrow y=9$.

   Therefore, $\frac{dy}{dx}=-\frac{\sqrt{9}}{\sqrt{1}}=-3$.

4. **B**

   Define $f(x)=\sin^2 x$ and $f'(x)=2\sin x\cos x$.

   $$\lim_{h\to 0}\frac{\sin^2\left(\frac{\pi}{4}+h\right)-\sin^2\left(\frac{\pi}{4}\right)}{h}=f'\left(\frac{\pi}{4}\right)$$

   Therefore, $f'\left(\frac{\pi}{4}\right)=2\sin\left(\frac{\pi}{4}\right)\cos\left(\frac{\pi}{4}\right)=2\left(\frac{1}{\sqrt{2}}\right)\left(\frac{1}{\sqrt{2}}\right)=1$.

5. **C**

   $f(x)=\frac{x^2}{x+1} \Rightarrow f'(x)=\frac{2x(x+1)-x^2}{(x+1)^2}=\frac{x(x-2)}{(x+1)^2}=0$

   At $x=0$, it has a local maximum. Because $f'(x)>0$ for $x<0$, and $f'(x)<0$ for $x>0$.

6. **C**

   $$\lim_{x\to 0}\frac{x}{\sin x+\tan x}=\lim_{x\to 0}\frac{x'}{(\sin x+\tan x)'}=\lim_{x\to 0}\frac{1}{\cos x+\sec^2 x}=\frac{1}{2}$$

7. **E**

   Since $x=e^t>0$ and $t=\ln x$, then $y=\ln x$.

8. **A**

   $y=e^{2x-1} \Rightarrow x=e^{2y-1} \Rightarrow 2y-1=\ln x \Rightarrow y=g(x)=\frac{\ln(x)+1}{2}$

   $g'(x)=\frac{1}{2x} \Rightarrow g'(e)=\frac{1}{2e}$

   Or

   $g(e)=k \leftrightarrow f(k)=e \Rightarrow e=e^{2x-1} \Rightarrow x=1$

   $f(x)=2e^{2x-1}$

Dr. John Chung's AP Calculus AB/BC

AP Calculus AB Test 3 Answers

Therefore, $g'(e) = \dfrac{1}{f'(1)} = \dfrac{1}{2e}$

9. **C**

$y = \arctan x \;\Rightarrow\; \dfrac{dy}{dx} = \dfrac{1}{1+x^2} \;\Rightarrow\; \left.\dfrac{dy}{dx}\right|_{x=1} = \dfrac{1}{2}$

$x = 1 \;\Rightarrow\; y = \tan^{-1} 1 \;\Rightarrow\; y = \dfrac{\pi}{4}$

The equation of the line tangent to the graph of $y = \arctan x$ at $x = 1$ is $y - \dfrac{\pi}{4} = \dfrac{1}{2}(x-1)$.

10. **E**

$f(x) = \dfrac{1}{4}x^4 - 2x^2 + 2 \;\Rightarrow\; f'(x) = x^3 - 4x = x(x+2)(x-2) = 0$ and $f''(x) = 3x^2 - 4$

At $x = 1$, $f'(1) = -3 < 0$ and $f''(1) = 3 - 4 = -1 < 0$: concave down.

(A) f is increasing. False
(B) f has a relative maximum. False
(C) f has a relative minimum. False
(D) f is discontinuous. False
(E) f is concave downward. True

11. **E**

$\dfrac{d}{dx}(e^x + 3^x) = e^x + 3^x \ln 3$

12. **C**

Area of $R = \displaystyle\int_0^1 (2 - \sin \pi x)\, dx = \left[2x + \dfrac{\cos \pi x}{\pi}\right]_0^1 = \left(2 + \dfrac{\cos \pi}{\pi}\right) - \left(0 + \dfrac{\cos 0}{\pi}\right) = 2 - \dfrac{2}{\pi}$

13. **A**

$\dfrac{dy}{dx} = \dfrac{dy/dt}{dx/dt} = \dfrac{2}{1/\sqrt{2t-1}} = 2\sqrt{2t-1} = 2x$

Or, eliminate parameter t.

$x^2 = 2t - 1 \;\Rightarrow\; 2t = x^2 + 1 \;\Rightarrow\; y = 2t = x^2 + 1$

Therefore, $\dfrac{dy}{dx} = 2x$.

14. **C**

$\dfrac{dy}{dx} = \dfrac{1}{x}$ and a point $(e, \ln e) = (e, 1)$. $\left.\dfrac{dy}{dx}\right|_{x=e} = \dfrac{1}{e}$

Slope of the normal to the graph of $y = \ln x$ at $x = e$ is $-e$. (Negative reciprocal)

Therefore, $y - 1 = -e(x - e) \;\Rightarrow\; y = -ex + e^2 + 1$.

15. **C**

$\displaystyle\int_0^4 \sqrt{x^2 - 4x + 4}\, dx = \int_0^4 \sqrt{(x-2)^2}\, dx = \int_0^4 |x - 2|\, dx = \int_0^2 -(x-2)\, dx + \int_2^4 (x-2)\, dx$

$= \left[2x - \dfrac{x^2}{2}\right]_0^2 + \left[\dfrac{x^2}{2} - 2x\right]_2^4 = 2 + 2 = 4$

Or, use the graph of $y = |x - 2|$.

AP Calculus AB Test 3 Answers

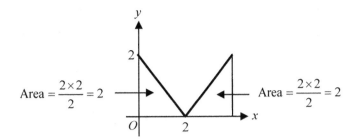

16. **D**

Amplitude of $f = \sqrt{3^2 + 4^2} = 5$

17. **B**

$$\int_3^4 \frac{1}{(x-1)(x-2)} dx = \int_3^4 \left(\frac{1}{x-2} - \frac{1}{x-1}\right) dx = \left[\ln|x-2| - \ln|x-1|\right]_3^4 = \left[\ln\left|\frac{x-2}{x-1}\right|\right]_3^4 = \ln\frac{4}{3}$$

18. **B**

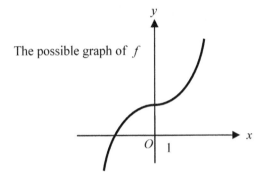

The possible graph of f

(A) f is not differentiable at $x = 0$. False $(f'(1) = 0)$

(B) A point of inflection exists at $x = 0$. True

(C) f is discontinuous at $x = 0$. False

(D) f has a relative minimum at $x = 0$. False

(E) $\lim_{x \to 0} f(x)$ does not exist. False

19. **C**

$x(t) = 8t - t^2 + 2 \Rightarrow v(t) = x'(t) = 8 - 2t = 0 \Rightarrow t = 4$

$d = \int_0^5 |v(t)| dt = \int_0^4 (8-2t) dt + \int_4^5 (2t-8) dt = \left[8t - t^2\right]_0^4 + \left[t^2 - 8t\right]_4^5 = 16 + 1 = 17$

20. **A**

I. $y = x^{\frac{2}{3}} \Rightarrow \frac{dy}{dx} = \frac{2}{3}x^{-\frac{1}{3}} = \frac{2}{3\sqrt[3]{x}}$: not differentiable at $x = 0$.

II. $y = x^{\frac{1}{3}} \Rightarrow \frac{dy}{dx} = \frac{1}{3}x^{-\frac{2}{3}} = \frac{1}{3\sqrt[3]{x^2}}$: not differentiable at $x = 0$.

III. $y = |x - 1|$: not differentiable at $x = 1$.

21. **C**

$f(x) = \frac{x^6 \sqrt{2x-1}}{(x+1)^2} \Rightarrow \ln f = \ln\left(\frac{x^6 \sqrt{2x-1}}{(x+1)^2}\right) \Rightarrow \ln f = 6\ln x + \frac{1}{2}\ln(2x-1) - 2\ln(x+1)$

Use logarithmic differentiation:

AP Calculus AB Test 3 Answers

$$\frac{f'}{f} = \frac{6}{x} + \frac{1}{2x-1} - \frac{2}{x+1} \Rightarrow f'(x) = f(x)\left(\frac{6}{x} + \frac{1}{2x-1} - \frac{2}{x+1}\right) = \frac{x^6\sqrt{2x-1}}{(x+1)^2}\left(\frac{6}{x} + \frac{1}{2x-1} - \frac{2}{x+1}\right)$$

$$f'(1) = \frac{1\sqrt{1}}{4}(6+1-1) = \frac{3}{2}$$

22. D

If $\lim_{x \to a} f(x) = f(a)$, then the graph of f is continuous at $x = a$.

I. f is continuous at $x = a$. **True**

II. $f'(a)$ exists. **Not always true** $(y = |x-a|)$

III. $\lim_{x \to a^+} f(x) = \lim_{x \to a^-} f(x)$ **True**

23. C

$$v(t) = \int \frac{1}{t^2} dt = -\frac{1}{t} + C \Rightarrow 2 = -\frac{1}{1} + C \Rightarrow C = 3$$

$$x(e) = 5 + \int_1^e \left(-\frac{1}{t} + 3\right) dt = 5[-\ln t + 3t]_1^e = 5 + (-1+3e) - (0+3) = 1+3e$$

24. D

$$V = \pi \int_3^9 y^2 \, dx = \pi \int_3^9 \left(\frac{x^2}{9} - 1\right) dx = \pi \left[\frac{x^3}{27} - x\right]_3^9 = 20\pi$$

25. B

$$f(x) = x^2 e^{-x} \Rightarrow f'(x) = 2xe^{-x} + x^2(-e^{-x}) = xe^{-x}(2-x) = 0$$

f is increasing for $0 < x < 2$, because $f'(x) > 0$ on the interval.

26. B

$$f'(x) = -\frac{2x}{x^2} = -\frac{2}{x} \Rightarrow f'(1) = -2$$

$f(1) = 2 - \ln 1 = 2$

The equation for the line tangent to the graph of f at $x = 1$ is

$y - 2 = -2(x-1) \Rightarrow y = -2x + 4$

27. D

Slopes are undefined at $y = 0$ and slopes are 0 at $x = -1$.

$$\frac{dy}{dx} = \frac{x+1}{y}$$

28. E

x	f	f'	g	g'
-8	4	2	-2	6
-6	6	4	2	9
-2	8	5	6	18
2	12	6	12	24

$A(x) = f(g(3x)) \Rightarrow A'(x) = f'(g(3x)) \cdot g'(3x) \cdot 3$

$A'(-2) = f'(g(-6)) \cdot g'(-6) \cdot 3 = f'(2)(9)(3) = (6)(9)(3) = 162$

AP Calculus AB Test 3 Answers

SECTION 1, PART B

76. **B**

    (A) $f'(-3.5) = 0$ **True**

    (B) $f'(-1)$ exists. **False** (Vertical line at $x = -1$)

    (C) $f'(4)$ does not exist. **True** $\left(f'(4^-) \neq f'(4^+)\right)$

    (D) f has a point of inflection at $x = -1$. **True** (f is continuous at $x = -1$ and $f''(-1)$ is undefined)

    (E) f has a relative maximum at $x = 7$. **True**

77. **B**

    The graph of f': Use your graphic calculator.

    $f(x) = x^3 - 3x^2 + 5 + \ln x \Rightarrow f'(x) = 3x^2 - 6x + \dfrac{1}{x}$.

    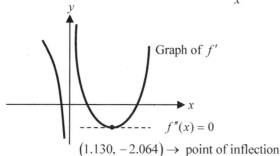

    $(1.130, -2.064) \rightarrow$ point of inflection

78. **C**

    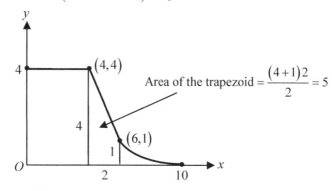

    Area of the trapezoid $= \dfrac{(4+1)2}{2} = 5$

    $g(10) - g(4) = \displaystyle\int_4^{10} f(x)\,dx = \int_4^6 f(x)\,dx + \int_6^{10} f(x)\,dx = 5 + 1.8 = 6.8$

79. **E**

    $f'(x) = e^{\sqrt{x}} \left(\sqrt{x}\right)' = \dfrac{e^{\sqrt{x}}}{2\sqrt{x}}$

    $f'(x) = \dfrac{e^{\sqrt{x}}}{2\sqrt{x}} = 5$

    Use your graphic calculator to find a point of intersection.

Dr. John Chung's AP Calculus AB/BC

AP Calculus AB Test 3 Answers

$y_1 = \dfrac{e^{\sqrt{x}}}{2\sqrt{x}}$ and $y_2 = 5$

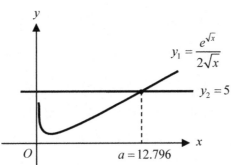

$a = 12.796$

80. B

Distance $= \sqrt{(x-3)^2 + (y-0)^2} = \sqrt{(x-3)^2 + (x^2+1)^2}$

Use your graphic calculator.
At $x = 0.735$, it has a minimum distance 2.739.
Therefore, the point on the graph is
$(0.735,\ 0.735^2 + 1) = (0.735,\ 1.540)$

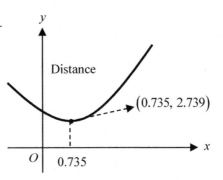

81. D

$f(x) = \dfrac{4x^2 - 16}{x^2 - 9} \Rightarrow f'(x) = \dfrac{8x(x^2-9) - (4x^2-16)(2x)}{(x^2-9)^2} = \dfrac{-40x}{(x^2-9)^2}$

$f'(x) = 0$ at $x = 0$ and $f'(x)$ is undefined at $x = 0$.

x		-3		0		3	
f'	+	Asym	+	0	−	Asym	−

- **Asym: Asymptote**

If $f'(x) > 0$, then f is increasing.

For $(-\infty, -3)$ or $(-3, 0)$, the graph of f is increasing.

82. E

Remember: $x(7) = x(-5) + \displaystyle\int_{-5}^{7} f(x)\,dx = 0 + 12.8 - 6.8 - 2.3 = 3.7$

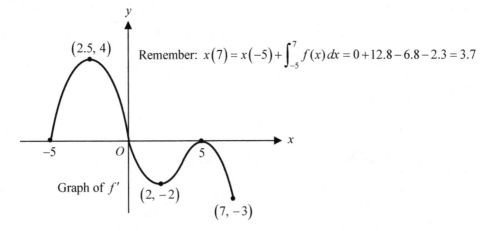

AP Calculus AB Test 3 Answers

$-5 \leq x < -2.5$	$f' > 0 \Rightarrow f$ is increasing $f'' > 0 \Rightarrow$ Concave up
$x = -2.5$	$f''(-2.5) = 0 \Rightarrow$ Point of inflection
$-2.5 < x < 0$	$f' > 0 \Rightarrow f$ is increasing $f'' < 0 \Rightarrow$ Concave down
$x = 0$	$f' = 0 \Rightarrow$ Local maximum
$0 < x < 2$	$f' < 0 \Rightarrow$ Decreasing $f'' < 0 \Rightarrow$ Concave down
$x = 2$	$f''(2) = 0 \Rightarrow$ Point of inflection
$2 < x < 5$	$f' < 0 \Rightarrow$ Decreasing $f'' > 0 \Rightarrow$ Concave up
$x = 5$	$f''(5) = 0 \Rightarrow$ Point of inflection
$5 < x \leq 7$	$f' < 0 \Rightarrow$ Decreasing $f'' < 0 \Rightarrow$ Concave down

83. A

Area: $A(x) = 2x(4 - x^3)$

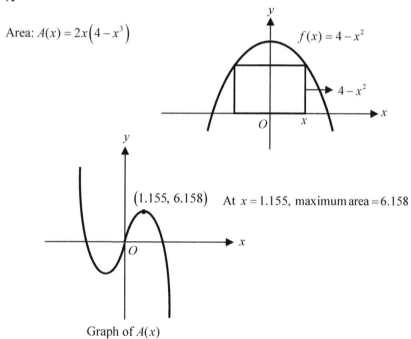

$(1.155, 6.158)$ At $x = 1.155$, maximum area $= 6.158$

Graph of $A(x)$

AP Calculus AB Test 3 Answers

84. C

$g'(x) = e^x - x^3$

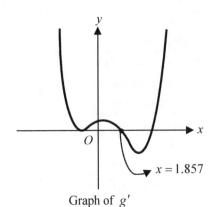

Graph of g'

$x = 1.857$

85. B

$f'(x) = \dfrac{-\cos x}{2\sqrt{1 - \sin x}}$

$f'(0) = \dfrac{-1}{2}$ and $f(0) = 1$

$y - 1 = -\dfrac{1}{2}(x - 0) \;\Rightarrow\; y = -\dfrac{1}{2}x + 1$

86. A

$V = \pi \displaystyle\int_1^4 (\ln x)^2\, dx \approx 8.158$

87. C

$\displaystyle\int_0^8 \dfrac{f(x-2)}{3}\, dx = \dfrac{1}{3}\int_{-2}^8 f(x)\, dx = \dfrac{1}{3}\left(\int_{-2}^6 f(x)\, dx + \int_6^8 f(x)\, dx\right) = \dfrac{1}{3}(15 - 3) = 4$

88. D

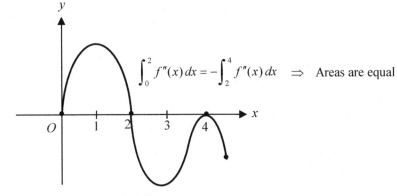

$\displaystyle\int_0^2 f''(x)\, dx = -\int_2^4 f''(x)\, dx \;\Rightarrow\;$ Areas are equal

Graph of f''

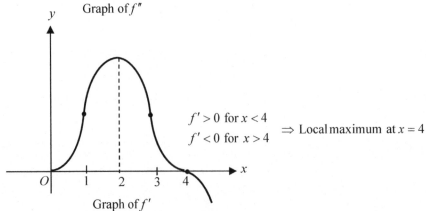

$f' > 0$ for $x < 4$
$f' < 0$ for $x > 4$ $\Rightarrow$ Local maximum at $x = 4$

Graph of f'

AP Calculus AB Test 3 Answers

89. **B**

$f''(x) = 2x - 1 \Rightarrow f'(x) = x^2 - x + C \Rightarrow f'(0) = -3 \Rightarrow C = -3$

$f'(x) = x^2 - x - 3 \Rightarrow f(x) = \dfrac{x^3}{3} - \dfrac{x^2}{2} - 3x + C$

Since $f(0) = 2 \Rightarrow C = 2$

Therefore, $f(x) = \dfrac{x^3}{3} - \dfrac{x^2}{2} - 3x + 2$

90. **C**

$L = \int_a^b \sqrt{1 + [f'(x)]^2} \, dx$

$x' = -e^{-y} \Rightarrow 1 + (x')^2 = 1 + e^{-2y}$

$L = \int_0^2 \sqrt{(1 + e^{2y})} \, dy \approx 2.221$

91. **A**

$x\left(\dfrac{\pi}{2}\right) = 3 + \int_0^{\pi/2} \left(e^{2x} \cos x\right) dx =$

At any time $t \geq 0$, the velocity of a particle traveling in a straight line is $v(t) = e^{2x} \cos x$. If the initial position is $x(0) = 3$, what is the position of the particle at time $t = \dfrac{\pi}{2}$?

(A) 7.229 (B) 4.147 (C) 1.283 (D) −1.325 (E) −4.147

92. **D**

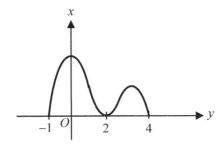

The graph of f is strictly increasing on the interval $[-1, 4]$.

For $-1 < x < 0$, the graph of f is increasing and concave up.

At $x = 0, 2, 3$, points of inflection.

AP Calculus AB Test 3 Answers

SECTION II, PART A

Question 1

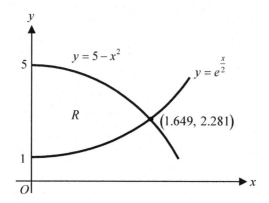

(a)
$$\text{Area} = \int_0^{1.649} \left(5 - x^2 - e^{x/2}\right) dx \approx 4.189$$

(b)

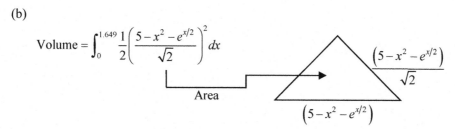

$$\text{Volume} = \int_0^{1.649} \frac{1}{2}\left(\frac{5 - x^2 - e^{x/2}}{\sqrt{2}}\right)^2 dx$$

(c)
$$\int_0^k \left(5 - x^2 - e^{x/2}\right) dx = \frac{4.189}{2}$$

(d)

Cylindrical Shell: $2\pi \int_0^{1.649} x\left(5 - x^2 - e^{x/2}\right) dx \approx 16.026$

Disk: $\pi \int_{2.281}^{5} (5 - y)\, dy + \pi \int_{1}^{2.281} (2\ln y)^2 \, dy \approx 16.026$

AP Calculus AB Test 3 Answers

Question 2

t (hours)	0	3	5	9	15
$G(t)$ (gallons)	220	260	320	350	440

(a)
$$G'(7) \approx \frac{G(9) - G(5)}{9 - 5} = \frac{350 - 320}{4} = 7.5 \text{ gallons per hour}$$

(b)

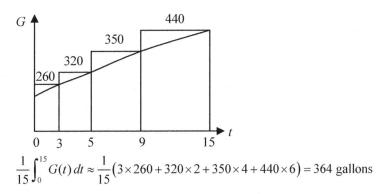

$$\frac{1}{15} \int_0^{15} G(t)\,dt \approx \frac{1}{15}(3 \times 260 + 320 \times 2 + 350 \times 4 + 440 \times 6) = 364 \text{ gallons}$$

Since $G(t)$ is strictly increasing on the interval $[0, 15]$, the answer is an overestimate.

(c)
$$\int_0^{15} G'(t)\,dt = G(15) - G(0) = 440 - 220 = 220 \text{ gallons: The amount of the water increased over the interval from } t = 0 \text{ to } t = 15 \text{ hours}$$

$G(0) + \int_0^{15} G'(t)\,dt$ is the amount of the water in the swimming pool at time $t = 15$.

(d)
$$\frac{1}{15}\int_0^{15}\left(220 + te^{t/5}\right)dt \approx 288.618 \text{ gallons}$$

(e)
$$G(20) = G(15) - \int_{15}^{20} G'(t)\,dt$$
$$G(20) = 440 - \int_{15}^{20}\left(\frac{50\ln t}{t}\right)dt = 398.978 \text{ gallons}$$

AP Calculus AB Test 3 Answers

SECTION II, PART B

Question 3

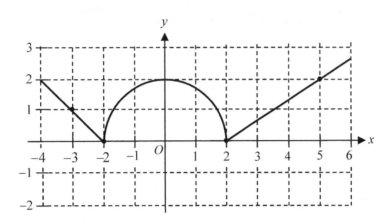

Graph of f

(a) Find the values of $g(0)$ and $g(5)$.

$g(0) = 3 + \int_2^0 f(t)\,dt = 3 - \int_0^2 f(t)\,dt = 3 - \pi$

$g(5) = 3 + \int_2^5 f(t)\,dt = 3 + \dfrac{3 \times 2}{2} = 6$

(b) Find the values of $g'(-2)$, $g''(-2)$. If the value does not exist, explain your reasoning.

$g'(2) = f(2) = 0$ but $g''(2) = f'(2)$ does not exist. Sharp corner

(c) Write an equation for the line tangent to the graph of g at $x = 5$.

$g'(5) = f(5) = 2$ and a point $(5, g(5)) = (5, 6)$

$y - 6 = 2(x - 5)$

(d) For $-4 < x < 6$, find all values of x for which the graph of g has a point of inflection.

The graph of g has a point of inflection at $x = -2, x = 0,$ and $x = 2$ because $g''(x)$ changes sign at each of these values.

(e) Find the average value of g' over the closed interval $[-3, 5]$.

$\dfrac{1}{5-(-3)} \int_{-3}^{5} g'(x)\,dx = \dfrac{1}{8} \int_{-3}^{5} f(x)\,dx = 2\pi + \dfrac{7}{2}$

AP Calculus AB Test 3 Answers

Question 4

(a) $\dfrac{dP}{200-P} = \dfrac{1}{5}dt \;\Rightarrow\; \displaystyle\int \dfrac{dP}{200-P} = \int \dfrac{1}{5}dt$

$-\ln(200-P) = \dfrac{1}{5}t + C_1 \;\Rightarrow\; \ln(200-P) = -\dfrac{1}{5}t + C_2 \;\Rightarrow\; 200 - P = Ce^{-\frac{t}{5}}$

$P = 200 - Ce^{-\frac{t}{5}} \;\rightarrow\;$ Initial condition $(0, 1000)$

$1000 = 200 - C \;\rightarrow\; C = -800$

Therefore, $P(t) = 200 + 800e^{-\frac{t}{5}}$.

(b) $\displaystyle\lim_{t \to \infty} 200 + 800e^{-\frac{t}{5}} = 200$

(c) At $t = 0$, $P = 1000$. Therefore, $\left.\dfrac{dP}{dt}\right|_{t=0} = \dfrac{1}{5}(200 - 1000) = -160$

An equation of the line tangent to the graph of P at $t = 0$ is

$y - 1000 = -160(t - 0) \;\Rightarrow\; y = -160t + 1000$

Another method: You can find the equation from the particular solution.

$P'(t) = -160e^{-\frac{t}{5}} \;\Rightarrow\; P'(0) = -160$

An equation of the line tangent to the graph of P at $t = 0$ is

$y - 1000 = -160(t - 0) \;\Rightarrow\; y = -160t + 1000$

(d) $P(1) \approx 1000 - 160(1) = 840$

(e) $\dfrac{d^2P}{dt^2} = -\dfrac{1}{5}\dfrac{dP}{dt} = -\dfrac{1}{5}\left(\dfrac{1}{5}(200-P)\right) = -\dfrac{1}{25}(200-P)$ and $P(1) = 840$

$\left.\dfrac{d^2P}{dt^2}\right|_{P=840} = -\dfrac{1}{25}(200-840) = \dfrac{-640}{-25} > 0$

The answer is an underestimate because the graph of P is concave up and decreasing over the interval.

AP Calculus AB Test 3 Answers

Question 5

The differential equation is given by $\dfrac{dy}{dx} = \dfrac{x-2}{y}$.

(a) Sketch a slope field for the given differential equation $\dfrac{dy}{dx} = \dfrac{x-2}{y}$ on the axis at the given points.

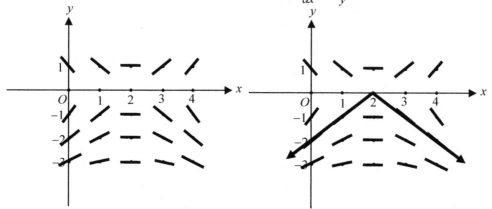

(b) Find the particular solution $y = f(x)$ to the differential equation $\dfrac{dy}{dx} = \dfrac{x-2}{y}$ with $f(0) = -2$.

$y\,dy = (x-2)\,dx \;\Rightarrow\; \dfrac{y^2}{2} = \dfrac{x^2}{2} - 2x + C_1 \;\Rightarrow\; y^2 = x^2 - 4x + C$

$y = \pm\sqrt{x^2 - 4x + C} \;\Rightarrow\;$ Initial condition $f(0) = -2$.

$-2 = -\sqrt{0+C} \;\Rightarrow\; C = 4$

Therefore, $y = -\sqrt{x^2 - 4x + 4} = -|x-2|$

Sketch the graph of $y = f(x)$ in the xy-plane provided above.

(c) Find the values of $\lim\limits_{x \to -2} f(x)$ and $f'(-2)$.

$\lim\limits_{x \to -2} f(x) = 0$

$f'(-2)$ does not exist.

(d) Does the graph of $y = f(x)$ have a relative maximum? Explain your reasoning.

Yes, it has a relative maximum at $x = 2$ because f is continuous and f' changes sign at $x = 2$.

AP Calculus AB Test 3 Answers

Question 6

(a) Find the velocity for each particle.

$$A'(t) = -4\left(\frac{\pi}{4}\right)\cos\left(\frac{\pi}{4}t\right) = -\pi\cos\left(\frac{\pi}{4}\right) \quad \text{and} \quad B'(t) = 3t^2 - 18t + 24$$

(b) For $0 \le t \le 8$, find all time t during which the two particles move in same directions.
For $4 \le t \le 6$, $A'(t)$ and $B'(t)$ have the same positive sign. They move in same direction.

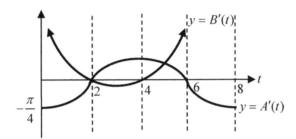

(c) For $0 \le t \le 8$, find all time t during which particle A is speeding up. Explain your reasoning.
If the velocity and the acceleration of the particle have the same sign, the speed is increasing.

$$a(t) = A''(t) = \frac{\pi^2}{4}\sin\left(\frac{\pi}{4}t\right)$$

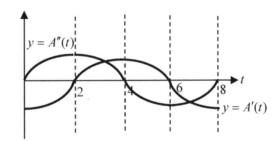

For $2 \le t \le 4$ and $6 \le t \le 8$, the particle is speeding up because A' and A'' have the same sign.

(d) Find the distance traveled by particle A from $t = 0$ to $t = 8$.

$$d = \int_0^8 |A'(t)|\, dt = 4\int_0^2 \left|-\pi\cos\left(\frac{\pi t}{4}\right)\right| dt = 4\pi\left[\frac{4}{\pi}\sin\left(\frac{\pi t}{4}\right)\right]_0^2 = 16\left(\sin\left(\frac{\pi}{2}\right) - \sin 0\right) = 16$$

(e) Find the average velocity of the particle B over the interval $0 \le t \le 4$.
The average velocity of the particle B over the interval $0 \le t \le 4$ is

$$\frac{1}{4}\int_0^4 (3t^2 - 18t + 24)\, dt = \frac{1}{4}\left[t^3 - 9t^2 + 24t\right]_0^4 = \frac{1}{4}(16 - 0) = 4$$

NO MATERIAL ON THIS PAGE

AP CALCULUS AB

TEST 4

AP Calculus AB Test 4   *Section 1*

AP CALCULUS AB

A CALCULATOR **CANNOT BE USED ON PART A OF SECTION 1**. A GRAPHIC CALCULATOR FROM THE APPROVED LIST **IS REQUIRED FOR PART B OF SECTION 1 AND FOR PART A OF SECTION II** OF THE EXAMINATION.

SECTION 1
Time: 1 hour and 45 minutes
All questions are given equal weight.
Percent of total grade – 50

Part A: 55 minutes, 28 multiple-choice questions
A calculator is NOT allowed.
Part B: 50 minutes, 17 multiple-choice questions
A graphing calculator is required.

CALCULUS AB
SECTION 1, PART A
Time— 55minutes
Number of questions— 28

A CALCULATOR MAY NOT BE USED ON THIS PART OF THE EXAMINATION.

Directions: Solve each of the following problems, using the available space for scratchwork. After examining the form of the choices, decide which is the best of the choices given and fill in the corresponding oval on the answer sheet. No credit will be given for anything written in the test book. Do not spend too much time on any one problem.

In this test:

(1) Unless otherwise specified, the domain of a function f is assumed to be the set of all real numbers x for which $f(x)$ is a real number.

(2) The inverse of a trigonometric function f may be indicated using the inverse function notation f^{-1} or with the prefix "arc" (e.g, $\sin^{-1} x = \arcsin x$)

AP Calculus AB Test 4

Section 1

1. A particle moves along the x-axis, so that, at any time $t \geq 0$, its acceleration is given by $a(t) = 6x - 12$. At time $t = 0$, the velocity of the particle is 9. For what values of t is the particle moving to the left?

   (A) $t \geq 0$ (B) $0 < t < 3$ (C) $1 < t < 3$ (D) $2 < t < 4$ (E) $t > 3$

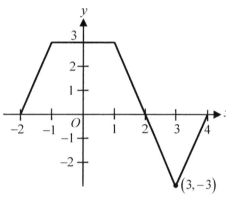

Graph of f'

2. The graph of f', the derivative of f is shown above. If $f(0) = -2$, then $f(4) =$

   (A) -0.5 (B) 1.0 (C) 1.5 (D) 1.8 (E) 2.0

3. $\int_0^2 \dfrac{x}{1+x^2}\,dx =$

   (A) $-\dfrac{5}{2}$ (B) -3 (C) $\dfrac{\ln 5}{2}$ (D) $\dfrac{e^3}{5}$ (E) $2\ln 5$

4. If $xy + x - y = 5$, then what is the value of $\dfrac{dy}{dx}$ when $x = 2$?

   (A) -4 (B) -2 (C) 1.5 (D) 2 (E) 4

5. If f is continuous on the closed interval $[a,b]$ and differentiable on the open interval (a,b), which of the following must be true?

   (A) f is increasing over the interval (a,b).
   (B) f is decreasing over the interval (a,b).
   (C) f has a relative maximum on $[a,b]$.
   (D) f has a point of inflection on (a,b).
   (E) There exists c such that $f(b) - f(a) = (b-a)f'(c)$ for $a < c < b$.

AP Calculus AB Test 4   Section 1

6. $\int_{1}^{2} t\sqrt{t-1}\, dt =$

(A) $-\infty$ (B) -2 (C) $\dfrac{1}{2}$ (D) $\dfrac{16}{15}$ (E) ∞

7. If $f(x) = \int_{0}^{x} (t \ln t)\, dt$, then $\lim\limits_{h \to 0} \dfrac{f(e+h) - f(e)}{h} =$

(A) $-\infty$ (B) $-e^2$ (C) e (D) $2e$ (E) ∞

8. Let f be the function given by $f(x) = \dfrac{1}{x-4}$. What is the equation for the line tangent to the graph of f at the point $(5, f(5))$?

(A) $y = -x + 6$
(B) $y = -2x + 9$
(C) $y = -3x + 6$
(D) $y = -3x + 9$
(E) $y = -3x + 10$

9. Let f be the differentiable function for all real numbers x. If $f'(a) = 0$ and $f''(a) > 0$, then which of the following must be true?

(A) The graph of f has a point of inflection at $x = a$.
(B) The graph of f is concave downward on the interval $(-\infty, \infty)$.
(C) The graph of f is concave upward on the interval $(-\infty, \infty)$.
(D) The graph of f has a relative maximum at $x = a$.
(E) The graph of f has a relative minimum at $x = a$.

10. If $f(x) = \cos^2(3x)$, then $f'\left(\dfrac{\pi}{9}\right) =$

(A) $\dfrac{-3\sqrt{3}}{2}$ (B) $-\dfrac{\sqrt{3}}{2}$ (C) $\dfrac{\sqrt{3}}{4}$ (D) $\dfrac{\sqrt{3}}{2}$ (E) $3\sqrt{3}$

AP Calculus AB Test 4   Section 1

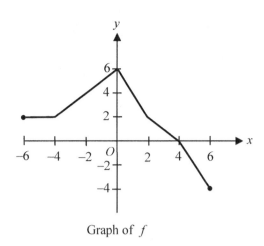

Graph of f

11. The graph of f is shown above. If $g(x) = \int_2^x f(t)\,dt$, which of the following values greatest?

(A) $g(-4)$ (B) $g(-2)$ (C) $g(0)$ (D) $g(4)$ (E) $g(6)$

---

12. Let f be differentiable function with the following properties.

(i) $f'(x) > 0$ on the interval $[0, 4]$.
(ii) $f''(x) > 0$ for all x.

Which of the following has the greatest value?

(A) Left Riemann sum approximation of $\int_0^4 f(x)\,dx$ with 10 subintervals of equal length

(B) Right Riemann sum approximation of $\int_0^4 f(x)\,dx$ with 10 subintervals of equal length

(C) Midpoint Riemann sum approximation of $\int_0^4 f(x)\,dx$ with 10 subintervals of equal length

(D) Trapezoidal sum approximation of $\int_0^4 f(x)\,dx$ with 10 subintervals of equal length

(E) $\int_0^4 f(x)\,dx$

---

13. If $f(x) = e^{2\ln(x+1)}$, then $f'(x) =$

(A) $e^{2\ln(x)}$ (B) $2\ln(x+1)$ (C) $\dfrac{2}{x+1}$ (D) $2x+2$ (E) $e^{2\ln(x+1)} \cdot \dfrac{1}{x+1}$

AP Calculus AB Test 4 — Section 1

14. If $f(x) = \begin{cases} \ln x & \text{for } 0 < x \leq 1 \\ x-1 & \text{for } 1 < x < \infty \end{cases}$, then $f'(1) =$

(A) 0   (B) 1   (C) 2   (D) e (E) nonexistent

---

15. $\dfrac{d}{dx}\left(\ln 5^{2x}\right) =$

(A) $\dfrac{1}{e}$ (B) 2   (C) $\ln 5$ (D) $2\ln 5$ (E) 5

---

16. $\displaystyle\int \cot x\, dx =$

(A) $\ln|\cos x| + C$ (B) $\ln|\sin x| + C$ (C) $\dfrac{1}{2}\cot^2 x + C$ (D) $\ln|\csc x| + C$ (E) $\ln|\sec x| + C$

---

17. If $e^{xy} = x$, then what is the value of $y'(1)$?

(A) -1 (B) 0   (C) 1   (C) $\dfrac{1}{e}$ (D) e

---

18. Let f be the function given by $f(x) = \cos^2 x + 2$ for $0 \leq x \leq \pi$. For what values of x does the graph of f have a point of inflection?

(A) $\dfrac{\pi}{4}$ only   (B) $\dfrac{\pi}{2}$ only   (C) $\dfrac{\pi}{4}$ and $\dfrac{\pi}{2}$ (D) $\dfrac{\pi}{2}$ and $\dfrac{3\pi}{4}$ (E) $\dfrac{\pi}{4}$ and $\dfrac{3\pi}{4}$

---

19. If the line $y = -2x + 3$ is tangent to the graph of $y = x^2 + 2x + k$, where k is a constant, what is the value of k?

(A) 0   (B) 2   (C) 4   (D) 6   (E) 7

AP Calculus AB Test 4   Section 1

20. What are all asymptotes of the graph of $y = \dfrac{x^2 + 3x + 2}{x^3 - 1}$?

   (A) $x = 1$ only

   (B) $x = 1$ and $y = 0$ only

   (C) $y = 0$ only

   (D) $x = 1, x = 2$ and $y = 0$

   (E) $x = \dfrac{1+\sqrt{3}}{2}, \dfrac{1-\sqrt{3}}{2}$, and $y = 0$

---

21. $\lim\limits_{x \to 2} \dfrac{x^5 - (2^5)}{10(x-2)} =$

   (A) 6   (B) 8   (C) 10   (D) 32   (E) nonexistent

---

22. Let g be a function given by $g(x) = x^6 - 10x^4 + 2x + 3$. What are the x-coordinates of points of inflection of the graph of g ?

   (A) 0 only   (B) 2 only   (C) 0 and 2 only   (D) -2 and 2 only   (E) $-2, 0, 2$

---

23. If $f(x) = \sin(\tan(3x))$, then $f'(x) =$

   (A) $f'(x) = \cos(\tan(3x))$
   (B) $f'(x) = 3\cos(\tan(3x))$
   (C) $f'(x) = 3\cos(\tan(3x))\sec^2 x$
   (D) $f'(x) = 3\cos(\tan(3x))\sec^2(3x)$
   (E) $f'(x) = \cos(3x)\sec^3(3x)$

---

24. A position of a moving particle is given by $x(t) = t^3 - 2t^2 + 4t + 2$. What is its average velocity for $1 \le t \le 4$?

   (A) 3.5   (B) 4   (C) 10.2   (D) 15   (E) 20.5

Dr. John Chung's AP Calculus AB/BC

Interval	Sign of f'	Sign of f''
$x < -3$	+	−
$-3 < x < 0$	−	+
$0 < x < 3$	−	−
$x > 3$	+	+

25. Let f be a continuous function over all real x. If the function f has properties shown above, then which of the following could be the graph of f?

(A)

(B)

(C)

(D)

(E)

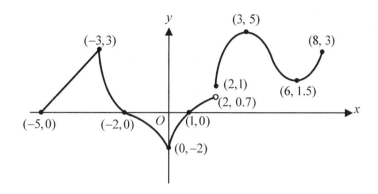

26. The graph of f is shown above. What are all critical numbers of the graph on the open interval $-5 < x < 8$?

(A) 3 and 6
(B) $-5, -2,$ and 1
(C) $-5, -2, 1, 3,$ and 6
(D) $-3, 0, 3,$ and 6
(E) $-3, 0, 3, 2,$ and 6

---

27. $\int x^2 e^x \, dx =$

(A) $\dfrac{x^3 e^x}{3} + C$

(B) $\dfrac{x^3}{3} + \dfrac{e^{2x}}{2} + C$

(C) $e^x(x^2 - 2x + 2) + C$

(D) $\dfrac{e^x(x^2 + 2)}{2} + C$

(E) $\dfrac{e^x(x^2 + 2x + 2)}{3} + C$

x	$f(x)$	$f'(x)$
3	−2	−8
−2	−6	−4

28. The polynomial function f has selected values of its first derivative f' given in the table above. If the function is differentiable and $g(x) = f^{-1}(x)$ for all x, then $g'(-2) =$

(A) $-\dfrac{1}{4}$ (B) $-\dfrac{1}{8}$ (C) $\dfrac{1}{8}$ (D) $\dfrac{1}{4}$ (E) $\dfrac{1}{2}$

END OF PART A OF SECTION 1

AP Calculus AB Test 4 *Section 1*

CALCULUS AB
SECTION 1, PART B
Time— 50minutes
Number of questions— 17

A GRAPHING CALCULATOR IS REQUIRED FOR SOME QUESTIONS ON THIS PART OF THE EXAMINATION.

Directions: Solve each of the following problems, using the available space for scratchwork. After examining the form of the choices, decide which is the best of the choices given and fill in the corresponding oval on the answer sheet. No credit will be given for anything written in the test book. Do not spend too much time on any one problem.

In this test:

(1) The exact numerical value of the correct answer does not always appear among the choices given. When this happens, select from among the choices the number that best approximates the exact numerical value.

(2) Unless otherwise specified, the domain of a function f is assumed to be the set of all real numbers x for which $f(x)$ is a real number.

(3) The inverse of a trigonometric function f may be indicated using the inverse function notation f^{-1} or with the prefix "arc" (e.g, $\sin^{-1} x = \arcsin x$)

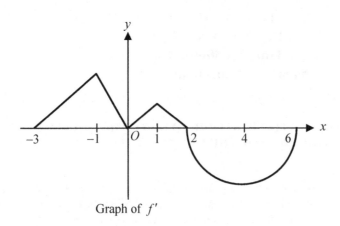

Graph of f'

76. The graph of the function f' on the closed interval $-3 \leq x \leq 6$ is shown above. How many points of inflection does the graph of f have on the interval $-3 < x < 6$?

(A) 1   (B) 2   (C) 3   (D) 4   (E) 5

77. The function f is defined by $f(x) = x^3 - 3x^2 + 5 + \ln x$. For what value of x does the graph of f change its concavity?

(A) 0   (B) 1.130   (C) 1.412   (D) 2.275   (E) 3.125

78. If $\int_{-5}^{2}(f(x)-2)dx = -28$ and $\int_{2}^{5} 2f(x)dx = 6$, what is the value of $\int_{-5}^{5}(f(x)+5)dx$?

(A) −27   (B) −17   (C) 17   (D) 23   (E) 39

79. Let f be the function defined as follows.

$$f(x) = \begin{cases} |x-2|+2, & x < 2 \\ x^2 + ax + b, & x \geq 2 \end{cases}$$

If f is both continuous and differentiable, what are the values of a and b?

(A) $a = 0, b = 1$
(B) $a = 2, b = -3$
(C) $a = -4, b = 8$
(D) $a = -3, b = 5$
(E) $a = -5, b = 8$

AP Calculus AB Test 4

Section 1

80. If $g(x) = \int_0^{\ln x} e^{2t} dt$, what is the equation of the line tangent to the graph of $y = g(x)$ at $x = 3$?

   (A) $y = 3x - 5$

   (B) $y = 3x + 10$

   (C) $y = -\dfrac{3}{2}x - 5$

   (D) $y = -3x + 5$

   (E) $y = -\dfrac{5}{2}x + 5$

---

81. A particle moves along the path given for time $t > 0$ by the parametric equations $x = e^{t^2}$ and $y = 3t^2$. What is the acceleration of the particle at time $t = 1$?

   (A) 4.723   (B) 17.378   (C) 31.918   (D) 35.902   (E) 54.518

---

82. The function f is continuous for $0 \leq x \leq 5$. If $f(0) = -3$ and $f(5) = 3$, which of the following statements must be true?

   (A) f has a relative maximum for $0 < x < 5$.
   (B) f has a point of inflection for $0 < x < 5$.
   (C) f is concave up for $0 < x < 5$.
   (D) f is differentiable for $0 < x < 5$.
   (E) There exists at least one c for $0 < c < 5$ such that $f(c) = 2$.

---

83. Let R be the region enclosed by the graphs of $y = e^x$, $y = (x-1)^2$, and the line $x = 2$. What is the area of R?

   (A) 5.722   (B) 7.125   (C) 8.654   (D) 12.761   (E) 12.912

---

x	0	4	6	7	10
$f(x)$	10	15	20	16	10

84. The function is continuous on the closed interval $[0, 10]$ and has selected values in the table above. What is the trapezoidal approximation of $\int_0^{10} f(x)\, dx$ using the four subintervals?

   (A) 138   (B) 142   (C) 145   (D) 148   (E) 156

AP Calculus AB Test 4   Section 1

85. Let f be the function given by $f(x) = x^3 - ax^2 + bx - 5$, where a and b are constants. If the graph of f has a relative maximum at $x = 1$, what is the smallest integer value of a?

   (A) 2   (B) 3   (C) 4   (D) 5   (E) 7

86. A particle moves on x-axis so that its position at any time $t \geq 0$ is given by $x(t) = 6t^2 - 24t + 22$. Find all values of t for which the speed of the particle is decreasing.

   (A) $0 < t < 2$
   (B) $2 < t < 3$
   (C) $3 < t < 4$
   (D) $4 < t < 6$
   (E) $6 < t < \infty$

87. $\lim\limits_{x \to \infty} \dfrac{\int_0^x \sqrt{t^2 + 5t + 3}\, dt}{x^2} =$

   (A) $\dfrac{1}{2}$ (B) 2   (C) $\dfrac{5}{2}$ (D) ∞ (E) undefined

88. Let f be the function with derivative given by $f(x) = \cos(e^{2x} - 10)$. How many points of inflection does f have on the interval $0 < x < 1$?

   (A) None   (B) 1   (C) 2   (D) 3   (E) 5

89. Let f be the function given by $f(x) = \int_0^{x^2} \sqrt{t} \sin\left(1 + \dfrac{t}{2}\right) dt$ for $0 \leq x \leq 3$. On which of the following intervals is f increasing?

   (A) $0 < x \leq 1$ only
   (B) $0 < x \leq 1.694$ only
   (C) $0 < x < 2.070$
   (D) $1.511 \leq x \leq 2.927$
   (E) $2.070 < x \leq 3$

AP Calculus AB Test 4 Section 1

90. Let $g(x)$ be an antiderivative of $x^2 \cos^2 x$. If $g(0) = 0$, then $g(5) =$

(A) 8.235   (B) 12.341   (C) 16.452   (D) 24.674   (E) 28.118

---

91. Let f be a differentiable function with $f(4) = -3$ and $f'(4) = 2$. If g is the function defined by $g(x) = (x^2 + 1)f(x)$, which of the following is an equation of the line tangent to the graph of g at the point $(4, g(4))$?

(A) $y + 45 = 6(x - 4)$
(B) $y - 51 = 10(x - 4)$
(C) $y + 45 = -6(x - 4)$
(D) $y + 51 = 10(x - 4)$
(E) $y - 45 = -6(x - 4)$

---

92. Let f be a function that is continuous on the closed interval $[-5, 5]$. If $f(-2) = -3$, $f(0) = 4$, and $f(3) = -5$, which of the following must be true?

   I. f has at least two zeros.
   II. f has at least one local maximum.
   III. The graph of f has at least one horizontal tangent.

(A) I only
(B) I and II only
(C) I and III only
(D) II and III only
(E) I, II, and III

END OF SECTION 1

NO MATERIAL ON THIS PAGE

AP Calculus AB Test 4 — Section II

CALCULUS AB

SECTION II
Time: 1 hour and 30 minutes
Percent of total grade – 50

Part A: 30 minutes, 2 problems
Part B: 60 minutes, 4 problems

Part A: A graphing calculator is required for some problems or parts of problems.
Part B: No calculator is allowed for these problems.

GENERAL INSTRUCTIONS FOR SECTION II PART A AND PART B

For each part of section II, you may wish to look over the problems before starting to work on them, since it is not expected that everyone will be able to complete all parts of all problems. All problems are given equal weight, but the parts of a particular problem are not necessary given equal weight.

- YOU SHOULD WRITE ALL WORK FOR EACH PART OF EACH PROBLEM IN THE SPACE PROVIDED FOR THAT PART IN THE TEST BOOKLET. Be sure to write clearly and legibly. If you make an error, you may save time by crossing it out rather than trying to erase it. Erased or cross-out work will not be graded.
- Show all work. Clearly label any functions, graphs, tables, or other object that you use. You will be graded on the correctness and completeness of your methods as well as your answers. Answers without supporting work may not receive credit.
- Justifications required that you give mathematical (noncalculator) reasons
- Your work must be expressed in standard mathematical notation rather than calculator syntax.
  For example, $\int_1^5 x^2 \, dx$ may not be written as $\text{fnInt}(X^2, X, 1, 5)$
- Unless otherwise specified, answers (numeric or algebraic) need not be simplified.
- If you use decimal approximations in calculations, you will be graded on accuracy. Unless otherwise specified, your final answers should be accurate to three places after the decimal point.
- Unless otherwise specified, the domain of a function f is assumed to be the set of all real numbers x for which $f(x)$ is a real number.

CALCULUS AB
SECTION II, PART A
Time — 30 minutes
Number of questions — 2

A Graphing calculator is required for some problems or parts of problems.

Question 1

A particle moves along the x-axis so that its velocity at time t is given by $v(t) = 2t \cdot \sin\left(\dfrac{\pi}{3}t\right)$. At time $t = 0$, the position of the particle is $x(0) = 1$.

(a) For $0 \leq t \leq 6$, find all time t during which the speed of the particle is decreasing. Explain your reasoning.

(b) For $0 < t < 6$, find the value of t for which the particle at rest.

(c) Write, but do not evaluate, an expression for the total distance traveled by the particle from time $t = 0$ to $t = 6$.

(d) Write, but do not evaluate, an expression for the average velocity of the particle over the interval $0 \leq t \leq 6$.

(e) Find the value of $x(3)$, the position of the particle at time $t = 3$.

AP Calculus AB Test 4 — Section II

Question 2

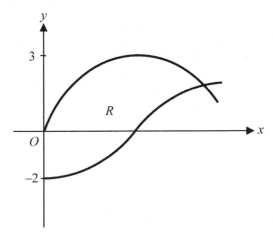

Let R be the region enclosed by the y-axis and the graphs of $y = 3\sin\left(\dfrac{\pi}{4}x\right)$ and $y = -2\cos\left(\dfrac{\pi}{3}x\right)$, as shown in the figure above.

(a) Find the area of R.

(b) Write, but do not evaluate, an expression for the volume of the solid generated when R is revolved about the y-axis.

(c) The region R is the base of a solid. For this region, each cross section perpendicular to the x-axis is a square. Find the volume of the solid.

END OF PART A OF SECTION II

AP Calculus AB Test 4 — Section II

CALCULUS AB
SECTION II, PART B
Time — 60minutes
Number of questions — 4

No calculator is required for these problems.

Question 3

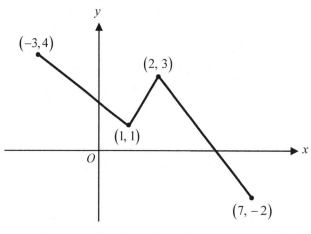

Graph of f

Let f be the continuous function defined on $[-3, 7]$ whose graph, consisting of three line segments is shown above. Let g be the function given by $g(x) = \int_1^x f(t)\,dt$.

(a) Find the values of $g(-3)$ and $g(7)$.

(b) Find the instantaneous rate of change of g at $x = 3$.

(c) Find the absolute maximum value of g on the closed interval $[-3, 7]$. Explain your reasoning.

(d) For $-3 < x < 7$, find all values of x coordinates of points of inflection of the graph of g. Explain your reasoning.

(e) Find the equation of the line tangent to the graph of g at $x = 2$.

AP Calculus AB Test 4 — Section II

Question 4

Consider the curve given by $x^2 - xy + y^2 = 12$.

(a) Find $\dfrac{dy}{dx}$.

(b) Find all points on the curve at which the line tangent to the curve is horizontal. Write an equation of the tangent line at each of these points.

(c) Find all points on the curve at which the line tangent line to the curve is vertical. Write an equation of the tangent line at each of these points.

(d) Find the value of $\dfrac{d^2y}{dx^2}$ at each of these points found in part (b). Does the curve have a relative maximum or a relative minimum at each of these points. Explain your reasoning.

AP Calculus AB Test 4

Question 5

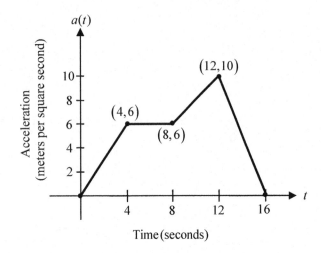

A particle moves along the x-axis. The acceleration $a(t)$ of the particle at time t, in meters per square second, is modeled by the piecewise-linear function defined by the graph above.

(a) Find $\int_0^{16} a(t)\,dt$. Explain the meaning of $\int_0^{16} a(t)\,dt$.

(b) Given that $v(t)$ is the velocity of the particle at time t and that $v(0) = 4$ meters/second, find the average rate of change of v over the interval $[0, 16]$. Indicate units of measure.

(c) For $0 \leq t \leq 16$, find the piecewise-defined function for $a(t)$.

(d) Given that $v(t)$ is the velocity of the particle at time t and that $v(0) = 12$, find the value of $v(8)$. Indicate units of measure.

Question 6

Consider the differential equation $\dfrac{dy}{dx} = -2xy^2$.

(a) On the axes provided below, sketch a slope field for the given differential equation at the points indicated.

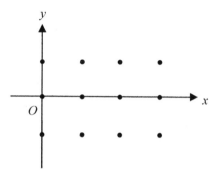

(b) Find the particular solution $y = f(x)$ to the differential equation with the initial condition $f(0) = 1$.

(c) Write an equation for the line tangent to the curve found in part (b) at $x = 2$.

(d) Find $\dfrac{d^2y}{dx^2}$ in terms of x and y.

(e) Find all values of x which the graph of f has a horizontal tangent and determine whether f has a local maximum or a local minimum. Justify your answer.

(f) Find the range of the solution found in part (b).

END OF EXAM

NO MATERIAL ON THIS PAGE

AP Calculus AB Test 4 Answers

SECTION 1, PART A

1. **C**

   $$v(t) = 9 + \int_0^t (6t - 12)\,dt = 9 + \left[3t^2 - 12t\right]_0^t = 3t^2 - 12t + 9$$
   $$\to v(t) = 3(t-1)(t-3) < 0 \implies v(t) < 0 \text{ for the interval } 1 < t < 3$$

2. **A**

   $$f(4) = -2 + \int_0^4 f'(t)\,dt = -2 + 4.5 - 3 = -0.5$$

   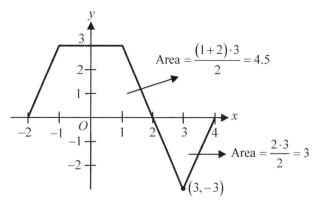

   Graph of f'

3. **C**

   $$\int_0^2 \frac{x}{1+x^2}\,dx = \frac{1}{2}\left[\ln(1+x^2)\right]_0^2 = \frac{1}{2}(\ln 5 - 0) = \frac{\ln 5}{2}$$

4. **A**

   If $xy + x - y = 5$, then what is the value of $\dfrac{dy}{dx}$ when $x = 2$?

   $$y + xy' + 1 - y' = 0 \implies (x-1)y' = -y - 1 \implies y' = \frac{-y-1}{x-1}$$

   When $x = 2$, $2y + 2 - y = 5 \implies y = 3$.

   $$\therefore y' = \frac{-3-1}{2-1} = -4$$

5. **E**

   Mean value theorem for derivative:
   There exists c such that $f(b) - f(a) = (b-a)f'(c)$ for $a < c < b$.

6. **D**

   $$\begin{cases} u = t - 1 \implies t = u + 1 \\ du = dt \end{cases}$$

   $$\int_1^2 t\sqrt{t-1}\,dt = \int_0^1 (u+1)\sqrt{u}\,du = \int_0^1 \left(u^{\frac{3}{2}} + u^{\frac{1}{2}}\right) du = \left[\frac{2}{5}u^{\frac{5}{2}} + \frac{2}{3}u^{\frac{3}{2}}\right]_0^1 = \frac{16}{15}$$

7. **C**

   $$\lim_{h \to 0} \frac{f(e+h) - f(e)}{h} = f'(e)$$
   $$f'(x) = x \ln x \implies f'(e) = e \ln e = e$$

AP Calculus AB Test 4 Answers

8. **A**

$f(5) = 1$ and $f'(x) = \dfrac{-1}{(x-4)^2} \Rightarrow f'(5) = -1$

The equation for the line tangent to the graph of f at the point $(5, f(5))$ is

$y - 1 = -(x-5) \Rightarrow y = -x + 6$

9. **E**

The graph of f has a relative minimum at $x = a$.

10. **A**

$f'(x) = 2\cos(3x)(-\sin(3x))3 \Rightarrow f'\left(\dfrac{\pi}{9}\right) = -6\cos\left(\dfrac{\pi}{3}\right)\sin\left(\dfrac{\pi}{3}\right) = -\dfrac{3\sqrt{3}}{2}$

11. **D**

(A) $g(-4) < 0$ (B) $g(-2) < 0$ (C) $g(0) < 0$ (D) $g(4) = 2$ (E) $g(6) = 2 - 4 = -2$

12. **B**

Right Riemann sum approximation of $\int_0^4 f(x)\,dx$ with 10 subintervals of equal length is greatest.

13. **D**

$f(x) = e^{2\ln(x+1)}$, then $f'(x) = e^{2\ln(x+1)}(2\ln(x+1))' = e^{2\ln(x+1)}\left(\dfrac{2}{x+1}\right) = \dfrac{2(x+1)^2}{(x+1)} = 2(x+1)$

14. **B**

If $f(x) = \begin{cases} \ln x & \text{for } 0 < x \le 1 \\ x-1 & \text{for } 1 < x < \infty \end{cases}$, then $f'(1) =$

$f(1) = \lim_{x \to 1}(x-1) = 0 \rightarrow f$ is continuous.

$(\ln x)' = \dfrac{1}{x} \rightarrow (\ln x)'_{x=1} = 1$ and $(x-1)'_{x=1} = 1$

Therefore, $f'(1) = 1$.

15. **D**

$\dfrac{d}{dx}\left(\ln 5^{2x}\right) = \dfrac{d}{dx}(2x \ln 5) = 2\ln 5$

16. **B**

$\int \cot x\, dx = \int \dfrac{\cos x}{\sin x}\, dx = \ln|\sin x| + C$

17. **C**

$\left(e^{xy}\right)' = 1 \rightarrow e^{xy}(xy)' = 1 \rightarrow e^{xy}(y + xy') = 1$

When $x = 1$, $e^y = 1 \rightarrow y = 0$

$e^{1(0)}(0 + y') = 1 \rightarrow y'(1) = 1$

18. **E**

$f'(x) = -2\cos x \sin x = -\sin 2x \Rightarrow f''(x) = -2\cos 2x = 0$

$2x = \dfrac{\pi}{2}, \dfrac{3\pi}{2}, \dfrac{5\pi}{2}\ldots \rightarrow x = \dfrac{\pi}{4}, \dfrac{3\pi}{4}, \ldots$

AP Calculus AB Test 4 Answers

19. **E**

Let the point of tangency be (a, b). $\left.\dfrac{dy_1}{dx}\right|_{x=a} = -2$ and

$\left.\dfrac{dy_2}{dx}\right|_{x=a} = 2x+2\big|_{x=a} = 2a+2 \Rightarrow 2a+2 = -2 \Rightarrow a = -2$

When $a = -2$, $b = f(-2) = -2(-2)+3 = 7$.

Since $y_1 = y_2$ at $x = -2$, $y_2(-2) = 7 \Rightarrow (-2)^2 + 2(-2) + k = 7 \Rightarrow k = 7$.

20. **B**

$y = \dfrac{x^2+3x+2}{x^3-1} = \dfrac{(x+1)(x+2)}{(x-1)(x^2+x+1)}$

Vertical asymptote: $x = 1$
Horizontal asymptote: $y = \lim\limits_{x \to \infty} f(x) = 0$

21. **B**

Define $f(x) = x^5$. $f'(x) = 5x^4$

$\lim\limits_{x \to 2} \dfrac{x^5 - (2^5)}{10(x-2)} = \dfrac{1}{10} \lim\limits_{x \to 2} \dfrac{x^5 - (2^5)}{x-2} = \dfrac{1}{10} f'(2)$

$f'(2) = \dfrac{5(2)^4}{10} = 8$

22. **D**

$g(x) = x^6 - 10x^4 + 2x + 3 \Rightarrow g'(x) = 6x^5 - 40x^3 + 2 \Rightarrow g''(x) = 30x^4 - 120x^2$

For points of inflection: $g''(x) = 0 \Rightarrow 30x^2(x-2)(x+2) = 0$

$g''(x)$ changes sign at each value of $x = 2$ and $x = -2$.

23. **D**

$f(x) = \sin(\tan(3x))$, then $f'(x) = \cos(\tan(3x))\sec^2(3x) \cdot 3$

24. **D**

Average velocity $= \dfrac{1}{4-1} \int_1^4 v(t)\,dt = \dfrac{x(4)-x(1)}{3} = \dfrac{45}{3} = 15$

25. **C**

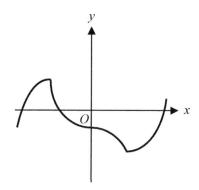

AP Calculus AB Test 4 Answers

26. **D**

Critical point: Let f be defined at c. If $f'(c) = 0$ or f' is undefined at c, then c is a critical point of f.

Therefore, the critical point are $x = -3, 0, 3,$ and 6.

27. **C**

$\int x^2 e^x \, dx = e^x \left(x^2 - 2x + 2 \right) + C$

Use Tabular Integration by Part:

+	x^2	e^x
−	$2x$	e^x
+	2	e^x
−	0	e^x

$\int x^2 e^x \, dx = x^2 e^x - 2xe^x + 2e^x + C$

28. **B**

$g(-2) = k \leftrightarrow f(k) = -2 \Rightarrow k = 3$

$g'(-2) = \dfrac{1}{f'(3)} = \dfrac{1}{-8}$

x	$f(x)$	$f'(x)$
3	−2	−8
−2	−6	−4

SECTION 1, PART B

76. **D**

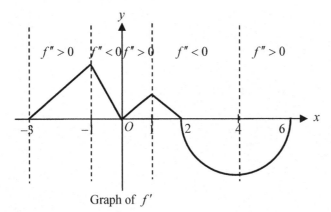

Graph of f'

$f''(x)$ changes sign at each value of $x = -1, 0, 1,$ and 4.

77. **B**

$f'(x) = 3x^2 - 6x + \dfrac{1}{x}$ for $x > 0$.

Use your calculator.

f has a point of inflection at $x = 1.130$.

Graph of f'

AP Calculus AB Test 4 Answers

78. E

$$\int_{-5}^{5}(f(x)+5)\,dx = \int_{-5}^{5}f(x)\,dx + \int_{-5}^{5}5\,dx \text{ and } \int_{-5}^{5}5\,dx = [5x]_{-5}^{5} = 50$$

$$\int_{-5}^{2}(f(x)-2)\,dx = \int_{-5}^{2}f(x)\,dx - 14 = -28 \implies \int_{-5}^{2}f(x)\,dx = -14$$

$$\int_{2}^{5}2f(x)\,dx = 6 \implies \int_{2}^{5}f(x)\,dx = 3$$

Therefore, $\int_{-5}^{5}(f(x)+5)\,dx = \int_{-5}^{2}f(x)\,dx + \int_{2}^{5}f(x)\,dx + 50 = -14 + 3 + 50 = 39$

79. E

$$f(x) = \begin{cases} |x-2|+2, & x<2 \\ x^2+ax+b, & x\geq 2 \end{cases}$$

Continuous: $f(2) = \lim_{x\to 2}|x-2|+2 \implies 4+2a+b=2 \implies 2a+b=-2$

Differentiable: $\begin{cases} x\geq 2, f'(x) = 2x+a \implies f'(2) = 4+a \\ x<2, f'(x) = (-x+4)' = -1 \end{cases} \implies 4+a=-1$

$\begin{cases} 4+a=-1 \\ 2a+b=-2 \end{cases} \implies a=-5 \text{ and } b=8$

80. A

$$g(x) = \int_{0}^{\ln x} e^{2t}\,dt \implies g'(x) = e^{\ln x} \cdot \frac{1}{x} \implies g'(3) = e^{\ln 3} = 3$$

$$g(3) = \int_{0}^{\ln 3} e^{2t}\,dt = \left[\frac{e^{2t}}{2}\right]_{0}^{\ln 3} = \frac{e^{2\ln 3}}{2} - \frac{1}{2} = \frac{9-1}{2} = 4$$

Therefore, the equation is $y-4 = 3(x-3) \implies y = 3x-5$

81. B

$x'(t) = 2te^{t^2} \implies x''(t) = 2e^{t^2} + 4t^2 e^{t^2}$

$y'(t) = 6t \implies y''(t) = 6$

$\begin{cases} x''(1) = 6e \\ y''(1) = 6 \end{cases}$

Acceleration = $\sqrt{(x''(t))^2 + (y''(t))^2} = \sqrt{36e^2+36} \approx 17.378$

A particle moves along the path given for time $t>0$ by the parametric equations $x = e^{t^2}$ and $y = 3t^2$. What is the acceleration of the particle at time $t=2$?

82. E

There exists at least one c for $0<c<5$ such that $f(c)=2$. Intermediate Theorem.

83. A

Area = $\int_{0}^{2}\left[e^x - (x-1)^2\right]dx \approx 5.722$

84. B

$\frac{1}{2}\left[(10+15)\cdot 4 + (20+15)\cdot 2 + (16+20)\cdot 1 + (10+16)\cdot 3\right] = 142$

AP Calculus AB Test 4 Answers

85. C

In order to have a relative maximum: $f'(1) = 0$ and $f''(1) < 0$

$f'(x) = 3x^2 - 2ax + b \Rightarrow f'(1) = 3 - 2a + b = 0 \Rightarrow 2a = 3 + b$

$f''(x) = 6x - 2a \Rightarrow f''(1) = 6 - 2a < 0 \Rightarrow 2a > 6 \Rightarrow 3 + b > 6 \Rightarrow b > 3$

Therefore, the smallest integer value of b is 4.

86. A

$x'(t) = 12t - 24$ and $x''(t) = 12$

Speed increasing: Since $a(t) = x''(t) = 12 > 0$, $v(t) = x'(t)$ must be negative.

Therefore, $12t - 24 < 0 \Rightarrow 0 < t < 2$

87. A

Type: $\dfrac{\infty}{\infty}$ (Use L'Hôpital's Rule)

$$\lim_{x\to\infty} \dfrac{\int_0^x \sqrt{t^2 + 5t + 3}\, dt}{x^2} = \lim_{x\to\infty} \dfrac{\dfrac{d}{dx}\int_0^x \sqrt{t^2 + 5t + 3}\, dt}{\dfrac{d(x^2)}{dx}} = \lim_{x\to\infty} \dfrac{\sqrt{x^2 + 5x + 3}}{2x} = \dfrac{1}{2}$$

88. C

$f'(x) = -2\sin(e^{2x} - 10) \cdot e^{2x}$

Use your graphic calculator.

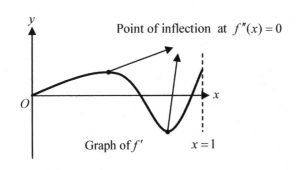

Point of inflection at $f''(x) = 0$

Graph of f' $x = 1$

89. C

$f'(x) = \sqrt{x^2} \sin\left(1 + \dfrac{x^2}{2}\right) \cdot 2x = 2x^2 \sin\left(1 + \dfrac{x^2}{2}\right)$

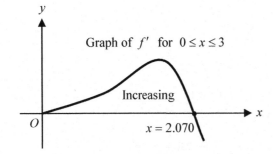

Graph of f' for $0 \le x \le 3$

Increasing

$x = 2.070$

90. C

$g(5) = g(0) + \int_0^5 x^2 \cos^2 x\, dx \approx 16.452$

91. D

$g'(x) = 2xf(x) + (x^2 + 1)f'(x) \Rightarrow g'(4) = 8f(4) + 17f'(4) = 10$

$g(4) = 17f(4) = 17(-3) = -51$

AP Calculus AB Test 4 Answers

$\begin{cases} g'(4) = 10 \\ \text{a point}\,(4, -51) \end{cases} \Rightarrow y - (-51) = 10(x-4) \Rightarrow y + 51 = 10(x-4)$

92. B

Note: The graph is just continuous. (It may not differentiable on the open interval $(-5, 5)$.)

I. f has at least two zeros. True (Intermediate Theorem)

II. f has at least one local maximum. True (It is true even for CUSP)

III. The graph of f has at least one horizontal tangent. False (Example would be CUSP, which has a vertical tangent. Another example would be sharp corner at which is not differentiable)

Compare) If the graph is differentiable on the open interval $(-5, 5)$, then the answer is (E).

SECTION II, PART A

Question 1

(a) $v(t) = 2t \sin\left(\dfrac{\pi}{3} t\right)$ and $a(t) = 2\sin\left(\dfrac{\pi t}{3}\right) + \dfrac{1}{3}\left(2\pi t \cos\left(\dfrac{\pi t}{3}\right)\right)$

Use your graphical calculator: Graph

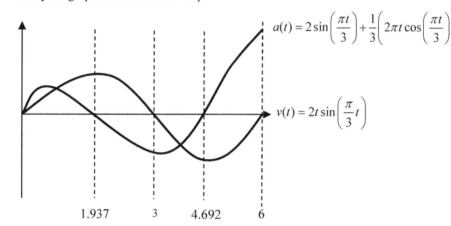

The speed of the particle is decreasing on the intervals $(1.937, 3)$ and $(4.692, 6)$ because the velocity and the acceleration have opposite sign.

(b) For $0 < t < 6$, find the value of t for which the particle at rest.

$v(t) = 0$ at time $t = 3.128$

(c) $d = \displaystyle\int_0^6 |v(t)|\,dt = \int_0^{3.128} 2t\sin\left(\dfrac{\pi t}{3}\right) dt + \int_{3.128}^6 -2t\sin\left(\dfrac{\pi t}{3}\right) dt$

(d) Average velocity $= \dfrac{1}{6}\displaystyle\int_0^6 \left(2t\sin\left(\dfrac{\pi t}{3}\right)\right) dt$

(e) $x(3) = 1 + \displaystyle\int_0^3 2t \sin\left(\dfrac{\pi t}{3}\right) dt \approx 6.730$

AP Calculus AB Test 4 Answers

Question 2

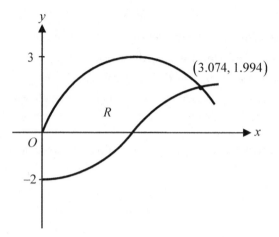

(a) Area $= \int_0^{3.074} \left(3\sin\left(\dfrac{\pi x}{4}\right) + 2\cos\left(\dfrac{\pi x}{4}\right) \right) dx \approx 6.525$

(b) Volume $= 2\pi \int_0^{3.074} x\left(3\sin\left(\dfrac{\pi t}{4}\right) + 2\cos\left(\dfrac{\pi t}{4}\right) \right) dx$

(c) Volume $= \int_0^{3.074} \left(3\sin\left(\dfrac{\pi x}{4}\right) + 2\cos\left(\dfrac{\pi x}{3}\right) \right)^2 dx \approx 16.611$

Question 3

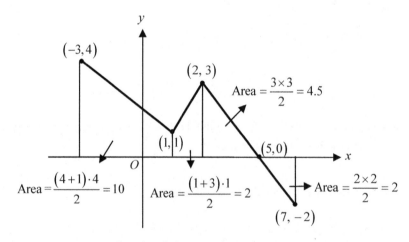

Graph of f

(a) $g(-3) = \int_1^{-3} f(t)\,dt = -\int_{-3}^1 f(t)\,dt = -\left(\dfrac{(1+4)4}{2}\right) = -10$

$g(7) = \int_1^7 f(t)\,dt = 2 + 4.5 - 2 = 4.5$

AP Calculus AB Test 4 Answers

(b) $g'(3) = f(3) = 2$

Find the equation of the line passing through the points $(2,3)$ and $(5,0)$.

$y = -x + 5$. If $x = 3$, then $y = 2$.

(c) The absolute maximum value of g is $g(5)$ because $g(x)$ has the greatest area at $x = 5$.

$g(5) = 2 + 4 = 6.5$

(d) The graph of g has a point of inflection at each of $x = 1$ and $x = 2$ because $g''(x)$ changes sign at each of these values.

(e) $g(2) = \int_1^2 f(t)\,dt = 2 \Rightarrow (2,2)$

$g'(2) = f(2) = 3$

The equation of the line tangent to the graph of g at $x = 2$ is $y - 2 = 3(x - 2)$

Question 4

Consider the curve given by $x^2 - xy + y^2 = 12$.

(a) Find $\dfrac{dy}{dx}$.

$2x - (y + xy') + 2yy' = 0 \Rightarrow y' = \dfrac{dy}{dx} = \dfrac{2x - y}{x - 2y}$

(b) $\dfrac{dy}{dx} = 0 \Rightarrow 2x - y = 0 \Rightarrow y = 2x$

Substitute $y = 2x$ into the equation of the curve.

$x^2 - x(2x) + (2x)^2 = 12 \Rightarrow 3x^2 = 12 \Rightarrow x^2 = 4 \Rightarrow x = \pm 2$

At $\begin{cases} x = 2 \Rightarrow \text{Tangent line } y = 4 \\ x = -2 \Rightarrow \text{Tangent line } y = -4 \end{cases}$

(c) $\dfrac{dy}{dx}$ is undefined at this point. $x = 2y$

$(2y)^2 - (2y)y + y^2 = 12 \Rightarrow y = \pm 2$

$\begin{cases} y = 2 \to \text{Tangent line } x = 4 \\ y = -2 \to \text{Tangent line } x = -4 \end{cases}$

(d) $\dfrac{d^2y}{dx^2} = \dfrac{(2 - y')(x - 2y) - (2x - y)(1 - 2y')}{(x - 2y)^2} = \dfrac{2(x - 2y) - (2x - y)}{(x - 2y)^2}$

For point $(2, 4)$, $\dfrac{d^2y}{dx^2} = \dfrac{2(x - 2y) - (2x - y)}{(x - 2y)^2}\bigg|_{(2,4)} = -\dfrac{1}{3} < 0 \Rightarrow$ Local maximum

For point $(-2, -4)$, $\dfrac{d^2y}{dx^2} = \dfrac{2(x - 2y) - (2x - y)}{(x - 2y)^2}\bigg|_{(-2,-4)} = \dfrac{1}{3} > 0 \Rightarrow$ Local minimum

AP Calculus AB Test 4 Answers

Question 5

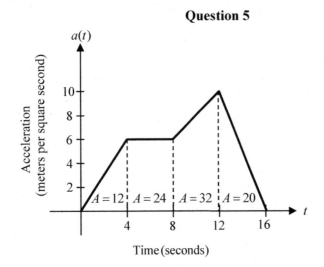

(a) $\int_0^{16} a(t)\,dt = 12 + 24 + 32 + 20 = 88$ meters per second

$\int_0^{16} a(t)\,dt$ is the particle's change in velocity in meters per second from $t = 0$ to $t = 16$ seconds.

(b) Average rate of change of v over the interval $[0, 16]$

$= \dfrac{v(16) - v(0)}{16} = \dfrac{1}{16}\int_0^{16} a(t)\,dt = \dfrac{88}{16} = 5.5$ meters/second^2

(c) For $0 \leq t \leq 8$, find the piecewise-defined function for $a(t)$.

$$a(t) = \begin{cases} y = \dfrac{3}{2}t & \text{for } 0 \leq t < 4 \\ y = 6 & \text{for } 4 \leq t < 8 \\ y = t - 2 & \text{for } 8 \leq t < 12 \\ y = -\dfrac{5}{2}t + 40 & \text{for } 12 \leq t \leq 16 \end{cases}$$

(d) $v(8) = v(0) + \int_0^8 f(t)\,dt = 12 + (12 + 24) = 48$ meters per second

AP Calculus AB Test 4 Answers

Question 6

(a)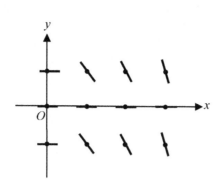

(b) $\dfrac{dy}{y^2} = -2x\,dx \;\to\; \int \dfrac{1}{y^2}\,dy = \int -2x\,dx \;\to\; -\dfrac{1}{y} = -x^2 + C$

Initial condition $(0,1) \;\to\; C = -1$

Therefore, $-\dfrac{1}{y} = -x^2 - 1 \;\to\; \dfrac{1}{y} = x^2 + 1 \;\to\; y = \dfrac{1}{x^2 + 1}$

(c) When $x = 2$, $y = \dfrac{1}{5}$.

$\left.\dfrac{dy}{dx}\right|_{(2,\frac{1}{5})} = -2(2)\left(\dfrac{1}{25}\right) = -\dfrac{4}{25}$

The equation is $y - \dfrac{1}{5} = -\dfrac{4}{25}(x-2)$

(d) $\dfrac{d^2y}{dx^2} = -2y^2 - 2x(2yy') = -2y^2 - 4xy(-2xy^2) = -2y^2 + 8x^2y^3$

(e) $\dfrac{dy}{dx} = -2xy^2 \;\Rightarrow\; \dfrac{dy}{dx} = 0$ at $x = 0$ or $y = 0$.

From part (b): $\begin{cases} x = 0 \;\Rightarrow\; y = 1 \\ y = 0 \;\Rightarrow\; x \text{ is undefined} \end{cases} \;\Rightarrow\; \left.\dfrac{d^2y}{dx^2}\right|_{(0,1)} = -2 < 0$

The graph of f has a local maximum at $x = 0$ because $f'(0) = 0$ and $f''(0) < 0$.

(f) The range is $0 < y \leq 1$.

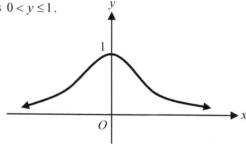

Graph of f'

NO MATERIAL ON THIS PAGE

Calculus BC Practice Test

NO MATERIAL ON THIS PAGE

AP CALCULUS BC

TEST 1

AP Calculus BC Test 1   *Section 1*

AP CALCULUS BC

A CALCULATOR **CANNOT BE USED ON PART A OF SECTION 1**. A GRAPHIC CALCULATOR FROM THE APPROVED LIST **IS REQUIRED FOR PART B OF SECTION 1 AND FOR PART A OF SECTION II** OF THE EXAMINATION.

SECTION 1
Time: 1 hour and 45 minutes
All questions are given equal weight.
Percent of total grade – 50

Part A: 55 minutes, 28 multiple-choice questions
A calculator is NOT allowed.
Part B: 50 minutes, 17 multiple-choice questions
A graphing calculator is required.

CALCULUS BC
SECTION 1, PART A
Time — 55 minutes
Number of questions — 28

A CALCULATOR MAY NOT BE USED ON THIS PART OF THE EXAMINATION.

Directions: Solve each of the following problems, using the available space for scratchwork. After examining the form of the choices, decide which is the best of the choices given and fill in the corresponding oval on the answer sheet. No credit will be given for anything written in the test book. Do not spend too much time on any one problem.

In this test:

(1) Unless otherwise specified, the domain of a function f is assumed to be the set of all real numbers x for which $f(x)$ is a real number.

(2) The inverse of a trigonometric function f may be indicated using the inverse function notation f^{-1} or with the prefix "arc" (e.g, $\sin^{-1} x = \arcsin x$)

AP Calculus BC Test 1

Section 1

1. If $f(x) = \cos(x^2 + 5)$, then $\lim_{h \to 0} \dfrac{f(x+h) - f(x)}{h} =$

   (A) $\sin(x^2 + 5)$ (B) $-\sin(x^2 + 5)$ (C) $-\sin(2x)$ (D) $-2x\sin(2x)$ (E) $-2x\sin(x^2 + 5)$

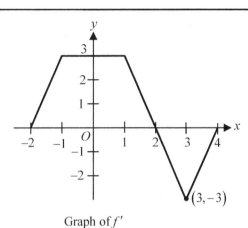

Graph of f'

2. The graph of f', the derivative of f, is shown above. If $f(4) = 8$, then $f(0) =$

   (A) 3   (B) 4.5   (C) 5   (D) 6.5   (E) 12.5

---

3. $\displaystyle\int \dfrac{e^{2\tan x}}{\cos^2 x}\,dx =$

   (A) $\dfrac{1}{2}e^{2\tan x} + C$ (B) $e^{\sec^2 x} + C$ (C) $\dfrac{e^{2\tan x}}{2\cos x \sin x}$ (D) $2e^{2\tan x}\sec^2 x$ (E) $e^{2\tan x} + C$

---

4. $\displaystyle\lim_{x \to 0} \dfrac{\cos x + 2x - e^{2x}}{\cos x - 1} =$

   (A) -3 (B) $-\dfrac{1}{2}$ (C) 0   (D) $\dfrac{1}{2}$ (E) 5

AP Calculus BC Test 1

Section 1

5. If the ratio test is applied to the series $\sum_{n=1}^{\infty} \frac{3^{n+1}}{(n+1)!}$, which of the following implies that the series converges?

   (A) $\lim_{n \to \infty} \frac{3}{n} < 1$

   (B) $\lim_{n \to \infty} \frac{9}{n+1} < 1$

   (C) $\lim_{n \to \infty} \frac{3}{n+1} > 1$

   (D) $\lim_{n \to \infty} \frac{n+1}{3} > 1$

   (E) $\lim_{n \to \infty} \frac{3}{n+2} < 1$

---

6. Let $y = g(x)$ be the solution to the differential equation $\frac{dy}{dx} = \frac{3x}{y}$ with the initial condition $g(1) = 3$. If Euler's method is used, what is the approximation for $g(2)$ starting at $x = 1$ with a step size of 0.5?

   (A) $\frac{5}{3}$ (B) $\frac{51}{14}$ (C) $\frac{29}{7}$ (D) 5.5   (E) 12.5

---

7. If $f(x) = \int_0^{x^2} (t \ln t) dt$, then $\lim_{h \to 0} \frac{f(e+h) - f(e)}{h} =$

   (A) $-\infty$ (B) e^2 (C) $2e^2$ (D) $4e^3$ (E) $6e^3$

AP Calculus BC Test 1 — Section 1

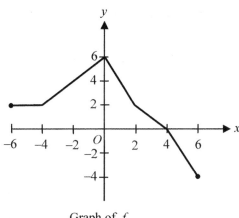

Graph of f

8. The graph of f is shown above. If $g(x) = \int_2^x f(t)\,dt$, which of the following values is the greatest?

   (A) $g(-4)$ (B) $g(-2)$ (C) $g(0)$ (D) $g(4)$ (E) $g(6)$

---

9. In the xy-plane, what is the slope of the line tangent to the graph of $x^2 - xy + y^2 = 3$ at the point $(1, -1)$?

   (A) $-\dfrac{2}{3}$ (B) $-\dfrac{1}{3}$ (C) 0 (D) 1 (E) $\dfrac{4}{3}$

---

10. $\dfrac{d}{dx}\int_x^{x+1}(t^2+t)\,dt =$

    (A) $x^2 + x$ (B) $x^2 + 3x + 1$ (C) $2x + 2$ (D) $x^2 - x + 1$ (E) $\dfrac{x^3}{3} + \dfrac{x^2}{2}$

---

11. If $y = e^{\ln(\cos x)}$, then $\dfrac{dy}{dx} =$

    (A) $-\sin x$ (B) $\sin x$ (C) $e^{\ln(\sin x)}$ (D) $\dfrac{\sin x}{\ln x}$ (E) $\dfrac{\cos x}{\ln x}$

---

12. Let R be the region between the graph of $y = e^x$, the x-axis, and the y-axis for $x \leq \ln 2$. What is the area of R?

    (A) $\dfrac{1}{4}$ (B) $\dfrac{1}{2}$ (C) 1 (D) $\dfrac{e^2}{2}$ (E) e^2

AP Calculus BC Test 1   Section 1

13. What are the all values of p for which $\sum_{n=1}^{\infty} \dfrac{1}{n^{3p}}$ converges?

   (A) $p<0$ (B) $0<p<1$ (C) $p>\dfrac{1}{3}$ (D) $p>1$ (E) $p>3$

14. If $f(x)=\ln(\cos x)$, then $f''\left(\dfrac{\pi}{3}\right)=$

   (A) -4 (B) $-\dfrac{1}{2}$ (C) $\dfrac{1}{2}$ (D) 2 (E) 4

15. If the line tangent to the graph of the function f at the point $(2, 5)$ passes through the point $(-1, 2)$, which of the following could be f?

   (A) $y=-\dfrac{1}{2}x^2+3x-1$

   (B) $y=-\dfrac{1}{2}x^2+3x+1$

   (C) $y=\dfrac{1}{2}x^2+3x-1$

   (D) $y=\dfrac{1}{2}x^2+3x-3$

   (E) $y=\dfrac{1}{2}x^2+3x-2$

16. If the arc length of the graph of $y=f(x)$ from 0 to 5 is given by $\int_0^5 \sqrt{1+4x^6}\,dx$, then which of the following could be the value of $f'(-2)$?

   (A) -20 (B) -16 (C) -3 (D) 3 (E) 5

17. The parametric equations of a curve are given by $x(t)=t^3-3t-3$ and $y(t)=\dfrac{t^2}{2}-5$. What is the slope of the line tangent to the graph of the curve at the point $(-1,-3)$?

   (A) $-\dfrac{3}{2}$ (B) $-\dfrac{1}{3}$ (C) 1 (D) $\dfrac{2}{9}$ (E) $\dfrac{3}{2}$

AP Calculus BC Test 1

Section 1

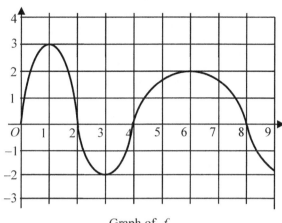

Graph of f

18. The graph of the function f is shown in the figure above. If $g(x) = 2x + \int_0^{3x} f(t)\,dt$, what is the value of $g'(2)$?

(A) 3
(B) 4
(C) 6
(D) 8
(E) 10

---

19. If a function f has Maclaurin series given by $f(x) = x^2 + \dfrac{x^3}{2!} + \dfrac{x^4}{3!} + \dfrac{x^5}{4!} + \cdots + \dfrac{x^{n+1}}{n!} + \cdots$, which of the following is an expression for $f(x)$?

(A) $e^x - 1$
(B) $x \sin x - 1$
(C) $x^2 \cos x - x$
(D) $xe^x - 1$
(E) $xe^x - x$

AP Calculus BC Test 1

Section 1

20. Let f be differentiable function with the following properties.

    (i) $f'(x) > 0$ on the interval $[0, 4]$.
    (ii) $f''(x) > 0$ for all x.

Which of the following has the greatest value?

(A) Left Riemann sum approximation of $\int_0^4 f(x)\,dx$ with 10 subintervals of equal length

(B) Right Riemann sum approximation of $\int_0^4 f(x)\,dx$ with 10 subintervals of equal length

(C) Midpoint Riemann sum approximation of $\int_0^4 f(x)\,dx$ with 10 subintervals of equal length

(D) Trapezoidal sum approximation of $\int_0^4 f(x)\,dx$ with 10 subintervals of equal length

(E) $\int_0^4 f(x)\,dx$

---

21. If $f(x) = e^{2\ln(x+1)}$, then $f'(x) =$

(A) $\dfrac{(x+1)^2}{\ln(x+1)}$ (B) $2\ln(x+1)$ (C) $\dfrac{2}{x+1}$ (D) $2x+2$ (E) $e^{2\ln(x+1)} \cdot \dfrac{1}{2\ln(x+1)}$

---

22. The graph of the function P is a logistic curve that satisfy $\dfrac{dP}{dt} = 0.002P(500 - P)$, where t is the time in years. If $P(0) = 200$, then $\lim\limits_{t \to \infty} P(t) =$

(A) 100   (B) 200   (C) 500   (D) 10000   (E) 25000

---

23. If the infinite series $\sum\limits_{n=1}^{\infty} \dfrac{n^2}{n^p + 2}$ converges, which of the following is all values of p?

(A) $0 < p < 1$ (B) $0 < p \leq 2$ (C) $p > 2$ (D) $p > 3$ (E) $p \geq 3$

AP Calculus BC Test 1

Section 1

24. Which of the following series converges?

I. $\sum_{n=1}^{\infty}\left(\frac{2^{n+1}}{3^n}\right)$

II. $\sum_{n=1}^{\infty}\frac{1}{\sqrt{n}}$

III. $\sum_{n=1}^{\infty}\frac{2e^n}{e^n+3}$

(A) I only (B) I and II only (C) I and III only (D) II and III only (E) I, II, and III

---

25. $\int x^2 \cos(2x)\,dx =$

(A) $\frac{x^2}{2}\sin(2x) + \frac{x}{2}\cos(2x) - \frac{1}{4}\sin(2x) + C$

(B) $\frac{x^2}{2}\sin(2x) - \frac{x}{2}\cos(2x) - \frac{1}{4}\sin(2x) + C$

(C) $x^2\sin(2x) + 2x\sin(2x) + C$

(D) $x^2\cos(2x) - x\cos(2x) + C$

(E) $-\frac{x^2}{2}\cos(2x) - \frac{x}{2}\sin(2x) + x\cos(2x) + C$

---

26. $\int \frac{2x}{x^2 - 5x + 6}\,dx =$

(A) $\ln|x^2 - 5x + 1| + C$

(B) $6\ln|x-3| - 4\ln|x-2| + C$

(C) $4\ln|x-2| - 6\ln|x-3| + C$

(D) $2\ln|x-2| - 3\ln|x-3| + C$

(E) $2\ln|x-2| + 3\ln|x-3| + C$

Dr. John Chung's AP Calculus AB/BC

AP Calculus BC Test 1

Section 1

27. If $g(x) = \int_0^{\sqrt{x}} te^{t^2}\, dt$, then which of the following is the equation of the line tangent to the graph of g at point $(\ln 4, g(\ln 4))$?

(A) $y - \dfrac{3}{2} = 2(x - \ln 4)$

(B) $y - 5 = \dfrac{2}{3}(x - \ln 4)$

(C) $y + 3 = 2(x - \ln 4)$

(D) $y - \dfrac{2}{3} = -2(x - \ln 4)$

(E) $y - 3 = 2(x - \ln 4)$

---

28. What is the coefficient of x^3 in the Taylor series for $\left(\dfrac{1}{x+1}\right)^2$ about $x = 0$?

(A) -6

(B) -4

(C) 2

(D) 4

(E) 6

END OF PART A OF SECTION 1

AP Calculus BC Test 1

Section 1

CALCULUS BC
SECTION 1, PART B
Time — 50minutes
Number of questions — 17

A GRAPHING CALCULATOR IS REQUIRED FOR SOME QUESTIONS ON THIS PART OF THE EXAMINATION.

Directions: Solve each of the following problems, using the available space for scratchwork. After examining the form of the choices, decide which is the best of the choices given and fill in the corresponding oval on the answer sheet. No credit will be given for anything written in the test book. Do not spend too much time on any one problem.

In this test:

(1) The exact numerical value of the correct answer does not always appear among the choices given. When this happens, select from among the choices the number that best approximates the exact numerical value.

(2) Unless otherwise specified, the domain of a function f is assumed to be the set of all real numbers x for which $f(x)$ is a real number.

(3) The inverse of a trigonometric function f may be indicated using the inverse function notation f^{-1} or with the prefix "arc" (e.g, $\sin^{-1} x = \arcsin x$)

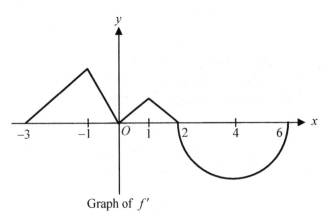

Graph of f'

76. The graph of the function f' on the closed interval $-3 \leq x \leq 6$ is shown above. How many points of inflection does the graph of f have on the interval $-3 < x < 6$?

(A) 1   (B) 2   (C) 3   (D) 4   (E) 5

---

77. If $f(x) = 3x - 5x^2 + 2x^3 - 7x^4 + \cdots$ is the Taylor series for the function f about $x = 0$, what is the value of $f'''(0) =$

(A) 2   (B) 6   (C) 12   (D) −2   (E) −12

---

78. The volume of a cube is increasing at a constant rate of 6 cubic inches per second. What is the rate of increase in the surface area of the cube when the volume of the cube is 50 cubic inches?

(A) 0.147 square inches/sec
(B) 1.125 square inches/sec
(C) 2.547 square inches/sec
(D) 3.012 square inches/sec
(E) 6.515 square inches/sec

---

79. Let f be the function given by $f(x) = \dfrac{(x-3)(x^2-1)}{x-k}$, where k is a constant. For what values of k is the function f continuous for all real numbers x?

(A) 3 only
(B) 1 and −1 only
(C) 3, 1, and −1 only
(D) All real numbers
(E) None

AP Calculus BC Test 1

Section 1

80. If $\frac{dy}{dx} = (1 - \ln y)^2$, then $\frac{d^2y}{dx^2} =$

(A) $-\frac{1}{y}$

(B) $-2(1 - \ln y)$

(C) $\frac{-2(1 - \ln y)}{y}$

(D) $\frac{-2(1 - \ln y)^2}{y}$

(E) $\frac{-2(1 - \ln y)^3}{y}$

---

81. Let f be a differentiable function defined by $f(x) = 2g(x) + 5$ and $g(0) = -10$. If $\int_0^3 g'(x)\,dx = 3$, then $f(3) =$

(A) -9 (B) -2 (C) 5 (D) 11 (E) 15

---

82. The Maclaurin series for e^x is given by $\sum_{n=0}^{\infty} \frac{x^n}{n!}$. Let $p(x)$ be the function given by the sum of the first three nonzero terms of the series. What is the maximum value of $|e^x - p(x)|$ for $-0.3 \le x \le 0.5$?

(A) 0.024 (B) 0.039 (C) 0.158 (D) 0.536 (E) 0.631

AP Calculus BC Test 1

Section 1

83. Maclaurin series for the function f is given by $f(x) = \sum_{n=0}^{\infty} \frac{(-1)^n (4x)^n}{n+1}$. Find the interval of convergence for f?

(A) $-4 \leq x \leq 4$

(B) $-\frac{1}{4} < x < \frac{1}{4}$

(C) $-\frac{1}{4} \leq x < \frac{1}{4}$

(D) $-\frac{1}{4} < x \leq \frac{1}{4}$

(E) $-\frac{1}{4} \leq x \leq \frac{1}{4}$

x	0	4	6	7	10
$f(x)$	10	15	20	16	10

84. The function is continuous on the closed interval $[0, 10]$ and has selected values in the table above. What is the trapezoidal approximation of $\int_0^{10} f(x)\, dx$ using the four subintervals?

(A) 138   (B) 142   (C) 145   (D) 148   (E) 156

85. Which of the following is an equation of the line tangent to the graph of $f(x) = x^3 + x - 1$ at the point where $f''(x) = 6$?

(A) $y = 4x - 3$
(B) $y = 4x - 3.5$
(C) $y = 4x + 0.5$
(D) $y = 4x + 1.4$
(E) $y = 8x - 2.5$

AP Calculus BC Test 1

86. A particle moves on x-axis so that its position at any time $t \geq 0$ is given by $x(t) = 6t^2 - 24t + 22$. Find all values of t for which the speed of the particle is decreasing.

(A) $0 < t < 2$
(B) $2 < t < 3$
(C) $3 < t < 4$
(D) $4 < t < 6$
(E) $6 < t < \infty$

---

87. $\lim\limits_{x \to \infty} \dfrac{\int_0^x \sqrt{t^2 + 5t + 3}\, dt}{x^2} =$

(A) $\dfrac{1}{2}$ (B) 2 (C) $\dfrac{5}{2}$ (D) ∞ (E) undefined

---

88. Let R be the region enclosed by the graph of $y = \sqrt{x-1}$, the line $x = 4$, and the x-axis. What is the volume of the solid generated by evolving region R about y-axis?

(A) 9.699
(B) 18.257
(C) 22.975
(D) 35.125
(E) 60.944

---

89. Let f be the function given by $f(x) = \int_0^{x^2} \sqrt{t} \sin\left(1 + \dfrac{t}{2}\right) dt$ for $0 \leq x \leq 3$. On which of the following intervals is f decreasing?

(A) $0 \leq x \leq 1$
(B) $0 \leq x \leq 1.694$
(C) $0.894 \leq x \leq 1.694$
(D) $1.511 \leq x \leq 2.927$
(E) $2.297 \leq x \leq 3$

AP Calculus BC Test 1

Section 1

90. Let f be the function given by $f(x) = \dfrac{1}{1-2x}$ with $f(0)=1$, $f'(0)=2$, $f''(0)=8$, and $f'''(0)=48$. Which of the following is the third degree Taylor polynomial for $g(x) = \dfrac{1}{1-2x}$ about $x=0$?

(A) $1 - 2x + 4x^2 - 8x^3$
(B) $1 + x + 4x^2 + 8x^3$
(C) $1 - 2x + 8x^2 - 48x^3$
(D) $1 + x + x^2 + x^3$
(E) $1 + 2x + 4x^2 + 8x^3$

---

91. Let f be a continuous function for $x > 0$ and let g be the function given by $g(x) = \int_1^x f(t)\,dt$. If $f'(1) = -3$ and $g'(1) = 3$, which of the following is the equation of the line tangent to the graph f at point $(1, f(1))$?

(A) $y = -3x + 6$
(B) $y = -x + 4$
(C) $y = 2x + 1$
(D) $y = 3x + 2$
(E) $y = 4x - 1$

---

92. Let f and g be the functions defined by $f(x) = \dfrac{1}{2}x + 2$ and $g(x) = -x^2 - 3$. If $K(x) = g(f(x) + x)$, what is the value of $K'(-2)$?

(A) $-\dfrac{3}{2}$
(B) $-\dfrac{1}{2}$
(C) $\dfrac{1}{2}$
(D) $\dfrac{3}{2}$
(E) 3

END OF SECTION 1

AP Calculus BC Test 1   Section II

CALCULUS BC

SECTION II
Time: 1 hour and 30 minutes
Percent of total grade – 50

Part A: 30 minutes, 2 problems
Part B: 60 minutes, 4 problems

Part A: A graphing calculator is required for some problems or parts of problems.
Part B: No calculator is allowed for these problems.

GENERAL INSTRUCTIONS FOR SECTION II PART A AND PART B

For each part of section II, you may wish to look over the problems before starting to work on them, since it is not expected that everyone will be able to complete all parts of all problems. All problems are given equal weight, but the parts of a particular problem are not necessary given equal weight.

- YOU SHOULD WRITE ALL WORK FOR EACH PART OF EACH PROBLEM IN THE SPACE PROVIDED FOR THAT PART IN THE TEST BOOKLET. Be sure to write clearly and legibly. If you make an error, you may save time by crossing it out rather than trying to erase it. Erased or cross-out work will not be graded.
- Show all work. Clearly label any functions, graphs, tables, or other object that you use. You will be graded on the correctness and completeness of your methods as well as your answers. Answers without supporting work may not receive credit.
- Justifications required that you give mathematical (noncalculator) reasons
- Your work must be expressed in standard mathematical notation rather than calculator syntax. For example, $\int_{1}^{5} x^2 dx$ may not be written as $\text{fnInt}(X^2, X, 1, 5)$
- Unless otherwise specified, answers (numeric or algebraic) need not be simplified.
- If you use decimal approximations in calculations, you will be graded on accuracy. Unless otherwise specified, your final answers should be accurate to three places after the decimal point.
- Unless otherwise specified, the domain of a function f is assumed to be the set of all real numbers x for which $f(x)$ is a real number.

AP Calculus BC Test 1

Section II

CALCULUS BC
SECTION II, PART A
Time — 30 minutes
Number of questions —2

A Graphing calculator is required for some problems or parts of problems.

Question 1

An object is moving along a curve in the *xy*-plane has position $(x(t), y(t))$ at time t with $\frac{dx}{dt} = \frac{\cos(t^2)}{e^t}$ and $\frac{dy}{dt} = 2t\sin(t)$ for $t \geq 0$. At time $t = 2$, the object is at position $(2, 4)$.

(a) Write an equation for the line tangent to the curve at (2, 4).

(b) Find the speed of the object at time $t = 2$. Is the horizontal movement of the object to the right or to the left at time $t = 2$. Explain your reasoning.

(c) Write an expression for the total distance traveled by the object from time $t = 2$ to $t = 5$. Do not evaluate.

(d) Find the value of $x(5)$.

(e) Find the acceleration vector of the object at time $t = 2$.

AP Calculus BC Test 1

Section II

Question 2

t (seconds)	0	4	8	12	16	20	24	28	32
$v(t)$ (meters per second)	0	16	30	40	60	65	60	55	30

A table of values for the velocity, $v(t)$ in meters per second, of an object at 4 second intervals of time t is given above.

(a) During which intervals of time is the acceleration of the object positive? Justify your answer.

(b) Find the value of $\int_0^{32} v'(t)\,dt$. What is the meaning of $\int_0^{32} v'(t)\,dt$? Find the average acceleration of the object over the interval $0 \leq t \leq 32$.

(c) Approximate $\int_0^{32} v(t)\,dt$ with a Riemann sum, using the midpoints of four subintervals of equal length. What is the meaning of $\int_0^{32} v(t)\,dt$?

(d) For $t \geq 32$, the acceleration a of the object is modeled by $a(t) = \dfrac{-\ln(100t^3)}{t}$. Based on the model, what is the velocity of the object at time $t = 36$?

END OF PART A OF SECTION II

AP Calculus BC Test 1

Section II

**CALCULUS BC
SECTION II, PART B
Time — 60 minutes
Number of questions — 4**

No calculator is required for these problems.

Question 3

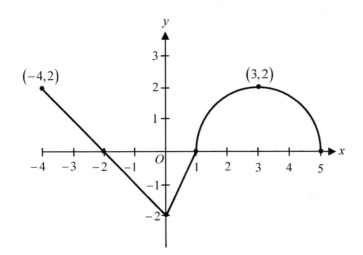

The graph of a function f defined on the closed interval $-4 \leq x \leq 5$ consists of a semicircle and two line segments as shown above. Let g be the function given by $g(x) = \int_0^x f(t)\,dt$.

a) Find the interval for which the graph of g is increasing.

b) Find $g(-2)$, $g(1)$, and $g(3)$.

c) Find all values of x on the open interval $(-4, 5)$ at which the graph of g have a relative maximum.

d) Find the x-coordinate of each point of inflection of the graph of g on the open interval $(-4, 5)$.

e) Write an equation for the line tangent to the graph of g at $x = 3$.

AP Calculus BC Test 1 *Section II*

Question 4

A lake is stocked with 1000 fish. If the population, at time t months, increases according to the logistic curve

$$\frac{dP}{dt} = \frac{1}{10^4} P(4000 - P).$$

Let $y = P(t)$ be the particular solution to the differential equation above with initial condition $P(0) = 1000$.

(a) Find the particular solution $y = P(t)$ with initial condition $P(0) = 1000$.

(b) Find $\lim_{t \to \infty} P(t)$ and $\lim_{t \to \infty} P'(t)$.

(c) For what value of y does the graph of P have a point of inflection?

(d) For what value of t is the population growing the fastest?

(e) Sketch possible solution curve through the point of inflection.

AP Calculus BC Test 1

Section II

Question 5

Let f be the function defined by $f(x) = e^{-x} + \cos x$.

(a) Write the fourth-degree Taylor polynomial for f about $x = 0$ and use it to approximate $f(1)$.

(b) Find the value of $f'(0)$, $f''(0)$, and $f^{(4)}(0)$.

(c) Let $P_4(x)$ be the fourth-degree Taylor polynomial for f about $x = 0$. If the fifth derivative of f satisfies the inequality $\left|f^{(5)}(x)\right| \leq 1.2$ for all x in the closed interval $\left[0, \frac{1}{2}\right]$, use the Lagrange error bound to prove that $\left|P_4\left(\frac{1}{2}\right) - f\left(\frac{1}{2}\right)\right| < \frac{1}{3000}$.

Question 6

Let f be the function whose derivative is given by $f'(x) = \dfrac{x^2 - 3}{x}$ with $f(2) = 5$.

(a) Write an equation of the line tangent to the graph of f at $x = 2$.

(b) Use the equation of the tangent line in part (a) to approximate $f(3)$.

(c) Use Euler's method, starting at $x = 2$ with two steps of equal size, to approximate $f(3)$.

(d) Use $f''(x)$ to prove that the approximation in part (c) is less than $f(3)$.

(e) Find $f(3)$.

END OF EXAM

AP Calculus BC Test 1 Answers

SECTION 1, PART A

1. **E**

   Since $\lim\limits_{h \to 0} \dfrac{f(x+h)-f(x)}{h} = f'(x)$, $f(x) = \cos(x^2+5)$ → $f'(x) = -2x\sin(x^2+5)$

2. **D**

   By the Fundamental Theorem of Calculus, $f(4) = f(0) + \int_0^4 f'(x)dx$.

   $\int_0^4 f'(t)dt = \dfrac{(1+2)\cdot 3}{2} - \dfrac{2\cdot 3}{2} = \dfrac{3}{2} = 1.5$ and $f(4) = 8$

   Therefore, $8 = f(0) + 1.5$ and $f(0) = 6.5$.

   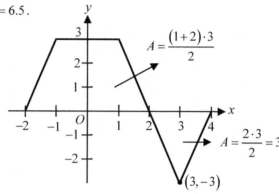

   Graph of f'

3. **A**

   $\int \dfrac{e^{2\tan x}}{\cos^2 x} dx = \int \dfrac{1}{\cos^2 x}\left(e^{2\tan x}\right)dx = \int \sec^2 x \left(e^{2\tan x}\right)dx$

   Use u-substitution.

   $\begin{cases} u = \tan x \\ du = \sec^2 x \, dx \end{cases}$ $\therefore \int \sec^2 x \left(e^{2\tan x}\right)dx = \int e^{2u} du = \dfrac{1}{2}e^{2u} + C$ → $\dfrac{1}{2}e^{2\tan x} + C$

4. **E**

   Type: $\dfrac{0}{0}$ → : We apply L'Hôpital's rule two times.

   $\lim\limits_{x\to 0} \dfrac{\cos x + 2x - e^{2x}}{\cos x - 1} = \lim\limits_{x\to 0} \dfrac{-\sin x + 2 - 2e^{2x}}{-\sin x} = \lim\limits_{x\to 0} \dfrac{-\cos x - 4e^{2x}}{-\cos x} = \dfrac{-1-4}{-1} = 5$

5. **E**

   We apply the ratio test to get the convergence of a series.

   Ratio $= \lim\limits_{n\to\infty} \left|\dfrac{3^{n+2}}{(n+2)!} \cdot \dfrac{(n+1)!}{3^{n+1}}\right| = \lim\limits_{n\to\infty} \left|\dfrac{3}{n+2}\right| < 1$

   If $\lim\limits_{n\to\infty} \dfrac{3}{n+2} < 1$, the series converges.

6. **C**

   $g(1.5) = g(1) + (0.5)\cdot g'(1)$ and $g'(1) = \dfrac{3(1)}{3} = 1$ at point $(1,3)$

   Therefore, $g(1.5) = 3 + 0.5(1) = 3.5$.

AP Calculus BC Test 1 Answers

$g(2) = g(1.5) + 0.5 \cdot g'(1.5)$ and $g'(1.5) = \dfrac{3(1.5)}{3.5} = \dfrac{4.5}{3.5} = \dfrac{9}{7}$ at point $(1.5, 3.5)$

Therefore, $g(2) = g(1.5) + (0.5)g'(1.5) = 3.5 + (0.5)\dfrac{9}{7} = \dfrac{7}{2} + \dfrac{9}{14} = \dfrac{58}{14} = \dfrac{29}{7}$.

7. **D**

We know that $\displaystyle\lim_{h \to 0} \dfrac{f(e+h) - f(e)}{h} = f'(e)$.

$f(x) = \displaystyle\int_0^{x^2} (t \ln t)\, dt \;\Rightarrow\; f'(x) = \dfrac{d}{dx}\left(\int_0^{x^2}(t \ln t)\, dt\right) = x^2 \ln x^2 \left(x^2\right)' = 2x^3 \ln x^2$

Therefore, $f'(e) = 2e^3 \ln e^2 = 4e^3 \ln e = 4e^3$.

8. **D**

The area of each section is shown on the graph.

(A) $g(-4) = -8 + (-10) + (-6) = -24$

(B) $g(-2) = -8 + (-10) = -18$

(C) $g(0) = -8$

(D) $g(4) = 2$

(E) $g(6) = 2 + (-4) = -2$

Therefore, $g(4)$ has the greatest value.

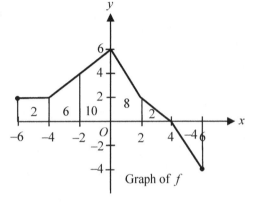

9. **D**

The slope of the tangent line is $\left.\dfrac{dy}{dx}\right|_{(1,-1)}$.

Apply implicit differentiation.

$x^2 - xy + y^2 = 3 \;\Rightarrow\; 2x - (y + xy') + 2yy' = 0 \;\Rightarrow\; y' = \dfrac{dy}{dx} = \dfrac{2x - y}{x - 2y}$

Therefore, $\left.\dfrac{dy}{dx}\right|_{(1,-1)} = \dfrac{2 - (-1)}{1 - 2(-1)} = \dfrac{3}{3} = 1$.

10. **C**

Since $\dfrac{d}{dx}\displaystyle\int_{k(x)}^{g(x)} f(t)\, dt = f(g) \cdot g' - f(k) k'$,

$\dfrac{d}{dx}\displaystyle\int_x^{x+1}(t^2 + t)\, dt = \left[(x+1)^2 + (x+1)\right] - \left[x^2 + x\right] = (x^2 + 3x + 2) - (x^2 + x) = 2x + 2$.

11. **A**

Since $y = e^{\ln(\cos x)} = \cos x$, then $\dfrac{dy}{dx} = -\sin x$.

12. **C**

$\text{Area} = \displaystyle\int_0^{\ln 2} e^x\, dx = \left[e^x\right]_0^{\ln 2} = e^{\ln 2} - e^0 = 2 - 1 = 1$

13. **C**

$\displaystyle\sum_{n=1}^{\infty} \dfrac{1}{n^{3p}}$: p-series test $\;\Rightarrow\; 3p > 1 \;\Rightarrow\; p > \dfrac{1}{3}$ for convergence.

AP Calculus BC Test 1 Answers

14. A

$$f'(x) = \frac{-\sin x}{\cos x} = -\tan x \text{ and } f''(x) = -\sec^2 x \Rightarrow f''\left(\frac{\pi}{3}\right) = -\frac{1}{\cos^2(\pi/3)} = -4$$

15. B

The slope of the tangent is $m = \dfrac{5-2}{2-(-1)} = 1$.

Therefore, the graph of f must pass through point $(2,5)$ and $f'(2)$ is equal to 1.

(B) $f(2) = -\dfrac{1}{2}(2)^2 + 3(2) + 1 = 5$ and $f'(x) = -x + 3 \Rightarrow f'(2) = -2 + 3 = 1$

16. B

The arc length of the graph of $y = f(x)$ from 0 to 5 is $\displaystyle\int_0^5 \sqrt{1 + [f'(x)]^2}\, dx = \int_0^5 \sqrt{1 + 4x^6}\, dx$.

Therefore, $[f'(x)]^2 = 4x^6$ or $f'(x) = 2x^3 \Rightarrow f'(-2) = -16$.

17. D

The slope of the tangent is $\left.\dfrac{dy}{dx}\right|_{(-1,-3)}$

When $y = -3$, the value of t can be obtained from the equation $y(t) = \dfrac{t^2}{2} - 5$.

$\dfrac{t^2}{2} - 5 = -3 \Rightarrow t^2 = 4 \Rightarrow t = 2$ We can see that $x(2) = 2^3 - 3(2) - 3 = -1$.

Therefore, $\left.\dfrac{dy}{dx}\right|_{(-1,-3)} = \left.\dfrac{dy}{dx}\right|_{t=2} = \left.\dfrac{dy/dt}{dx/dt}\right|_{t=2} = \left.\dfrac{t}{3t^2 - 3}\right|_{t=2} = \dfrac{2}{9}$.

18. D

$g'(x) = 2 + \dfrac{d}{dx}\displaystyle\int_0^{3x} f(t)\, dt$

Since $g'(x) = 2 + f(3x)(3x)' = 2 + 3f(3x)$, then $g'(2) = 2 + 3f(6) = 2 + 3(2) = 8$.

19. E

We remember that $e^x = 1 + x + \dfrac{x^2}{2!} + \dfrac{x^3}{3!} + \dfrac{x^4}{4!} + \cdots + \dfrac{x^n}{n!} + \cdots$.

Multiply both sides of the equation by x.

$xe^x = x + x^2 + \dfrac{x^3}{2!} + \dfrac{x^4}{3!} + \dfrac{x^5}{4!} + \cdots + \dfrac{x^{n+1}}{n!} + \cdots$

Therefore, $xe^x - x = x^2 + \dfrac{x^3}{2!} + \dfrac{x^4}{3!} + \dfrac{x^5}{4!} + \cdots + \dfrac{x^{n+1}}{n!} + \cdots \Rightarrow f(x) = xe^x - x$.

20. B

Right Riemann sum approximation of $\displaystyle\int_0^4 f(x)\, dx$ is the greatest because the graph of f is increasing and concave up on the closed interval $[0, 4]$.

21. D

Since $f(x) = e^{\ln(x+1)^2} = (x+1)^2$, $f'(x) = 2(x+1) \Rightarrow f'(x) = 2x + 2$.

AP Calculus BC Test 1 Answers

22. C

$$\frac{dP}{dt} = 0.002P(500-P) \rightarrow \frac{dP}{dt} = P\left(1 - \frac{P}{500}\right)$$

$k = 1$, $m = 500$

Since carrying capacity is 500, then $\lim_{t \to \infty} P(t) = 500$.

Or

Using the values of k and m, we can build up the equation $P(t) = \dfrac{500}{1 + Ae^{-t}}$.

Therefore, $\lim_{t \to \infty} P(t) = \lim_{t \to \infty} \dfrac{500}{1 + Ae^{-t}} = 500$. You can get the value of A from the initial condition $P(0) = 200$. $A = 1.5$.

23. D

$$\sum_{n=1}^{\infty} \frac{n^2}{n^p + 2} < \sum_{n=1}^{\infty} \frac{n^2}{n^p} = \sum_{n=1}^{\infty} \frac{1}{n^{p-2}}$$ Comparison Test and p-Series Test, the series converges if

$p - 2 > 1 \rightarrow p > 3$.

24. A

I. $\sum_{n=1}^{\infty} \left(\dfrac{2^{n+1}}{3^n}\right)$: Ratio test or Geometric Series test $\rightarrow$ Geometric convergence

$\lim_{n \to \infty} \left|\dfrac{2^{n+2}}{3^{n+1}} \cdot \dfrac{3^n}{2^{n+1}}\right| = \lim_{n \to \infty} \left|\dfrac{2}{3}\right| < 1$ or $\sum_{n=1}^{\infty}\left(\dfrac{2^{n+1}}{3^n}\right) = 2\sum_{n=1}^{\infty}\left(\dfrac{2}{3}\right)^n \rightarrow$ Geometric convergence

II. $\sum_{n=1}^{\infty} \dfrac{1}{\sqrt{n}} = \sum_{n=1}^{\infty} \dfrac{1}{n^{\frac{1}{2}}}$: p-Series Test $\rightarrow$ $p < 1$: Divergence

III. $\sum_{n=1}^{\infty} \dfrac{2e^n}{e^n + 3}$: nth term divergence test: $\lim_{n \to \infty} \dfrac{2e^n}{e^n + 3} = 2$: Divergence

25. A

Use Tabular Integration by Part.

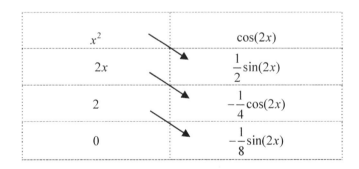

$\int x^2 \cos(2x)\, dx = \dfrac{x^2}{2}\sin(2x) + \dfrac{x}{2}\cos(2x) - \dfrac{1}{4}\sin(2x) + C$

AP Calculus BC Test 1 Answers

26. B

Partial fraction: $\dfrac{2x}{x^2-5x+6} = \dfrac{2x}{(x-3)(x-2)} = \dfrac{A}{(x-3)} + \dfrac{B}{(x-2)} = \dfrac{A(x-2)+B(x-3)}{(x-3)(x-2)}$

At $x=3$, $A=6$
At $x=2$, $B=-4$

$\displaystyle\int \dfrac{2x}{x^2-5x+6}\,dx = \int\left(\dfrac{6}{x-3} - \dfrac{4}{x-2}\right)dx = 6\ln|x-3| - 4\ln|x-2| + C$

27. D

Since $g(x) = \displaystyle\int_0^{\sqrt{x}} te^{t^2}\,dt = \left[\dfrac{1}{2}e^{t^2}\right]_0^{\sqrt{x}} = \dfrac{1}{2}e^x - \dfrac{1}{2}$, then $g'(x) = \dfrac{1}{2}e^x$ and $g'(\ln 4) = \dfrac{1}{2}e^{\ln 4} = 2$.

The value of $g(\ln 4) = \dfrac{1}{2}e^{\ln 4} - \dfrac{1}{2} = \dfrac{4}{2} - \dfrac{1}{2} = \dfrac{3}{2}$.

The equation of the tangent is $y - \dfrac{3}{2} = 2(x - \ln 4)$.

28. B

Since $f(x) = \dfrac{1}{x+2} = (x+2)^{-2}$, $f'(x) = -2(x+1)^{-3}$, $f''(x) = 6(x+1)^{-4}$ and $f'''(x) = -24(x+1)^{-5}$.

We know that $f'''(0) = -24$.

From the Taylor series, $\left(\dfrac{1}{x+1}\right)^2 = f(0) + f'(0)\cdot x + \dfrac{f''(0)}{2!}x^2 + \dfrac{f'''(0)}{3!}x^3 + \cdots$

Therefore, the coefficient of x^3 is $\dfrac{f'''(0)}{3!} = \dfrac{-24}{6} = -4$.

Or

Remember: Taylor series for $\dfrac{1}{x+1}$ about $x=0$ is

If $\dfrac{1}{x+1} = 1 - x + x^2 - x^3 + x^4 + \cdots$,

then, $\left(\dfrac{1}{x+1}\right)^2 = (1 - x + x^2 - x^3 + x^4 - x^5 + \cdots)(1 - x + x^2 - x^3 + x^4 - x^5 + \cdots)$

Search x^3 term. Therefore, $2(1)(-x^3) + 2(-x)(x^2) = -4x^3$.

AP Calculus BC Test 1 Answers

SECTION 1, PART B

76. D

The graph of f has a point of inflection at each of these values $x = -1, 0, x = 1$, and $x = 4$ because f'' changes sign at these values.

The possible graph of f is as follows.

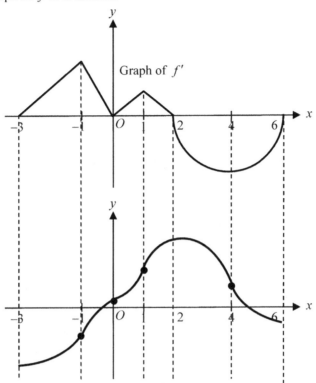

77. C

Since $\dfrac{f'''(0)}{3!}x^3 = 2x^3$, $\dfrac{f'''(0)}{3!} = 2$ or $f'''(0) = 12$.

78. E

Define: $x =$ Length of the cube

The volume of the cube is $V = x^3$ and the surface area is $A = 6x^2$. We know that $x = V^{\frac{1}{3}}$.

$$A = 6x^2 = 6 \cdot \left(V^{\frac{1}{3}}\right)^2 = 6V^{2/3} \Rightarrow \dfrac{dA}{dt} = 4V^{-\frac{1}{3}} \cdot \dfrac{dV}{dt} \text{ and } \dfrac{dV}{dt} = 6 \text{ is given.}$$

Therefore, $\dfrac{dA}{dt} = 4(50)^{-\frac{1}{3}}(6) = 6.5146 \approx 6.515$.

79. E

$$f(x) = \dfrac{(x-3)(x^2-1)}{x-k} = \dfrac{(x-3)(x+1)(x-1)}{(x-k)}$$

If $k = 3, -1, 1 \Rightarrow$ then the graph of f has a POD. (Point of discontinuity)

If $k \neq 3, -1, 1 \Rightarrow$ then the graph of f has an essential discontinuity at $x = k$.

AP Calculus BC Test 1 Answers

80. E

$$\frac{dy}{dx} = (1-\ln y)^2, \text{ then } \frac{d^2y}{dx^2} = 2(1-\ln y)\left(-\frac{y'}{y}\right) = 2(1-\ln y)\left(-\frac{(1-\ln y)^2}{y}\right) = \frac{-2(1-\ln y)^3}{y}$$

81. A

$f(x) = 2g(x) + 5$ and $f(3) = 2g(3) + 5$ with $g(0) = -10$. We need to find $g(3)$.

$$g(3) = g(0) + \int_0^3 g'(x)dt = -10 + 3 = -7$$

Therefore, $g(3) = 2(-7) + 5 = -9$.

82. A

The first three nonzero terms of the series is $1 + x + \frac{x^2}{2}$.

$$|e^x - p(x)| = \left|e^x - 1 - x - \frac{x^2}{2}\right|$$

Draw the graph using your graphing calculator: $y = e^x - 1 - x - \frac{x^2}{2}$

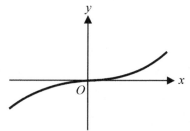

Graph of f is increasing on the interval $[-0.3, 0.5]$.

At $x = -0.3 \rightarrow y = -0.004 \Rightarrow |-0.004| = 0.004$

At $x = 0.5 \rightarrow 0.023721 \approx 0.024 \rightarrow$ maximum actual error

Remember. It is not asking Lagrange error bound. The Lagrange error must be larger than the actual error.

83. D

Ratio Test: $\displaystyle\lim_{n\to 0}\left|\frac{(-1)^{n+1}(4x)^{n+1}}{n+2} \times \frac{n+1}{(-1)^n(4x)^n}\right| = \lim_{n\to 0}\left|\frac{(-1)(4x)(n+1)}{n+2}\right| = |4x| < 1$

$|x| < \frac{1}{4} \rightarrow$ Interval of convergence is $-\frac{1}{4} < x < \frac{1}{4}$

We apply end point convergence test:

At $x = -\frac{1}{4} \rightarrow \displaystyle\sum_{n=0}^{\infty}\frac{(-1)^n(-1)^n}{n+1} = \sum_{n=0}^{\infty}\frac{1}{n+1} = \frac{1}{1} + \frac{1}{2} + \frac{1}{3} + \cdots$: Harmonic series $\rightarrow$ Divergence

At $x = \frac{1}{4} \rightarrow \displaystyle\sum_{n=0}^{\infty}\frac{(-1)^n(1)^n}{n+1} = \sum_{n=0}^{\infty}\frac{(-1)^n}{n+1} = \frac{1}{1} - \frac{1}{2} + \frac{1}{3} - $: Alternate Harmonic series $\rightarrow$ Convergence

Therefore, the interval of convergence is $-\frac{1}{4} < x \leq \frac{1}{4}$.

AP Calculus BC Test 1 Answers

84. **B**

We know that the area of a trapezoid is $A = \dfrac{(b_1 + b_2)h}{2}$.

The area of the trapezoids is $\dfrac{1}{2}\left[(10+15)\cdot 4 + (20+15)\cdot 2 + (16+20)\cdot 1 + (10+16)\cdot 3\right] = 142$.

85. **A**

Since $f'(x) = 3x^2 + 1$ and $f''(x) = 6x = 6$, the value of x is 1.

And also $f(1) = 1^3 + 1 - 1 = 1$ and $f'(1) = 4$ which is the slope of the tangent.

Therefore, the equation of the tangent is $y - 1 = 4(x-1)$ or $y = 4x - 3$.

86. **A**

If the velocity and the acceleration have the opposite sign, the particle's speed is decreasing.
We know that $x'(t) = 12t - 24$ and $x''(t) = 12$. The acceleration is positive.

If $v(t) = x'(t)$ is negative, the speed is decreasing.

Therefore, $12t - 24 < 0$ or $0 < t < 2$.

87. **A**

Since $\lim\limits_{x \to \infty} \dfrac{\int_0^x \sqrt{t^2 + 5t + 3}\, dt}{(x^2)} = \dfrac{\infty}{\infty}$, we apply L'Hopital's rule.

$\lim\limits_{x \to \infty} \dfrac{\dfrac{d}{dx}\int_0^x \sqrt{t^2+5t+3}\,dt}{(x^2)'} = \lim\limits_{x \to \infty} \dfrac{\sqrt{x^2+5x+3}}{2x} = \lim\limits_{x \to \infty} \sqrt{\dfrac{x^2+5x+3}{4x^2}} = \dfrac{1}{2}$.

88. **E**

The volume of cylindrical Shell is $2\pi \int_1^4 xy\, dx$, where $y = \sqrt{x-1}$

Therefore, $V = 2\pi \int_1^4 \left(x\sqrt{x-1}\right) dx \approx 60.944$.

We can find the volume using the washer method.

Washer: $\pi \int_0^{\sqrt{3}} \left(4^2 - (y^2+1)^2\right) dx \approx 60.944$

89. **E**

$f'(x) = \sqrt{x^2}\sin\left(1 + \dfrac{x^2}{2}\right)\cdot 2x = 2x^2 \sin\left(1 + \dfrac{x^2}{2}\right)$

$f'(x)$ is negative on the interval $(2.070, 3]$, choice (E) belongs to the interval.

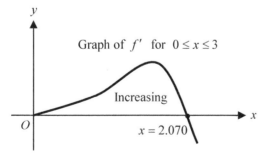

Graph of f' for $0 \le x \le 3$

Increasing

$x = 2.070$

Dr. John Chung's AP Calculus AB/BC

AP Calculus BC Test 1 Answers

90. A

The third degree Taylor polynomial is $P_3(x) = f(0) + f'(0)x + \dfrac{f''(0)}{2!}x^2 + \dfrac{f'''(0)}{3!}x^3$.

Substitute all values given into the polynomial.

Therefore, $P_3(x) = 1 + 2x + \dfrac{8}{2!}x^2 + \dfrac{48}{3!}x^3$ or $P_3(x) = 1 + 2x + 4x^2 + 8x^3$.

91. A

The equation of the line tangent to the graph of f at point $(1, f(1))$ is $y - f(1) = f'(1)(x - 1)$.

Since $g'(x) = f(x)$, then $g'(1) = f(1) = 3$. $f'(1) = -3$ is given.

Therefore, the equation of the line tangent to the graph of f at point $(1, 3)$ is

$y - 3 = -3(x - 1) \Rightarrow y = -3x + 6$

92. E

We know that $K'(x) = g'(f(x) + x)(f(x) + x)' = g'(f(x) + x)(f'(x) + 1)$ and

$K'(-2) = g'(f(-2) - 2)(f'(-2) + 1)$.

Because $f'(x) = \dfrac{1}{2}$ and $g'(x) = -2x$, $f(-2) = 1$, $f'(-2) = \dfrac{1}{2}$, and $g'(-1) = 2$.

Therefore, $K'(-2) = g'(-1)\left(\dfrac{1}{2} + 1\right) = 2\left(\dfrac{3}{2}\right) = 3$.

SECTION II

Question 1

(a) $m = \dfrac{dy}{dx} = \dfrac{dy/dt}{dx/dt} = \dfrac{2t \sin t}{\cos(t^2)/e^t} \Rightarrow \left.\dfrac{dy}{dx}\right|_{t=2} = \dfrac{4 \sin 2}{\cos 4/e^2} = \dfrac{4e^2 \sin 2}{\cos 4} \approx -41.116$

Therefore, the equation of the tangent is $y - 4 = -41.16(x - 2)$.

(b) The speed S, at time $t = 2$ is $S = \sqrt{\left(\dfrac{dx}{dt}\right)^2 + \left(\dfrac{dy}{dt}\right)^2} dt = \sqrt{\left(\dfrac{\cos(t^2)}{e^t}\right)^2 + (2t \sin t)^2}$.

Therefore, $\left.S\right|_{t=2} = \sqrt{\left(\dfrac{\cos 4}{e^2}\right)^2 + (4 \sin 2)^2} \approx 3.638$.

The horizontal velocity is $\left.\dfrac{dx}{dt}\right|_{t=2} = \dfrac{\cos 4}{e^2} = -.088 < 0$ which is negative.

The object moves to the left because the horizontal velocity is negative.

(c) $d = \int_2^5 \sqrt{\left(\dfrac{dx}{dt}\right)^2 + \left(\dfrac{dy}{dt}\right)^2} dt = \int_2^5 \sqrt{\left(\dfrac{\cos(t^2)}{e^t}\right)^2 + (2t \sin t)^2} \, dt$

AP Calculus BC Test 1 Answers

(d) $x(5) = x(2) + \int_2^5 \frac{\cos(t^2)}{e^t} dt = 2 + \int_2^5 \frac{\cos(t^2)}{e^t} dt = 2.01554 \approx 2.016$

(e) $a_x = \frac{d^2x}{dt^2} = \frac{-\sin t^2 \cdot 2t \cdot e^t - \cos t^2 \cdot e^t}{e^{2t}}$ and $a_x\big|_{t=2} = \frac{-\sin 4 \cdot (4) \cdot e^2 - \cos 4 \cdot e^2}{e^4} \approx 0.498$

$a_y = \frac{d^2y}{dt^2} = 2\sin t + 2t\cos t$ and $a_y\big|_{t=2} = 2\sin 2 + 4\cos 2 \approx 0.154$

The acceleration vector at time $t = 2$ is $\langle 0.498, 0.154 \rangle$.

Question 2

(a) Acceleration is $\frac{v(b) - v(a)}{b - a}$ on the interval $a \le t \le b$.

During $0 \le t \le 20$, the acceleration is positive because velocity is increasing on the interval $[0, 20]$.

(b) $\int_0^{32} v'(t)\, dt = v(32) - v(0) = 32 - 0 = 32$ is the change in velocity from time $t = 0$ to $t = 32$.

Average acceleration $= \frac{1}{32 - 0} \int_0^{32} v'(t)\, dt = \frac{v(32) - v(0)}{32} = \frac{30}{32} = \frac{15}{16}$ meters/square seconds

(c)
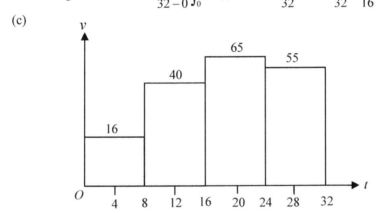

Riemann sum $= 8(16 + 40 + 65 + 55) = 1408$ m^2

We know that $\int_0^{32} v(t)\, dt = x(32) - x(0)$, where $x(t)$ is position function.

Therefore, $\int_0^{32} v(t)\, dt$ is the change in position of the object from time $t = 0$ to $t = 32$.

(d) For $t \ge 32$, the acceleration a of the object is modeled by $a(t) = \frac{-\ln(100t^3)}{t}$. Based on the model, what is the velocity of the object at time $t = 36$?

$v(36) = v(32) + \int_{32}^{36} \left(\frac{-\ln(100t^3)}{t} \right) dt = 30 + \int_{32}^{36} \left(\frac{-\ln(100t^3)}{t} \right) dt \approx 28.212$ meters/ second

AP Calculus BC Test 1 Answers

Question 3

a) From $g(x) = \dfrac{d}{dx}\int_0^x f(t)dt$ we get $g'(x) = f(x)$.

We know that the graph of g is increasing when $g'(x)$ is positive.

Therefore, the graph of g is increasing on the intervals $[-4, -2)$ and $(1, 5)$.

b) $g(-2) = \int_0^{-2} f(t)\,dt = -\int_{-2}^0 f(t)\,dt = -\left(-\dfrac{2\times 2}{2}\right) = 2$, $g(1) = \int_0^1 f(t)\,dt = -\dfrac{1\times 2}{2} = -1$

$g(3) = \int_0^3 f(t)\,dt = \int_0^1 f(t)\,dt + \int_1^3 f(t)\,dt = -1 + \pi = \pi - 1$

c) Because $\begin{cases} x < -2,\ g'(x) > 0 \\ x > -2,\ g'(x) < 0 \end{cases}$, the graph of g has a relative maximum at $x = -2$.

d) The graph of g have points of inflection at $x = 0, 3$.

At $x = 0$, $g'(x)$ changes sign from negative to positive.

At $x = 3$, $g'(x)$ changes sign from positive to negative.

e) The equation of the tangent is $y - y_1 = m(x - x_1)$.

Since $m = g'(3) = f(3) = 2$ and $g(3) = \pi - 1$, the equation of the tangent is $y - (\pi - 1) = 2(x - 3)$.

Question 4

(a) We know that the particular solution of the differential equation $\dfrac{dP}{dt} = \dfrac{k}{m}P(m - P)$ is

$$P(t) = \dfrac{m}{1 + Ae^{-kt}}.$$

From the differential equation

$\dfrac{dP}{dt} = \dfrac{1}{10000}P(4000 - P) = \dfrac{4000}{10000}P\left(1 - \dfrac{P}{4000}\right) \to \dfrac{dP}{dt} = \dfrac{2}{5}P\left(1 - \dfrac{P}{4000}\right)$

Carrying capacity is 4000 and $k = 0.4$.

Therefore, the particular solution is $P(t) = \dfrac{4000}{1 + Ae^{-0.4t}}$.

Initial condition: $1000 = \dfrac{4000}{1 + Ae^0}$ → We obtain $A = 3$.

Thus $P(t) = \dfrac{4000}{1 + 3e^{-0.4t}}$

(b) $\lim\limits_{t\to\infty} P(t) = \lim\limits_{t\to\infty} \dfrac{4000}{1 + 3e^{-0.4t}} = 4000$

$\lim\limits_{t\to\infty} P'(t) = \lim\limits_{P\to 4000} \dfrac{1}{10000} P(4000 - P) = 0$

(c) Graph of $P(t)$ has a point of inflection at $P = \dfrac{4000}{2} = 2000$. (Half of carrying capacity)

Proof) In order to find a point of inflection, we have to find $\dfrac{d^2 P}{dt^2}$.

AP Calculus BC Test 1 Answers

$$\frac{dP}{dt} = \frac{1}{10000}P(4000-P) = \frac{1}{10000}(4000P - P^2)$$

$$\frac{d^2P}{dt^2} = \frac{1}{10000}\left(4000\frac{dP}{dt} - 2P\frac{dP}{dt}\right) = \frac{1}{10000}\frac{dP}{dt}(4000 - 2P) = 0$$

$\rightarrow P = 2000$ Since $\frac{dP}{dt} \neq 0$ on $1000 < P < 4000$

Therefore, $P = 2000$ on the interval $1000 \leq P \leq 4000$.

(d) At the point of inflection the population is growing the fastest.

$$2000 = \frac{4000}{1 + 3e^{-0.4t}} \rightarrow 1 + 3e^{-0.4t} = 2 \rightarrow e^{-0.4t} = \frac{1}{3} \rightarrow -0.4t = -\ln 3$$

$$t = \frac{\ln 3}{0.4} = \frac{5\ln 3}{2} \text{ Months}$$

(e) Sketch possible solution curve through the point of inflection.

Point of inflection is at $\left(\frac{5\ln 3}{2}, 20\right)$.

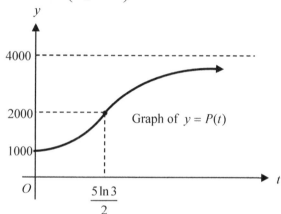

Question 5

(a) $e^x = 1 + x + \frac{x^2}{2!} + \frac{x^3}{3!} + \frac{x^4}{4!} + \frac{x^5}{5!} + \cdots$ and $\cos x = 1 - \frac{x^2}{2!} + \frac{x^4}{4!} - \frac{x^6}{6!} + \cdots$

$e^{-x} = 1 - x + \frac{x^2}{2!} - \frac{x^3}{3!} + \frac{x^4}{4!} - \frac{x^5}{5!} + \cdots$ and $\cos x = 1 - \frac{x^2}{2!} + \frac{x^4}{4!} - \frac{x^6}{6!} + \cdots$

Therefore, the fourth degree Taylor polynomial is $P_4(x) = 2 - x - \frac{x^3}{6} + \frac{x^4}{12}$.

$$P_4\left(\frac{1}{2}\right) = 2 - \frac{1}{2} - \frac{(1/2)^3}{6} + \frac{(1/2)^4}{12} = \frac{95}{64}$$

(b) From the Taylor series for f

$$f = f(0) + f'(0)x + \frac{f''(0)}{2!}x^2 + \frac{f'''(0)}{3!}x^3 + \frac{f^{(4)}(0)}{4!}x^4 + \cdots = 2 - x - \frac{x^3}{6} + \frac{x^4}{12} + \cdots$$

Compare the coefficients of the expression.

AP Calculus BC Test 1 Answers

$f'(0) = -1$

$\dfrac{f''(0)}{2!} = 0$ or $f''(0) = 0$

$\dfrac{f^{(4)}(0)}{4!} = \dfrac{1}{12} \Rightarrow f^{(4)}(0) = 2$

(c) Lagrange error $= |P_4(0.5) - f(0.5)| \leq \max\limits_{0 \leq z \leq 0.5} |f^{(5)}(z)| \dfrac{1}{5!} x^5 = \dfrac{1.2}{120}\left(\dfrac{1}{2}\right)^5 = \dfrac{1}{3200} < \dfrac{1}{3000}$

Question 6

(a) The slope of the tangent is $f'(2) = \dfrac{1}{2}$ and the y-coordinate at $x = 2$ is $f(2) = 5$.

Therefore, the equation of the tangent is $y - 5 = 0.5(x - 2)$.

(b) $y - 5 = 0.5(x - 2)$ or $y = 0.5x + 4$

Therefore, $f(3) \approx 0.5(3) + 5 = 5.5$.

(c) $f(2.5) \approx 5 + 0.5 \times f'(2) = 5 + 0.5 \times \dfrac{(2^2 - 3)}{2} = 5.25 \qquad f'(2.5) = \dfrac{2.5^2 - 3}{2.5} = 1.3$

$f(3) \approx 5.25 + 0.5 \times 1.3 = 5.9$

(d) Use $f''(x)$ to prove that the approximation in part (c) is less than $f(3)$.

$f''(x) = \dfrac{2x \cdot x - (x^2 - 3)}{x^2} = \dfrac{x^2 + 3}{x^2} > 0$ for $x > 0$.

The approximation is less than $f(3)$ because the graph of f is concave up for $x > 0$.

(e) Find $f(3)$.

$f(3) = 5 + \displaystyle\int_2^3 \left(\dfrac{x^2 - 3}{x}\right) dx = 5 + \int_2^3 \left(x - \dfrac{3}{x}\right) dx = 5 + \left[\dfrac{x^2}{2} - 3\ln x\right]_2^3$

$= 5 + \left(\dfrac{9}{2} - 3\ln 3\right) - (2 - 3\ln 2) = \dfrac{15}{2} - 3\ln 3 + 3\ln 2$

AP CALCULUS BC

TEST 2

AP Calculus BC Test 2 Section 1

AP CALCULUS BC

A CALCULATOR **CANNOT BE USED ON PART A OF SECTION 1**. A GRAPHIC CALCULATOR FROM THE APPROVED LIST **IS REQUIRED FOR PART B OF SECTION 1 AND FOR PART A OF SECTION II** OF THE EXAMINATION.

SECTION 1
Time: 1 hour and 45 minutes
All questions are given equal weight.
Percent of total grade – 50

Part A: 55 minutes, 28 multiple-choice questions
A calculator is NOT allowed.
Part B: 50 minutes, 17 multiple-choice questions
A graphing calculator is required.

CALCULUS BC
SECTION 1, PART A
Time — 55 minutes
Number of questions — 28

A CALCULATOR MAY NOT BE USED ON THIS PART OF THE EXAMINATION.

Directions: Solve each of the following problems, using the available space for scratchwork. After examining the form of the choices, decide which is the best of the choices given and fill in the corresponding oval on the answer sheet. No credit will be given for anything written in the test book. Do not spend too much time on any one problem.

In this test:

(1) Unless otherwise specified, the domain of a function f is assumed to be the set of all real numbers x for which $f(x)$ is a real number.

(2) The inverse of a trigonometric function f may be indicated using the inverse function notation f^{-1} or with the prefix "arc" (e.g, $\sin^{-1} x = \arcsin x$)

AP Calculus BC Test 2

Section 1

1. The position of a particle moving in the xy-plane at any time t is given by the parametric equations $x = \sin(2t)$ and $y = \cos t$. What is the acceleration vector of the particle at time $t = \frac{\pi}{6}$?

   (A) $\left\langle -\sqrt{3}, -\frac{\sqrt{3}}{2} \right\rangle$

   (B) $\left\langle -2\sqrt{3}, \frac{\sqrt{3}}{2} \right\rangle$

   (C) $\left\langle -2\sqrt{3}, -\frac{\sqrt{3}}{2} \right\rangle$

   (D) $\left\langle -4\sqrt{3}, -\frac{\sqrt{3}}{2} \right\rangle$

   (E) $\left\langle -4\sqrt{3}, \frac{\sqrt{3}}{2} \right\rangle$

---

2. $\int \frac{\cos x}{\sin^2 x} dx =$

   (A) $\sin x + C$ (B) $-\cos x + C$ (C) $-\csc x + C$ (D) $\sec x + C$ (E) $\tan x + C$

---

3. The function f given by $f(x) = (x^2 - 3)e^x$ has a local minimum at $x =$

   (A) -3 (B) -1 (C) 1 (D) 2 (E) 3

---

4. $\lim\limits_{x \to \pi/2} \frac{1 - \sin x}{\cos x} =$

   (A) -1 (B) 0 (C) $\frac{1}{2}$ (D) 2 (E) nonexistent

AP Calculus BC Test 2 Section 1

5. Which of the following sequences converge?

   I. $\left\{\dfrac{2n-1}{3n}\right\}$

   II. $\left\{\dfrac{\ln n}{n}\right\}$

   III. $\left\{\dfrac{e^n}{n^2}\right\}$

   (A) I only   (B) II only   (C) I and II only   (D) I and III only   (E) I, II, and III

---

6. Which of the following is the length of the path described by the parametric equations $x = \sin(2t)$ and $y = \cos(2t)$ from $t = 0$ to $t = \pi$?

   (A) $\dfrac{\pi}{2}$ (B) π (C) 4.5   (D) 6   (E) 2π

$$f(x) = \begin{cases} x^2 - 2 & \text{if } x \leq 2 \\ ax + b & \text{if } x > 2 \end{cases}$$

7. Let f be the function defined above. Find the value of a and b so that the function is differentiable for all real numbers x.

   (A) $a = -4, b = -6$
   (B) $a = -2, b = -4$
   (C) $a = 4, b = -4$
   (D) $a = 4, b = -6$
   (E) $a = -6, b = 4$

---

8. Which of the following is the slope of the line normal to the graph $x^2 + xy = 3$ at $x = 1$?

   (A) -4 (B) $-\dfrac{1}{4}$ (C) 1   (D) $\dfrac{1}{4}$ (E) 4

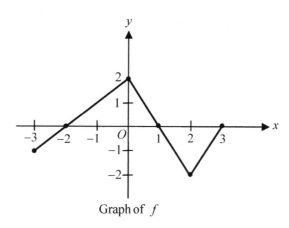

Graph of f

9. Let f be the piecewise linear function in the figure above. If $g(x) = \int_0^x f(t)\,dt$, which of the following values is the greatest?

(A) $g(-3)$ (B) $g(-2)$ (C) $g(0)$ (D) $g(1)$ (E) $g(3)$

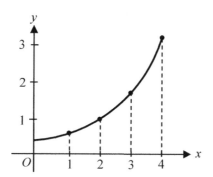

10. The graph of the function is shown above for $0 \le x \le 4$. Which of the following has the greatest value ?

(A) Left Riemann sum approximation with using 4 equal subintervals

(B) Midpoint Riemann sum approximation using 4 equal subintervals

(C) Right Riemann sum approximation using 4 equal subintervals

(D) Trapezoidal sum approximation using 4 equal subintervals

(E) $\int_0^4 f(x)\,dx$

AP Calculus BC Test 2

Section 1

11. Euler's method is used with a step size of 0.5 to approximate for $y(1)$. If $\frac{dy}{dx} = x + 2y$ and $y(0) = -2$, What is the approximation for $y(1)$?

   (A) −7.75 (B) −3.5 (C) 0.75 (D) 1.5 (E) 2

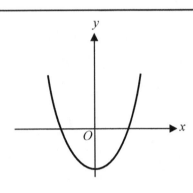

Graph of f'

12. The graph of f', the derivative of f, is shown above. Which of the following could be the graph of f?

(A)   (B)   (C)

(D)   (E)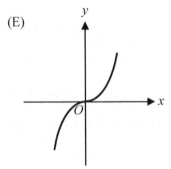

AP Calculus BC Test 2

Section 1

13. If $f(x) = 4^{1/x}$, then $f'(2) =$

(A) $-\ln 2$ (B) -1 (C) $\ln 2$ (D) 2 (E) $2\ln 2$

---

x	0	2	4	6
$f''(x)$	-2	0	3	-1

14. The function f is continuous and differentiable for all real x and selected values of its second derivative f'' are shown in the table above. Which of the following statements must be true?

(A) f is increasing on the interval $(2, 3)$.
(B) f is increasing on the interval $(0, 2)$.
(C) f has a local maximum at $x = 4$.
(D) The graph of f has a point of inflection at $x = 2$.
(E) The graph of f has at least two points of inflection in the interval $(0, 6)$.

---

15. A particle moves along the x-axis so that its acceleration at any time t is $a(t) = 2\cos t$. If the initial velocity of the particle at time $t = 0$ is -1, at what time t during the interval $0 \leq t \leq \pi$ is the particle farthest to the right?

(A) $\dfrac{\pi}{12}$ (B) $\dfrac{\pi}{6}$ (C) $\dfrac{\pi}{2}$ (D) $\dfrac{2\pi}{3}$ (E) $\dfrac{5\pi}{6}$

---

16. $\displaystyle\int x\sqrt{x+1}\, dx =$

(A) $\arcsin x + c$

(B) $\dfrac{x^2}{2} + \dfrac{1}{\sqrt{x+1}} + C$

(C) $\dfrac{2(x+1)^{1/2}}{5} - \dfrac{2(x+1)^{-1/2}}{3} + C$

(D) $\dfrac{2(x+1)^{5/2}}{5} - \dfrac{2(x+1)^{3/2}}{3} + C$

(E) $\dfrac{5(x+1)^{2/5}}{2} - \dfrac{3(x+1)^{2/3}}{2} + C$

AP Calculus BC Test 2   Section 1

17. What are the all values of x for which the series $\displaystyle\sum_{n=1}^{\infty} \frac{5^{n+1}}{\left(x^2+1\right)^n}$ converges?

    (A) $x < -2$ only
    (B) $-2 \leq x \leq 2$
    (C) $-2 < x < 2$
    (D) $x < -2$ or $x > 2$
    (E) $x \leq -2$ or $x \geq 2$

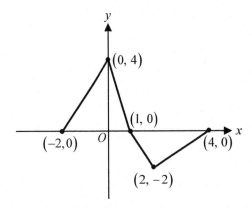

18. The graph of the function f is shown above. Let g be the function defined by $g(x) = \displaystyle\int_{-2}^{x} f(t)\,dt$. Find the equation of the line tangent to the graph of g at $x = 2$.

    (A) $y = -2x + 3$
    (B) $y = -2x + 5$
    (C) $y = -2x + 9$
    (D) $y = 2x + 3$
    (E) $y = 2x + 5$

19. Find all asymptotes of the graph of $y = \dfrac{1+e^x}{1-e^x}$ in the xy-plane.

    (A) $x = 0$ only
    (B) $x = 0$ and $y = -1$ only
    (C) $y = -1$ only
    (D) $y = -1$ and $y = 1$ only
    (E) $x = 0$, $y = -1$, and $y = 1$

20. Let g be a differentiable function, and let f be the function defined by $f(x) = \dfrac{g(x^2-5)}{12}$. Which of the following could be equal to $g'(4)$?

(A) $f'(3)$ (B) $2f'(3)$ (C) $2f'(4)$ (D) $4f'(3)$ (E) $4f'(4)$

---

21. What is the sum of the series $1 + (\ln 5) + \dfrac{(\ln 5)^2}{2!} + \dfrac{(\ln 5)^3}{3!} + \dfrac{(\ln 5)^4}{4!} + \cdots + \dfrac{(\ln 5)^n}{n!} + \cdots$?

(A) 1 (B) e (C) e^5 (D) 5 (E) $\dfrac{1}{1-\ln 5}$

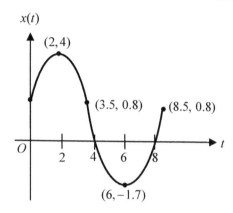

22. A particle moves along a straight line. The graph of the particle's position $x(t)$ is shown above for $0 \le t \le 8.5$ and the graph has a point of inflection at $(3.5, 0.8)$. For what value of t is the velocity of the particle decreasing on the open interval $(0, 8.5)$?

(A) $2 < t < 4$
(B) $2 < t < 6$
(C) $2 < t < 3.5$ only
(D) $0 < t < 3.5$
(E) $3.5 < t < 6$

AP Calculus BC Test 2 — Section 1

23. Which of the following is equal to the area of the region inside the curve $r = 3\sin\theta$ and outside the curve $r = \sin\theta$?

(A) $2\int_0^{\pi} \sin^2\theta \, d\theta$

(B) $8\int_0^{\pi/2} \sin^2\theta \, d\theta$

(C) $16\int_0^{\pi} \sin\theta \, d\theta$

(D) $16\int_0^{\pi/2} \sin^2\theta \, d\theta$

(E) $8\int_0^{2\pi} \sin^2\theta \, d\theta$

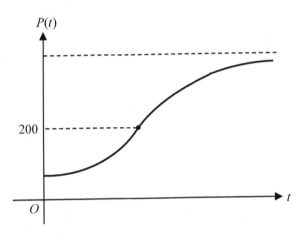

24. The graph of a logistic growth that is modeled by a function P is shown in the figure above. If the graph of P has a point of inflection at $P = 200$, which of the following could be the differential equation for P?

(A) $\dfrac{dP}{dt} = 2P - 400P^2$

(B) $\dfrac{dP}{dt} = 0.1P - 400P^2$

(C) $\dfrac{dP}{dt} = 0.2P - 0.0005P^2$

(D) $\dfrac{dP}{dt} = 0.2p - 0.001P^2$

(E) $\dfrac{dP}{dt} = 0.2P - 0.002P^2$

AP Calculus BC Test 2

25. If Taylor series for $\sin x$ about $x = 0$ is $x - \dfrac{x^3}{3!} + \dfrac{x^5}{5!} - \dfrac{x^7}{7!} + \cdots + \dfrac{(-1)^n x^{2n+1}}{(2n+1)!} + \cdots$, which of the following is the coefficient of x^8 in the Taylor series for $\cos(x^2)$ about $x = 0$?

(A) $-\dfrac{1}{9!}$ (B) $-\dfrac{1}{7!}$ (C) $\dfrac{1}{24}$ (D) $\dfrac{1}{6}$ (E) $\dfrac{1}{2}$

---

26. Let f be a function such that $f''(x) = 12x + 8$. If the graph of f is tangent to the line $2x - y = 2$ at the point $(0, -2)$, what is the average value of $f(x)$ on the closed interval $[-1, 1]$

(A) $-\dfrac{4}{3}$ (B) $-\dfrac{2}{3}$ (C) 2   (D) 4   (E) 8

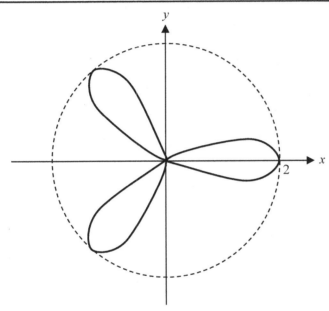

27. Which of the following expression gives the area of one petal enclosed by the polar curve $r = 2\cos 3\theta$ shown in the figure above?

(A) $\dfrac{1}{2} \displaystyle\int_{-\pi/2}^{\pi/2} \cos^2(3\theta)\, d\theta$

(B) $4 \displaystyle\int_{0}^{\pi/3} \cos^2(3\theta)\, d\theta$

(C) $4 \displaystyle\int_{0}^{\pi/6} \cos^2(3\theta)\, d\theta$

(D) $2 \displaystyle\int_{0}^{\pi/6} \cos^2(3\theta)\, d\theta$

(E) $\dfrac{1}{2} \displaystyle\int_{0}^{\pi/6} \cos^2(3\theta)\, d\theta$

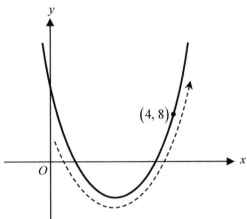

28. In the *xy*-plane, a particle moves to the right along the parabola $y = x^2 - 3x + 4$. If the particle moves with a constant speed of $\sqrt{104}$ units per second, which of the following is the velocity vector of the particle when the particle is at point $(4, 8)$?

(A) $\langle 2, 10 \rangle$ (B) $\langle 2, \sqrt{26} \rangle$ (C) $\langle 6, 2\sqrt{17} \rangle$ (D) $\langle 9, 2\sqrt{14} \rangle$ (E) $\langle 10, 2 \rangle$

END OF PART A OF SECTION 1

AP Calculus BC Test 2

Section 1

CALCULUS BC
SECTION 1, PART B
Time — 50minutes
Number of questions — 17

A GRAPHING CALCULATOR IS REQUIRED FOR SOME QUESTIONS ON THIS PART OF THE EXAMINATION.

Directions: Solve each of the following problems, using the available space for scratchwork. After examining the form of the choices, decide which is the best of the choices given and fill in the corresponding oval on the answer sheet. No credit will be given for anything written in the test book. Do not spend too much time on any one problem.

In this test:

(1) The exact numerical value of the correct answer does not always appear among the choices given. When this happens, select from among the choices the number that best approximates the exact numerical value.

(2) Unless otherwise specified, the domain of a function f is assumed to be the set of all real numbers x for which $f(x)$ is a real number.

(3) The inverse of a trigonometric function f may be indicated using the inverse function notation f^{-1} or with the prefix "arc" (e.g, $\sin^{-1} x = \arcsin x$)

AP Calculus BC Test 2

76. For what value of k does the series $\sum_{n=1}^{\infty} \dfrac{k^n}{2^{2n+1}}$ converge?

    (A) $-2 \le k \le 2$ only
    (B) $-8 < k < 8$
    (C) $-\dfrac{1}{2} \le k \le \dfrac{1}{2}$
    (D) $-4 < k < 4$
    (E) $-\dfrac{1}{4} < k < \dfrac{1}{4}$

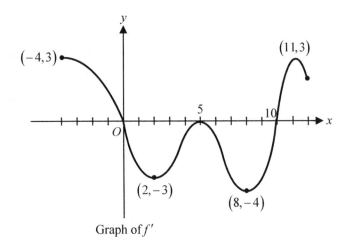

Graph of f'

77. The graph of f', the derivative of f, is shown above for $-4 \le x \le 12$. On what intervals is the graph of f concave upward?

    (A) $-4 < x < 0$ only
    (B) $0 < x < 5$ and $5 < x < 10$
    (C) $-4 < x < 0$ and $2 < x < 8$
    (D) $2 < x < 5$ only
    (E) $2 < x < 5$ and $8 < x < 11$

---

78. Let R be the region enclosed by the graph of $y = -x^2 + 4x - 3$, and the x-axis. What is the volume of the solid generated when R is rotated about the y-axis?

    (A) 16.755   (B) 18.003   (C) 20.152   (D) 23.568   (E) 25.925

AP Calculus BC Test 2

79. Let f be the function defined by $f(x) = e^x \cos x$ for $0 \leq x \leq 2\pi$. Which of the following is the intervals on which f is decreasing?

   (A) $0 < x < 1.57$
   (B) $0 < x < 3.14$
   (C) $0.785 < x < 3.927$
   (D) $0.785 < x < 5.498$
   (E) $1.571 < x < 4.712$

---

80. A parametric equations are given by $x = 2t^3 - t^2$ and $y = t^3 - 8t$. Which of the following is the equation for the line tangent to the curve at the point where $t = 2$?

   (A) $x - 5y = 52$
   (B) $x - 5y = 26$
   (C) $2x - 5y = 26$
   (D) $2x - 4y = 13$
   (E) $3x - 4y = 26$

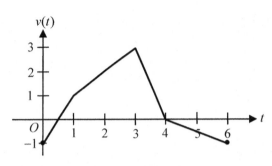

81. The graph of the velocity $v(t)$ of a particle moving along the x-axis is shown above. At time $t = 0$, the particle is at the origin. Which of the following could be the graph of the position, $x(t)$, of the particle for $0 \le t \le 6$?

(A)

(B)

(C)

(D)

(E)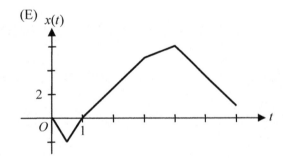

82. The velocity of a particle moves along a path in the xy-plane is given by the parametric equations $\frac{dx}{dt} = e^t \sin t$ and $\frac{dy}{dt} = e^t \cos t$. Which of the following is the distance traveled by the particle along the path from $t = 1$ to $t = 2$?

(A) e
(B) $e^2 - 1$
(C) $e(e-1)$
(D) $e^2 + 1$
(E) $e^2 + e + 1$

---

83. If the series $\sum_{n=3}^{\infty} \frac{1}{n^p}$ converges, where $p \geq 0$, which of the following statements must be true?

(A) $\sum_{n=3}^{\infty} \frac{n}{n^p}$ converges.

(B) $\sum_{n=3}^{\infty} \frac{n^2}{n^p}$ converges.

(C) $\sum_{n=3}^{\infty} \frac{1}{n^p \ln(n)}$ converges.

(D) $\sum_{n=3}^{\infty} \frac{1}{n^p e}$ diverges.

(E) $\sum_{n=3}^{\infty} \frac{1}{n^p n^2}$ diverges.

AP Calculus BC Test 2 — Section 1

84. A particle moves on the x-axis with velocity given by $v(t) = t^3 - 7t^2 + 14t - 8$ for $0 \le t \le 5$. Which of the following must be true?

   I. The velocity of the particle is strictly decreasing on the interval $(2, 3)$.
   II. The acceleration of the particle is strictly decreasing on the interval $(2, 3)$.
   III. The particle is speeding up on the interval $(2, 3)$.

(A) I only
(B) II only
(C) III only
(D) I and II only
(E) I and III only

---

85. The position of a particle at any time $t \ge 0$ is given by the parametric equation $x(t) = 2t + 1$ and $y = \dfrac{x^3}{3} + 3x^2 - 1$. Which of the following is the value of $\dfrac{dy}{dt}$ at time $t = 2$?

(A) 80
(B) 90
(C) 100
(D) 110
(E) 120

---

86. Let R be the region enclosed by the graphs of $y = e^{-x}$, $y = \sqrt{x}$, and the line $y = 0$. If cross sections of the solid perpendicular to the x-axis are semicircles, what is the volume of the solid?

(A) 0.035
(B) 1.045
(C) 1.153
(D) 2.546
(E) 3.123

AP Calculus BC Test 2   Section 1

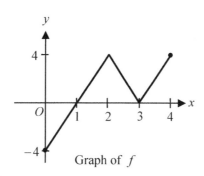

Graph of f

87. Let g be the function given by $g(x) = \int_2^x f(t)\,dt$ and let $P(x)$ be the second-degree Taylor polynomial for g about $x=1$. The graph of $f(x)$ on the interval $[0,4]$ is shown above. Which of the following is the value of $P(1.5)$?

(A) -3.5
(B) -2.5
(C) -1.5
(D) 1.50
(E) 2.10

---

88. The Taylor series for $f(x)$ about $x=0$ is given by $f(x) = 1 + x^2 + \dfrac{x^4}{2!} + \dfrac{x^6}{3!} + \cdots + \dfrac{(x^2)^n}{n!} + \cdots$.

What is the point of intersection of the graphs $y = f(x)$ and the line $x = \dfrac{1}{2}$?

(A) $\left(\dfrac{1}{2}, 1.284\right)$

(B) $\left(\dfrac{1}{2}, 1.648\right)$

(C) $\left(\dfrac{1}{2}, 1.895\right)$

(D) $\left(\dfrac{1}{2}, 2.252\right)$

(E) $\left(\dfrac{1}{2}, 1.315\right)$

AP Calculus BC Test 2

Section 1

89. Let f be the differentiable function for all values of x. If $f(x) > 0$ and the graph of f is increasing for all values of x, and $g(x) = \int_0^x f(t)\,dt$, which of the following must be true?

    I. The graph of g is increasing over the x-axis.
    II. The graph of g is continuous over the x-axis.
    III. The graph of g is concave up over the x-axis.

(A) I only   (B) II only   (C) I and II only   (D) I and III only   (E) I, II, and III

---

90. Let g be the function given by $g(x) = \int_{0.5}^{x^2} e^{-\sqrt{t}} \sin\left(\dfrac{1}{t}\right) dt$ for $0.5 \le x \le 2$. At which of the following value of x does g have a point of inflection on the graph of g?

(A) 0.564
(B) 0.601
(C) 0.720
(D) 0.750
(E) 0.814

---

91. Let f be the function given by $f(x) = \ln x$. If the tangent line to the graph of f at $x = e$ is used to approximate $f(2e)$, which of the following is the value of the error resulting from this tangent line approximation?

(A) 0.005
(B) 0.012
(C) 0.032
(D) 0.307
(E) 0.405

---

92. Let $F(x) = \int_0^x \sin(t)\,dt$ for $x \ge 0$. Use the trapezoidal rule with four equal subintervals of the closed interval $[0, 1]$ to approximate $F(1)$.

(A) 0.141
(B) 0.281
(C) 0.302
(D) 0.457
(E) 0.704

END OF SECTION 1

AP Calculus BC Test 2 *Section II*

CALCULUS BC

SECTION II
Time: 1 hour and 30 minutes
Percent of total grade – 50

Part A: 30 minutes, 2 problems
Part B: 60 minutes, 4 problems

Part A: A graphing calculator is required for some problems or parts of problems.
Part B: No calculator is allowed for these problems.

GENERAL INSTRUCTIONS FOR SECTION II PART A AND PART B

For each part of section II, you may wish to look over the problems before starting to work on them, since it is not expected that everyone will be able to complete all parts of all problems. All problems are given equal weight, but the parts of a particular problem are not necessary given equal weight.

- YOU SHOULD WRITE ALL WORK FOR EACH PART OF EACH PROBLEM IN THE SPACE PROVIDED FOR THAT PART IN THE PINK TEST BOOKLET. Be sure to write clearly and legibly. If you make an error, you may save time by crossing it out rather than trying to erase it. Erased or cross-out work will not be graded.
- Show all work. Clearly label any functions, graphs, tables, or other object that you use. You will be graded on the correctness and completeness of your methods as well as your answers. Answers without supporting work may not receive credit.
- Justifications required that you give mathematical (noncalculator) reasons
- Your work must be expressed in standard mathematical notation rather than calculator syntax. For example, $\int_{1}^{5} x^2 \, dx$ may not be written as $\text{fnInt}(X^2, X, 1, 5)$
- Unless otherwise specified, answers (numeric or algebraic) need not be simplified.
- If you use decimal approximations in calculations, you will be graded on accuracy. Unless otherwise specified, your final answers should be accurate to three places after the decimal point.
- Unless otherwise specified, the domain of a function f is assumed to be the set of all real numbers x for which $f(x)$ is a real number.

CALCULUS BC
SECTION II, PART A
Time — 30 minutes
Number of questions — 2

A Graphing calculator is required for some problems or parts of problems.

Question 1

100 feet

As shown in the figure above, water is entering from a water pipe during 20 minutes into a cylindrical tank that is initially empty. The cylindrical tank has a radius 5 feet and the height of the water in the cylindrical tank is changing at the rate of $2\sqrt{t} \cdot \sin(0.05t)$ feet per minute.

(a) Write an expression involving an integral that gives the volume of the water in the tank as a function of time t.

(b) At time $t = 10$, what is the height of the water in the tank?

(c) At what rate is the volume of the water in the tank changing at time $t = 10$?

(d) Find the average rate of change in the height of the water and the average rate of change in the volume of the water in the tank over the 10 minutes time period.

(e) At the same time, water is entering from the other pipe into a tank identical to the one as shown above. The height of the water in this tank is changing at the rate of $0.002t^3$ feet per minute. Let R_1 and R_2 be the rate of change in the height of the water each tank given by $R_1(t) = 2\sqrt{t} \cdot \sin(0.05t)$ and $R_2(t) = 0.002t^3$ and let $F(t)$ be the function defined by $F(t) = R_1(t) - R_2(t)$. Is there some time t, $5 < t < 20$, such that the heights of water in the two cylindrical tanks are changing at the same rate? Apply the Intermediate Value Theorem to the function F on the interval $5 < t < 20$, and justify your answer.

Question 2

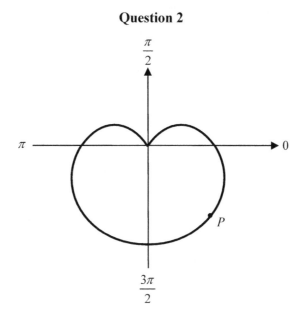

The graph of polar curve $r = 2(1-\sin\theta)$, where $0 \le \theta \le 2\pi$, is shown above. The polar coordinates (r, θ) of a point are related to the rectangular coordinates (x, y) of the point.

(a) Express x and y in terms of θ, and find $\dfrac{dy}{dx}$.

(b) Find the area in the second quadrant enclosed by the x-axis and the graph of r.

(c) If the x-coordinate of point P on the curve in the second quadrant is -1, then what is the y-coordinate of point P.

END OF PART A OF SECTION II

AP Calculus BC Test 2

Section II

CALCULUS BC
SECTION II, PART B
Time — 60minutes
Number of questions —4

No calculator is required for these problems.

Question 3

A particle moves in the xy-plane for $0 \le t \le 10$, in seconds. Its velocity, in meters per seconds, is modeled by a semicircle and the piecewise-linear function whose graph is shown above.

(a) Find the acceleration of the particle at time $t = 5.5$ seconds.

(b) Find the values of $\int_0^{10} v(t)dt$ and explain the meaning of $\int_0^{10} v(t)dt$.

(c) Find the value of $\int_0^{10} |v(t)|dt$ and explain the meaning of $\int_0^{10} |v(t)|dt$.

(d) At what time the particle changes its direction?

Question 4

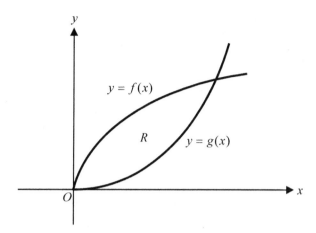

The shaded region, R, is bounded by the graphs of f and g, as shown in the figure above. Let f and g be the functions given by $f(x) = 8\sqrt{x}$ and $g(x) = x^2$.

(a) Find the area of R.

(b) Find the volume of the solid generated when R is revolved about y-axis.

(c) The base of a solid is the region R. Each cross section of the solid perpendicular to the y-axis is an equilateral triangle. Write the expression involving integral that gives the volume of the solid. Do not evaluate.

(d) The line tangent to the graph of f at $x = 1$ is parallel to the line tangent to the graph of g at point Q. Find the coordinates of point Q.

AP Calculus BC Test 2

Section II

Question 5

The number of a certain type of bacteria increases at a rate proportional to the difference 500 and the number present. There are 100 bacteria at time $t=0$. If $P(t)$ is the number of bacteria at time t minutes, then

$$\frac{dP}{dt} = \frac{1}{10}(500 - P).$$

(a) If $y = P(t)$ is the particular solution to the differential equation at time $t \geq 0$, then find $y = P(t)$.

(b) Find the value of $\lim\limits_{t \to \infty} P(t)$.

(c) Is the graph of $y = P(t)$ is concave down? Explain your reasoning.

(d) Find $\dfrac{d^2P}{dt^2}$ in terms of P. Is the value of $\dfrac{d^2P}{dt^2}$ always positive? Explain your reasoning.

Question 6

Let f be the function given by $f(x) = \dfrac{1}{1+2x}$ and the Maclaurin series for $\dfrac{1}{1+x}$ is

$$\frac{1}{1+x} = 1 - x + x^2 - x^3 + x^4 - x^5 + \cdots + (-1)^n x^n + \cdots.$$

(a) Write the first four nonzero terms and the general term of the Maclaurin series for f.

(b) Find the radius and interval of convergence for the Maclaurin series for f.

(c) Let $P_3(x)$ be the third-degree Taylor polynomial for g about $x = 0$. If $g(x) = \int_0^x f(t)\,dt$, then find the value of $P_3\left(\dfrac{1}{4}\right)$.

(d) Find the value of $g^{(4)}(0)$.

(e) Let $P_3(x)$ be the third-degree Taylor polynomial for g about $x = 0$. Use the Lagrange error bound to show that $\left| P_3\left(\dfrac{1}{4}\right) - g\left(\dfrac{1}{4}\right) \right| < \dfrac{1}{100}$.

END OF EXAM

No material on this page

AP Calculus BC Test 2 Answers   Section 1

SECTION 1, PART A

1. **C**

    The acceleration vector is $\langle a_x, a_y \rangle = \langle x''(t), y''(t) \rangle$.

    $x'(t) = 2\cos(2t) \;\to\; x''(t) = -4\sin(2t) \;\to\; x''\left(\dfrac{\pi}{6}\right) = -4\sin\left(\dfrac{\pi}{3}\right) = -2\sqrt{3}$

    $y'(t) = -\sin t \;\to\; y''(t) = -\cos t \;\to\; y''\left(\dfrac{\pi}{6}\right) = -\cos\left(\dfrac{\pi}{6}\right) = -\dfrac{\sqrt{3}}{2}$

    Therefore, $\langle a_x, a_y \rangle = \left\langle -2\sqrt{3}, -\dfrac{\sqrt{3}}{2} \right\rangle$.

2. **C**

    We know that $\dfrac{d}{dx}(\csc x) = -\csc x \cot x$.

    $\displaystyle \int \dfrac{\cos x}{\sin^2 x}\,dx = \int \left(\dfrac{1}{\sin x} \cdot \dfrac{\cos x}{\sin x}\right) dx = \int \csc x \cdot \cot x \, dx = -\csc x + C$

    Or, you can use u-substitution.

    $\begin{cases} u = \sin x \\ du = \cos x\, dx \end{cases}$

    $\displaystyle \int \dfrac{\cos x}{\sin^2 x}\,dx = \int \dfrac{du}{u^2} = -\dfrac{1}{u} + C \;\to\; -\dfrac{1}{\sin x} + C = -\csc x + c$

3. **C**

    We need the critical points of $f'(x) = 0$.

    $f'(x) = 2xe^x + (x^2 - 3)e^x = e^x(x^2 + 2x - 3) = e^x(x+3)(x-1) = 0$ where $(e^x > 0)$

    The critical points are $x = 1$ and $x = -3$.

    At $x = 1$, f' changes in sign from $-$ to $+$. Therefore, f has a local minimum at $x = 1$.

4. **B**

    Since $\dfrac{1 - \sin(\pi/2)}{\cos(\pi/2)} = \dfrac{0}{0}$, We apply L'Hopital's rule.

    $\displaystyle \lim_{x \to \pi/2} \dfrac{1 - \sin x}{\cos x} = \lim_{x \to \pi/2} \dfrac{-\cos x}{-\sin x} = \dfrac{-\cos\left(\dfrac{\pi}{2}\right)}{-\sin\left(\dfrac{\pi}{2}\right)} = \dfrac{0}{1} = 0$

5. **C**

    I. $\left\{\dfrac{2n-1}{3n}\right\} \;\to\; \displaystyle\lim_{n \to \infty} \dfrac{2n-1}{3n} = \dfrac{2}{3}$: The sequence converges to $\dfrac{2}{3}$.

    II. $\left\{\dfrac{\ln n}{n}\right\} \;\to\; \displaystyle\lim_{n \to \infty} \dfrac{\ln n}{n} = \lim_{n \to \infty} \dfrac{1/n}{1} = 0$ (L'Hopital's rule): The sequence converges to 0.

    III. $\left\{\dfrac{e^n}{n^2}\right\} \;\to\; \displaystyle\lim_{n \to \infty} \dfrac{e^n}{n^2} = \lim_{n \to \infty} \dfrac{e^n}{2n} = \lim_{n \to \infty} \dfrac{e^n}{2} = \infty$ (Apply L'Hopital's rule two times)

6. **E**

    $d = \displaystyle\int_0^\pi \sqrt{(x')^2 + (y')^2}\, dt = \int_0^\pi \sqrt{(2\cos 2t)^2 + (-2\sin 2t)^2}\, dt = \int_0^\pi \sqrt{4\cos^2(2t) + 4\sin^2(2t)}\, dt$

AP Calculus BC Test 2 Answers

$$= \int_0^\pi 2\,dt = [2t]_0^\pi = 2\pi$$

7. **D**

$$f(x) = \begin{cases} x^2 - 2 & \text{if } x \leq 2 \\ ax + b & \text{if } x > 2 \end{cases}$$

The graph of f is continuous at $x = 2$. We know that $f(2) = 2^2 - 2 = 4 - 2 = 2$.

Therefore, $\lim_{x \to 2} ax + b = 2 \rightarrow 2a + b = 2. \cdots (1)$

$$f'(x) = \begin{cases} 2x & \text{if } x \leq 2 \\ a & \text{if } x > 2 \end{cases}$$

Because the function is differentiable at $x = 2$, the derivatives of two piece-wise functions must be equal. Therefore, $2(2) = a$ or $a = 4$. From (1), $2a + b = 2 \rightarrow 2(4) + b = 2 \rightarrow b = -6$

8. **D**

The slope of the line normal to the graph is negative reciprocal of the slope of the tangent at $x = 1$.

$$2x + y + xy' = 0 \rightarrow y' = \frac{-2x - y}{x}$$

At $x = 1$, $1 + y = 3 \rightarrow y = 2$.

Therefore, $\left.\dfrac{dy}{dx}\right|_{(1,2)} = \dfrac{-2-2}{1} = -4$: The slope of the line normal to the graph is $\dfrac{1}{4}$.

9. **D**

(A) $g(-3) = \int_0^{-3} f(t)\,dt = -\int_{-3}^0 f(t)\,dt = -(-0.5 + 2) = -1.5$

(B) $g(-2) = \int_0^{-2} f(t)\,dt = -\int_{-2}^0 f(t)\,dt = -2$ (C) $g(0) = \int_0^0 f(t)\,dt = 0$

(D) $g(1) = \int_0^1 f(t)\,dt = 1$ (E) $g(3) = \int_0^3 f(t)\,dt = 1 - 2 = -1$

10. **C**

Because the graph of the function is strictly increasing and concave up, the right Riemann sum approximation has the greatest value.

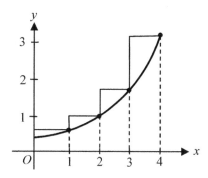

11. **A**

At point $(0, -2)$, $y'(0) = 0 + 2(-2) = -4$

$y(0.5) = -2 + 0.5(-4) = -4$

At point $(0.5, -4)$, $y'(0.5) = 0.5 + 2(-4) = -7.5$

Therefore, $y(1) = -4 + 0.5(-7.5) = -7.75$

Dr. John Chung's AP Calculus AB/BC

AP Calculus BC Test 2 Answers Section 1

12. D

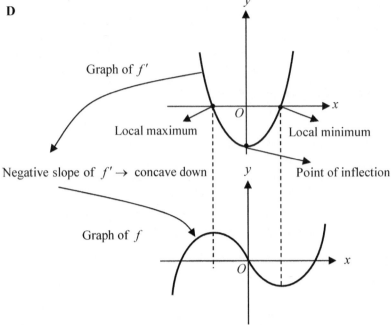

13. A

We know that $\dfrac{d}{dx}\left(a^{u(x)}\right) = a^{u(x)} \cdot u'(x) \cdot \ln a$.

Therefore, $f(x) = 4^{1/x} \;\to\; f'(x) = 4^{1/x} \cdot \ln 4 \cdot \left(-\dfrac{1}{x^2}\right)$ or

$f'(2) = 4^{1/2} \cdot \ln 4 \cdot \left(-\dfrac{1}{4}\right) = -\dfrac{1}{2}\ln 4 = -\dfrac{1}{2}\ln 2^2 = -\ln 2$.

14. E

(A) f is increasing on the interval $(2, 4)$. Not always true.

(B) f is increasing on the interval $(0, 2)$. Not always true.

(C) f has a local maximum at $x = 4$. Not always true.

(D) The graph of f has a point of inflection at $x = 2$. Not always true.

(E) The graph of f has at least two points of inflection in the interval $(0, 6)$. True

By Intermediate Theorem, there must be at least two x-intercepts in the interval $(0, 6)$.

15. E

$v(t) = \int 2\cos t \, dt = 2\sin t + C \;\to\; v(0) = 2\sin 0 + C = -1 \;\to\; C = -1$

$v(t) = 2\sin t - 1 = 0 \;\to\; t = \dfrac{\pi}{6}, \dfrac{5\pi}{6}$

AP Calculus BC Test 2 Answers   Section 1

The particle is farthest to the right at time $t = \frac{5\pi}{6}$. At $t = \frac{5\pi}{6}$, the particle changes its direction.

16. D

We change variable using u.

$$\begin{cases} u = x+1, \quad x = u-1 \\ du = dx \end{cases}$$

$$\int x\sqrt{x+1}\, dx = \int (u-1)\sqrt{u}\, dx = \int \left(u^{\frac{3}{2}} - u^{\frac{1}{2}} \right) du = \frac{2}{5}u^{\frac{5}{2}} - \frac{2}{3}u^{\frac{3}{2}} + C$$

Therefore, $\int x\sqrt{x+1}\, dx = \dfrac{2(x+1)^{\frac{5}{2}}}{5} - \dfrac{2(x+1)^{\frac{3}{2}}}{3} + C$

17. D

We apply ratio test as follows.

$$\lim_{n\to\infty} \left| \frac{5^{n+2}}{(x^2+1)^{n+1}} \cdot \frac{(x^2+1)^n}{5^{n+1}} \right| = \lim_{n\to\infty} \left| \frac{5}{(x^2+1)} \right| = \frac{5}{x^2+1}$$

For the interval of convergence, we solve the inequality $\dfrac{5}{(x^2+1)} < 1$.

From the inequality, $5 < x^2 + 1 \to x^2 - 4 > 0 \to x > 2$ or $x < 2$ which is interval of convergence. Now we apply end point test.

At $x = 2$, $\sum_{n=1}^{\infty} \dfrac{5^{n+1}}{(5)^n} = \sum_{n=1}^{\infty} 5 = 5 + 5 + 5 + \cdots \to$ The series diverges.

At $x = -2$, $\sum_{n=1}^{\infty} \dfrac{5^{n+1}}{(5)^n} = \sum_{n=1}^{\infty} 5 = 5 + 5 + 5 + \cdots \to$ The series diverges.

Therefore, the interval of convergence is $x > 2$ or $x < -2$.

18. C

The equation of the tangent is $y - g(2) = m(x - 2)$.

Therefore, $g(2) = \int_{-2}^{2} f(t)\, dt = \dfrac{3\times 4}{2} - \dfrac{1\times 2}{2} = 5$

and $m = g'(2) = f(2)$, which is going to be $m = -2$.
Therefore, the equation of the line tangent to the graph is
$y - 5 = -2(x-2)$ or $y = -2x + 9$.

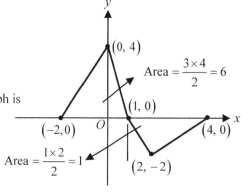

19. E

If you have a function which has a term e^x like $y = \dfrac{1+e^x}{1-e^x}$, you have to check both directions

$\left(\lim_{x\to +\infty} f(x) \text{ and } \lim_{x\to -\infty} f(x) \right)$ when you need to find the horizontal asymptotes.

1) Vertical asymptote: $1 - e^x = 0 \to e^x = 1 \to x = 0$

2) Horizontal asymptote: $\lim_{x\to\infty} \dfrac{1+e^x}{1-e^x} = -1$ and $\lim_{x\to-\infty} \dfrac{1+e^x}{1-e^x} = \dfrac{1+e^{-\infty}}{1-e^{-\infty}} = 1$

AP Calculus BC Test 2 Answers

Section 1

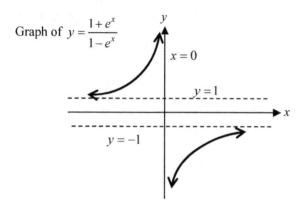

Graph of $y = \dfrac{1+e^x}{1-e^x}$

20. B

$f'(x) = \dfrac{g'(x^2 - 5)2x}{12}$ → $g'(x^2 - 5) = \dfrac{12 f'(x)}{2x}$ or $g'(x^2 - 5) = \dfrac{6 f'(x)}{x}$

From $g'(x^2 - 5) = g'(4)$, $x^2 - 5 = 4$ → $x^2 = 9$ → $x = 3$ or -3.

Therefore,

at $x = 3$, $g'(4) = \dfrac{6 f'(3)}{3} = 2 f'(3)$ and at $x = -3$, $g'(4) = \dfrac{6 f'(-3)}{-3} = -2 f'(-3)$.

21. D

We know that Taylor series of e^x is $e^x = 1 + x + \dfrac{x^2}{2!} + \dfrac{x^3}{3!} + \dfrac{x^4}{4!} + \cdots$.

When we substitute $\ln 5$ into the series, we will get

$e^{\ln 5} = 1 + (\ln 5) + \dfrac{(\ln 5)^2}{2!} + \dfrac{(\ln 5)^3}{3!} + \dfrac{(\ln 5)^4}{4!} + \cdots + \dfrac{(\ln 5)^n}{n!} + \cdots$.

Therefore, the sum of the series is equal to $e^{\ln 5} = 5$.

---

22. D

Slope of the line tangent to the curve is decreasing on the interval $(0, 3.5)$.

For example, the slope of the line tangent is decreasing as follows on the interval $(0, 3.5)$.

$4 \to 3 \to 2 \to 1 \to 0 \to -1 \to -2 \to -3 \to -4$

We can see that the number is decreasing.

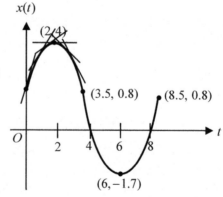

AP Calculus BC Test 2 Answers   Section 1

23. B

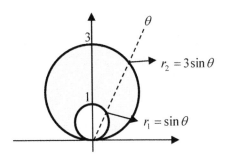

We apply the formula $A = \dfrac{1}{2}\int_{\theta_1}^{\theta_2} r_2^{\,2}\,d\theta - \dfrac{1}{2}\int_{\theta_1}^{\theta_2} r_1^{\,2}\,d\theta = \dfrac{1}{2}\int_{\theta_1}^{\theta_2}\left(r_2^{\,2} - r_1^{\,2}\right)d\theta$ for the area of a polar curve.

Because both curves are symmetric with respect to the y-axis, we can work with the right half-plane. Since the region that is shaded lies between the radial lines $\theta = 0$ and $\theta = \pi$, the area of common region is

$$\text{Area} = 2\left(\dfrac{1}{2}\right)\int_0^{\frac{\pi}{2}}\left(r_2^{\,2} - r_1^{\,2}\right)d\theta = \int_0^{\frac{\pi}{2}}\left(9\sin^2\theta - \sin^2\theta\right)d\theta = 8\int_0^{\frac{\pi}{2}}\sin^2\theta\, d\theta.$$

24. C

We know that the carrying capacity of a logistic growth is twice the value of P at which the graph has a point of inflection.
Carrying capacity, m, is $m = 2 \times 200 = 400$.

$\dfrac{dP}{dt} = kP\left(1 - \dfrac{P}{m}\right)$

Only choice (C) has the carrying capacity of 400.
At $k = 0.2$

(C) $\dfrac{dP}{dt} = 0.2P - 0.0005P^2 = 0.2P\left(1 - \dfrac{P}{400}\right)$

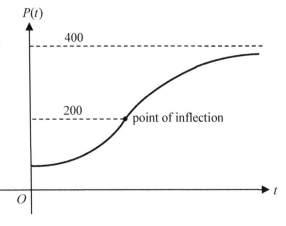

25. C

Because $\cos x = (\sin x)' = 1 - \dfrac{x^2}{2!} + \dfrac{x^4}{4!} - \dfrac{x^6}{6!} + \cdots$, it follows that

$\cos(x^2) = 1 - \dfrac{x^4}{2!} + \dfrac{x^8}{4!} - \cdots = 1 - \dfrac{x^4}{2!} + \dfrac{x^8}{4!} - \dfrac{x^{12}}{6!}\cdots.$

Therefore, the coefficient of x^8 is $\dfrac{1}{4!} = \dfrac{1}{24}$.

26. B

The average value of $f(x)$ on the closed interval $[-1, 1] = \dfrac{1}{1-(-1)}\int_{-1}^{1} f(x)\,dx = \dfrac{1}{2}\int_{-1}^{1} f(x)\,dx$

From the equation of the line tangent $y = 2x - 2$, we know that $f'(0) = 2$ and $f(0) = -2$.

Therefore, $f'(x) = \int(12x + 8)\,dx = 6x^2 + 8x + C$ and $f'(0) = 2$. We got $C = 2$.

$f(x) = \int\left(6x^2 + 8x + 2\right)dx = 2x^3 + 4x^2 + 2x + C$ and $f(0) = -2 \;\to\; C = -2$.

Now we got $f(x) = 2x^3 + 4x^2 + 2x - 2$.

The average value of $f(x)$ on the interval $[-1, 1]$ is

AP Calculus BC Test 2 Answers Section 1

$$\frac{1}{2}\int_{-1}^{1}(2x^3+4x^2+2x-2)dx = \frac{1}{2}\left[\frac{1}{2}x^4+\frac{4}{3}x^3+x^2-2x\right]_{-1}^{1} = -\frac{2}{3}.$$

27. C

From the equation $r = 2\cos 3\theta$, we can find the angles as follows.

$$0 = 2\cos 3\theta \rightarrow 3\theta = \pm\frac{\pi}{2} \rightarrow \theta = \pm\frac{\pi}{6}.$$

We can see that the right petal is traced as θ increases from $-\frac{\pi}{6}$ to $\frac{\pi}{6}$, the area is

$$\text{Area} = \frac{1}{2}\int_{-\pi/6}^{\pi/6}(2\cos 3\theta)^2 d\theta = \int_{0}^{\pi/6}(2\cos 3\theta)^2 d\theta = 4\int_{0}^{\pi/6}(\cos 3\theta)^2 d\theta$$

28. A

Since the speed of a particle is $s = \sqrt{\left(\frac{dx}{dt}\right)^2+\left(\frac{dy}{dt}\right)^2}$,

$$s = \sqrt{\left(\frac{dx}{dt}\right)^2+\left(\frac{dy}{dt}\right)^2} = \sqrt{104} \text{ or } \left(\frac{dx}{dt}\right)^2+\left(\frac{dy}{dt}\right)^2 = 104. \text{ ----(1)}$$

From the equation $y = x^2 - 3x + 4$, we got that $\frac{dy}{dt} = 2x\frac{dx}{dt} - 3\frac{dx}{dt}$.

At $x = 4$, $\frac{dy}{dt} = 2(4)\frac{dx}{dt} - 3\frac{dx}{dt}$ or $\frac{dy}{dx} = 5\left(\frac{dx}{dt}\right)$. Substitute this into equation (1).

Therefore, $\left(\frac{dx}{dt}\right)^2 + \left(5\frac{dx}{dt}\right)^2 = 104$ or $26\left(\frac{dx}{dt}\right)^2 = 104$.

Therefore, $\left(\frac{dx}{dt}\right)^2 = 4$ or $\frac{dx}{dt} = \pm 2$. from the graph we know that $\frac{dx}{dt}$ is positive at $x = 4$ because the

particle moves to the right. Since $\frac{dx}{dt} = 2$, then $\frac{dy}{dt} = 5\frac{dx}{dt} = 5(2) = 10$.

SECTION 1, PART B

76. D

From $\sum_{n=1}^{\infty}\frac{k^n}{2^{2n+1}} = \frac{1}{2}\sum_{n=1}^{\infty}\frac{k^n}{2^{2n}} = \frac{1}{2}\sum_{n=1}^{\infty}\frac{k^n}{4^n} = \frac{1}{2}\sum_{n=1}^{\infty}\left(\frac{k}{4}\right)^n$, we can see that the series is a geometric series

with a common ratio $\frac{k}{4}$.

If $\left|\frac{k}{4}\right| < 1$, then the geometric series converges. Therefore, $\left|\frac{k}{4}\right| < 1$ or $-4 < k < 4$.

On geometric series convergence, you don't need the end point test.
Or
When we apply the ratio test, we need the end point test as follows.

$$r = \lim_{n\to\infty}\left|\frac{k^{n+1}}{2^{2(n+1)+1}}\cdot\frac{2^{2n+1}}{k^n}\right| = r = \lim_{n\to\infty}\left|\frac{k}{4}\right| \rightarrow \left|\frac{k}{4}\right| < 1 \rightarrow |k| < 4 \rightarrow -4 < k < 4.$$

End point test:

At $k = 4$, $\sum_{n=1}^{\infty}\frac{4^n}{2^{2n+1}} = \sum_{n=1}^{\infty}\frac{2^{2n}}{2\cdot 2^{2n}} = \sum_{n=1}^{\infty}\frac{1}{2} = \frac{1}{2}+\frac{1}{2}+\frac{1}{2}+\cdots$. The series diverges.

432

AP Calculus BC Test 2 Answers

Section 1

At $k = -4$, $\sum_{n=1}^{\infty} \frac{(-4)^n}{2^{2n+1}} = \sum_{n=1}^{\infty} \frac{(-1)^n 4^n}{2 \cdot 2^{2n}} = \sum_{n=1}^{\infty} \frac{(-1)^n 2^{2n}}{2 \cdot 2^{2n}} = \sum_{n=1}^{\infty} (-1)^n \frac{1}{2} = -\frac{1}{2} + \frac{1}{2} - \frac{1}{2} + \cdots$. The series diverges.

Therefore, the interval of convergence is $-4 < k < 4$.

77. E

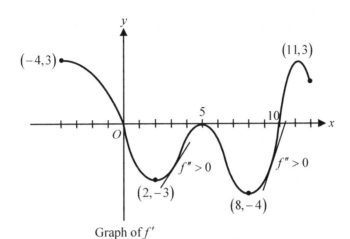

Graph of f'

We know that $f''(x)$ is the slope of the line tangent on the graph of f'.

For the graph concave up, $f''(x)$ should be positive. The slope of tangent line on the curve is positive on the interval $2 < x < 5$ and $8 < x < 11$.

78. A

In this case, Cylindrical Shell method is useful.

Since $V = 2\pi \int_a^b xy \, dx$ and $y = -x^2 + 4x - 3$,

Volume $= 2\pi \int_1^3 x(-x^2 + 4x - 3) dx \approx 16.755$

79. C

If $f'(x) > 0$, then the graph of f is increasing.

Therefore, $f'(x) = e^x \cos x - e^x \sin x = e^x (\cos x - \sin x)$ and the graph of f' is as follows.

$f'(x)$ is negative for $0.785 < x < 3.927$.

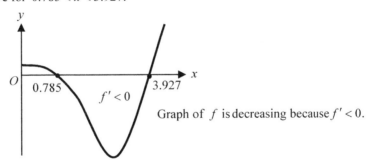

Graph of f is decreasing because $f' < 0$.

AP Calculus BC Test 2 Answers — Section 1

80. A

We know that $\dfrac{dx}{dt} = 6t^2 - 2t$ and $\dfrac{dy}{dt} = 3t^2 - 8$.

Because slope $m = \dfrac{dy}{dx} = \dfrac{dy/dt}{dx/dt}$, $\left. m = \dfrac{3t^2 - 8}{6t^2 - 2t} \right|_{t=2} = \dfrac{3(2)^2 - 8}{6(2)^2 - 2(2)} = \dfrac{4}{20} = \dfrac{1}{5}$.

Next, find $x(2)$ and $y(2)$.

$x(2) = 2(2)^3 - (2)^2 = 12$ and $y(2) = 2^3 - 8(2) = -8$

Therefore, the equation of the tangent is $y - (-8) = \dfrac{1}{5}(x - 12) \rightarrow 5y + 40 = x - 12 \rightarrow x - 5y = 52$.

81. C

From the graph of v, the derivative of $x(t)$, we can see the following properties.

1) For $0 < t < 0.5$ and $4 < t < 6$, the graph of $x(t)$ is decreasing.
2) For $0.5 < t < 4$, the graph of $x(t)$ is increasing.
3) For $-1 < t < 3$, the graph of $x(t)$ is concave up.
4) At $x = 3$, the graph of $x(t)$ has a point of inflection.
5) For $3 < t < 6$, the graph of $x(t)$ is concave down because the slope of v is negative.

82. C

We apply distance formula $d = \displaystyle\int_a^b \sqrt{\left(\dfrac{dx}{dt}\right)^2 + \left(\dfrac{dy}{dt}\right)^2}\, dt$.

$d = \displaystyle\int_1^2 \sqrt{(e^t \sin t)^2 + (e^t \cos t)^2}\, dt = \int_1^2 \sqrt{e^{2t}\sin^2 t + e^{2t}\cos^2 t}\, dt = \int_1^2 \sqrt{e^{2t}(\sin^2 t + \cos^2 t)}\, dt$

$= \displaystyle\int_1^2 \sqrt{e^{2t}}\, dt = \int_1^2 e^t\, dt = \left[e^t\right]_1^2 = e^2 - e = e(e-1)$

83. C

Because the series $\displaystyle\sum_{n=3}^{\infty} \dfrac{1}{n^p}$ converges, p must be greater than 1. ($p > 1$)

(A) $\displaystyle\sum_{n=3}^{\infty} \dfrac{n}{n^p} = \sum_{n=3}^{\infty} \dfrac{1}{n^{p-1}} \rightarrow p - 1 > 0$: False because the series diverges when $0 < p - 1 < 1$.

(B) $\displaystyle\sum_{n=3}^{\infty} \dfrac{n^2}{n^p} = \sum_{n=3}^{\infty} \dfrac{1}{n^{p-2}} \Rightarrow p - 2 > -1$: False because the series diverges when $0 < p - 2 < 1$.

(C) $\displaystyle\sum_{n=3}^{\infty} \dfrac{1}{n^p \ln(n)} = \sum_{n=3}^{\infty} \dfrac{1}{n^p \ln(n)}$: True because of the comparison test as follows.

$\displaystyle\sum_{n=3}^{\infty} \dfrac{1}{n^p \ln(n)} = \dfrac{1}{3^p \ln 3} + \dfrac{1}{3^p \ln 4} + \dfrac{1}{3^p \ln 5} + \cdots < \sum_{n=3}^{\infty} \dfrac{1}{n^p \ln(3)} = \dfrac{1}{\ln 3} \sum_{n=3}^{\infty} \dfrac{1}{n^p}$: The series converges.

(D) $\displaystyle\sum_{n=3}^{\infty} \dfrac{1}{n^p e} = \dfrac{1}{e} \sum_{n=3}^{\infty} \dfrac{1}{n^p}$: The series converges because $p > 1$.

(E) $\displaystyle\sum_{n=3}^{\infty} \dfrac{1}{n^p n^2} = \sum_{n=3}^{\infty} \dfrac{1}{n^{p+2}}$: The series converges because $p + 2 > 3$.

AP Calculus BC Test 2 Answers

84. E

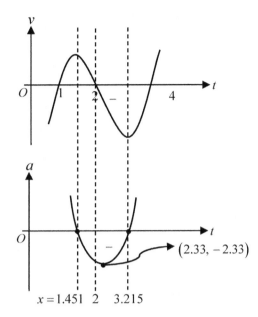

In the graph of v, the graph of $v(t)$ is decreasing on the interval $(2,3)$.

In the graph of a, the graph of $a(t)$ is decreasing and then increasing after the turning point on the interval $(2,3)$.

In the graphs of v and a, the particle is speeding up on the interval $(2,3)$ because they both have the same signs on that interval.

85. D

From the equation $y = \dfrac{x^3}{3} + 3x^2 - 1$, $\dfrac{dy}{dt} = x^2 \dfrac{dx}{dt} + 6x \dfrac{dx}{dt}$ where $\dfrac{dx}{dt} = 2$.

Therefore, $\dfrac{dy}{dt} = 2x^2 + 12x$.

When $t = 2$, $x(2) = 4 + 1 = 5$.

Therefore, $\left. \dfrac{dy}{dt} \right|_{t=2} = 2(5)^2 + 12(5) = 110$.

86. A

Draw the graphs of $y = e^{-x}$ and $y = \sqrt{x}$
using your graphing calculator and find
the point of intersection.
Since the radius of the semicircle is

$r = \dfrac{e^{-x} - \sqrt{x}}{2}$, the volume is

$V = \dfrac{1}{2} \int_0^{0.426} \left[\pi \left(\dfrac{e^{-x} - \sqrt{x}}{2} \right)^2 \right] dx = 0.034768 \cdots \approx 0.035$.

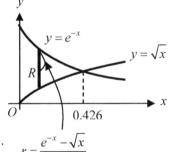

87. C

The second-degree Taylor polynomial for g about $x = 1$ is $P(x) = g(1) + g'(1)(x-1) + \dfrac{g''(1)}{2!}(x-1)^2$.

We need to find $g(1), g'(1),$ and $g''(1)$.

AP Calculus BC Test 2 Answers

$g(1) = \int_2^1 f(x)dx = -\int_1^2 f(x)dx = -\left(\frac{1 \cdot 4}{2}\right) = -2$

$g'(1) = f(1) = 0$ because $g'(x) = f(x)$.

$g''(1) = 4$ because the slope of the graph of f at $x = 1$ is 4.

Therefore, $P(x) = -2 + 0 \cdot (x-1) + \frac{4}{2!}(x-1)^2 = -2 + 2(x-1)^2$ and $P(1.5) = -2 + 2(1.5-1)^2 = -1.5$.

88. A

We know that the Taylor series for e^x about $x = 0$ is $e^x = 1 + x + \frac{x^2}{2!} + \frac{x^3}{3!} + \frac{x^4}{4!} + \cdots$.

It is followed that $e^{x^2} = 1 + x^2 + \frac{x^4}{2!} + \frac{x^6}{3!} + \cdots + \frac{(x^2)^n}{n!} + \cdots$.

Now we will find the point of intersection of $y = e^{x^2}$ and $x = \frac{1}{2}$. That is, by substitution,

$y = e^{\left(\frac{1}{2}\right)^2} \approx 1.284$. Therefore, the point of intersection is $\left(\frac{1}{2}, 1.284\right)$.

89. E

The graph of g is increasing over the x-axis because $g'(x) = f(x) > 0$. True

The graph of g is continuous over the x-axis because the graph of f is continuous. True

The graph of g is always concave up over the x-axis. True

g'' is the slope of the line tangent to the graph of f. The slope at any point on the curve is positive, because the graph of f is increasing over the x-axis.

Therefore, the graph of g is concave up over the x-axis.

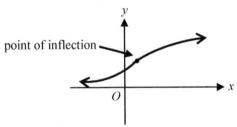

The graph of f could possibly be concave up and concave down in the figure as shown above. But the graph of g is concave up because $g''(x) > 0$ over the x-axis.

90. E

Since $g'(x) = \frac{d}{dx}\int_0^{x^2} e^{-\sqrt{t}}\sin\left(\frac{1}{t}\right)dt = e^{-\sqrt{x^2}}\sin\left(\frac{1}{x^2}\right) \cdot 2x$,

we got that $g'(x) = 2xe^{-x}\sin\left(\frac{1}{x^2}\right)$. Draw the graph of g' using a graphing calculator.

We can find the point of inflection on the graph of g' such that $g''(x) = 0$.

Therefore, we have a point of inflection at $x = 0.814$.

AP Calculus BC Test 2 Answers   Section 1

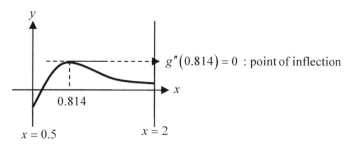

Graph of g'

91. **D**

$f'(x) = \dfrac{1}{x}$ or $f'(e) = \dfrac{1}{e}$ is the slope of the tangent. We need to find the y-coordinate of the point on the curve at $x = e$. Therefore, $f(e) = 1$, that is $(e, 1)$.

Therefore, the equation of tangent is $y - 1 = \dfrac{1}{e}(x - e)$ or $y - 1 = \dfrac{1}{e}x - 1$.

At $x = 2e$, $y = \dfrac{1}{e}(2e) = 2$.

Now the error resulting from the tangent line approximation is
Error $= |\ln(2e) - 2| = 0.30685\cdots \approx 0.307$

92. **D**

We know that $F(1) = \displaystyle\int_0^1 \sin t\, dt$ is the area under the curve on the closed interval $[0, 1]$.

We apply the formula for the area of trapezoidal approximation.

Area $= \dfrac{h}{2}\left[y_0 + 2(y_1 + y_2 + y_3) + y_4\right]$

where $h = \dfrac{1-0}{4} = 0.25$, $y_0 = \sin 0$, $y_1 = \sin(0.25)$, $y_2 = \sin(0.5)$, $y_3 = \sin(0.75)$, and $y_4 = \sin(1)$.

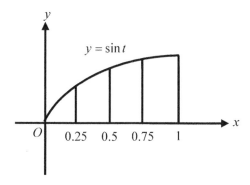

Area $F(1) \approx \dfrac{0.25}{2}\left[0 + 2(\sin 0.25 + \sin 0.5 + \sin 0.75) + \sin 1\right] = 0.4573009376 \approx 0.457$

Compare with $F(1) = \displaystyle\int_0^1 \sin t\, dt = 0.459698\cdots \approx 0.459$.

We know that the trapezoidal approximation is under estimation because the graph of $y = \sin t$ is concave down on the interval $[0, 1]$.

Dr. John Chung's AP Calculus AB/BC

AP Calculus BC Test 2 Answers

SECTION II, PART A

Question 1

(a) The volume of the water in the tank at time t is $W = \pi r^2 h(t)$.

Because $\dfrac{dh}{dt} = 2\sqrt{t} \cdot \sin(0.05t)$, $h(t) = \int_0^t 2\sqrt{t} \cdot \sin(0.05t)\, dt$.

Therefore,
$$W = \pi r^2 h = 25\pi h(t) = 25\pi \int_0^t 2\sqrt{t} \cdot \sin(0.05t)\, dt \text{ ft}^3$$

(b) $h(10) = \int_0^{10} 2\sqrt{t} \cdot \sin(0.05t)\, dt \approx 12.359 \text{ ft}$

(c) $\dfrac{dW}{dt} = \pi r^2 \dfrac{dh}{dt} = 25\pi \cdot 2\sqrt{t} \cdot \sin(0.05t)$

$\left.\dfrac{dW}{dt}\right|_{t=10} = 25\pi \cdot 2\sqrt{10} \cdot \sin(0.05 \times 10) \approx 238.145 \text{ ft}^3/\text{minute}$

(d) Average rate of change in the height $= \dfrac{h(10) - h(0)}{10} = \dfrac{1}{10}\int_0^{10} 2\sqrt{t} \cdot \sin(0.05t)\, dt \approx 1.236 \text{ ft/minute}$

Average rate of change in the volume of the water is
$$\dfrac{W(10) - W(0)}{10} = \dfrac{1}{10}\int_0^{10} 25\pi \cdot 2\sqrt{t} \cdot \sin(0.05t)\, dt = 97.066 \text{ ft}^3/\text{minute}$$

(e) We know that $R_1(t)$ is the rate of change in the height of the first tank and $R_2(t)$ is the rate of change in the height of the second tank.

For some time t, $F(t) = 2\sqrt{t} \cdot \sin(0.05t) - 0.002t^3 = 0$ such that $R_1(t) = R_2(t)$.

We apply the Intermediate Value Theorem.
$$\begin{cases} F(5) = 2\sqrt{5} \cdot \sin(0.05 \times 5) - 0.002(5^3) \approx 0.856 > 0 \\ F(20) = 2\sqrt{20}\sin(0.05 \times 20) - 0.002(20^3) \approx -8.474 < 0 \end{cases}$$

Because $F(t)$ is continuous on the closed interval $[5, 20]$ and 0 is between $F(5)$ and $F(20)$, there is some time c in $(5, 20)$ such that $F(c) = 0$.

Intermediate Value Theorem:
If f is continuous on the closed interval $[a,b]$ and k is any number between $f(a)$ and $f(b)$, then there is at least one number c in (a,b) such that $f(c) = k$.

Question 2

(a) $\begin{cases} x = r\cos\theta \to x = 2(1-\sin\theta)\cos\theta \\ y = r\sin\theta \to y = 2(1-\sin\theta)\sin\theta \end{cases}$, where $r = 2(1-\sin\theta)$.

$$\dfrac{dy}{dx} = \dfrac{dy/d\theta}{dx/d\theta} = \dfrac{2\cos\theta - 4\sin\theta\cos\theta}{-2\sin\theta - 2(\cos^2\theta - \sin^2\theta)} = \dfrac{2\cos\theta - 2\sin(2\theta)}{-2\sin\theta - 2\cos(2\theta)} = \dfrac{\cos\theta - \sin(2\theta)}{-\sin\theta - \cos(2\theta)}$$

AP Calculus BC Test 2 Answers — Section 1

(b) If $0 = 2(1 - \sin\theta)$, then $\sin\theta = 1$ or $\theta = \dfrac{\pi}{2}$.

Therefore, Area $= \dfrac{1}{2}\int_{\pi/2}^{\pi} r^2 d\theta = \dfrac{1}{2}\int_{\pi/2}^{\pi}[2(1-\sin\theta)]^2 d\theta \approx 0.712$.

(c) Draw the graphs of $y_1 = 2(1 - \sin\theta)\cos\theta$ and $y_2 = -1$ and find the value of θ at the point of intersection. We will get $\theta = 2.683$ in the interval $\left(\dfrac{\pi}{2}, \pi\right)$.

Now find the y-coordinate of point P from the equation $y = 2(1 - \sin\theta)\sin\theta$.

Therefore, $y = 2(1 - \sin 2.683)(\sin 2.683) \approx 0.493$

Question 3

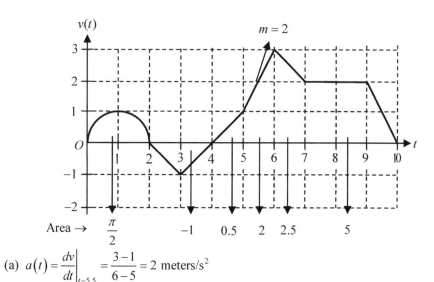

(a) $a(t) = \dfrac{dv}{dt}\bigg|_{t=5.5} = \dfrac{3-1}{6-5} = 2$ meters/s^2

Dr. John Chung's AP Calculus AB/BC

AP Calculus BC Test 2 Answers — Section 1

(b) $\int_0^{10} v(t)\,dt = \dfrac{\pi}{2} - 1 + 0.5 + 2 + 2.5 + 5 = \dfrac{\pi}{2} + 9$ meters

$\int_0^{10} v(t)\,dt$ is the change in position from time $t=0$ to $t=10$.

(c) $\int_0^{10} |v(t)|\,dt = \dfrac{\pi}{2} + 1 + 0.5 + 2 + 2.5 + 5 = \dfrac{\pi}{2} + 11$

$\int_0^{10} |v(t)|\,dt$ is the total distance, in meters, that the particle traveled over the first 10 seconds.

(d) At $t=2$ and $t=4$, the particle changes its direction because $v(t)=0$ at each time of these values.

Question 4

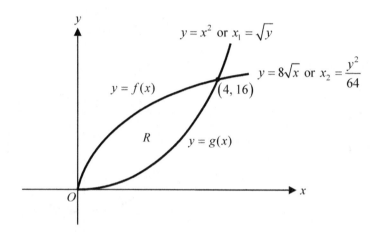

(a) Find the point of intersection of the graphs $y = 8\sqrt{x}$ and $y = x^2$.

Solve the equation $8\sqrt{x} = x^2 \;\to\; 64x = x^4 \;\to\; x(64 - x^3) = 0 \;\to\; x=0$ or $x=4$.

$$\text{Area} = \int_0^4 \left(8\sqrt{x} - x^2\right)dx = \left[\dfrac{16x^{3/2}}{3} - \dfrac{x^3}{3}\right]_0^4 = \dfrac{16(4)^{3/2} - 4^3}{3} = \dfrac{16(2^2)^{3/2} - 4^3}{3} = \dfrac{64}{3}$$

(b) Washer: $V = \pi \displaystyle\int_0^{16}\left[\left(\sqrt{y}\right)^2 - \left(\dfrac{y^2}{64}\right)^2\right]dy = \pi\int_0^{16}\left(y - \dfrac{y^4}{64^2}\right)dy = \pi\left[\dfrac{y^2}{2} - \dfrac{y^5}{5\cdot 64^2}\right]_0^{16} = 76.8\pi$

Shell: $V = 2\pi \displaystyle\int_0^4 x\left(8\sqrt{x} - x^2\right)dx = 2\pi\int_0^4 \left(8x^{3/2} - x^3\right)dx = 2\pi\left[\dfrac{16}{5}x^{5/2} - \dfrac{x^4}{4}\right]_0^4$

$\quad = 2\pi\left[\dfrac{16}{5}(2^2)^{\frac{5}{2}} - \dfrac{4^4}{4}\right] = 76.8\pi$

(c) The length of the base of a equilateral triangle is $x_1 - x_2 = \sqrt{y} - \dfrac{y^2}{64}$ and the area of the triangle is

$$A(y) = \dfrac{1}{2}\left(\sqrt{y} - \dfrac{y^2}{64}\right)^2 \sin 60 = \dfrac{1}{2}\left(\sqrt{y} - \dfrac{y^2}{64}\right)^2 \dfrac{\sqrt{3}}{2} = \dfrac{\sqrt{3}}{4}\left(\sqrt{y} - \dfrac{y^2}{64}\right)^2.$$

Therefore, volume of the solid is

AP Calculus BC Test 2 Answers Section 1

$$V = \int_0^{16} A(y)\, dy = \int_0^{16} \frac{\sqrt{3}}{4}\left(\sqrt{y} - \frac{y^2}{64}\right)^2 dy.$$

(d) Since $f'(x) = \dfrac{4}{\sqrt{x}}$, the slope of the tangent at $x = 1$ is $f'(1) = 4$.

The slope of the tangent to the graph of g at $x = k$ is $g'(k) = 2k$ where $g'(x) = 2x$.

Because these two line are parallel, $f'(1) = g'(k)$ or $4 = 2k \rightarrow k = 2$.

Because $g(2) = 4$, the coordinates of point P is $(2, 4)$.

Question 5

(a) If $y = P(t)$ is the particular solution to the differential equation at time $t \geq 0$, then find $y = P(t)$.

$$\frac{dP}{500-P} = \frac{1}{10} dt \;\Rightarrow\; \int \frac{dP}{500-P} = \int \frac{1}{10} dt \;\Rightarrow\; -\ln|500-P| = \frac{1}{10} t + C_1$$

The solution is $500 - P = A e^{-\frac{1}{10} t} \;\rightarrow\; P = 500 - A e^{-\frac{1}{10} t}$

$P = 500 - A e^{-\frac{t}{10}}$. With initial condition $(0, 100)$. $100 = 500 - A e^0 \;\rightarrow\; A = 400$

Therefore, $P(t) = 500 - 400 e^{-\frac{t}{10}}$.

(b) $\lim\limits_{t \to \infty} \left(500 - 400 e^{-\frac{t}{10}} \right) = 500$

(c) $P(t) = 500 - 400 e^{-\frac{t}{10}} \;\rightarrow\; P'(t) = 40 e^{-\frac{t}{10}} \;\rightarrow\; P''(t) = -4 e^{-\frac{t}{10}}$

Since $P''(t) < 0$ over all t, the graph of $y = P(t)$ is concave down.

(d) Find $\dfrac{d^2 P}{dt^2}$ in terms of P. Is the value of $\dfrac{d^2 P}{dt^2}$ always positive? Explain your reasoning.

$$\frac{dP}{dt} = \frac{1}{10}(500 - P) \;\rightarrow\; \frac{d^2 P}{dt^2} = -\frac{1}{10} \cdot \frac{dP}{dt} = -\frac{1}{10}\left(\frac{500-P}{10}\right) = -\frac{500-P}{100}$$

Since $100 \leq P < 500$, $\dfrac{d^2 P}{dt^2} = \dfrac{P - 500}{100} < 0$. The value of $\dfrac{d^2 P}{dt^2}$ is always negative.

Question 6

(a) Replace x with x^2 as follows.

$$f(x) = \frac{1}{1+2x} = 1 - (2x) + (2x)^2 - (2x)^3 + \cdots + (-1)^n (2x)^n + \cdots$$

(b) $\lim\limits_{n \to \infty} \left| \dfrac{a_{n+1}}{a_n} \right| = \lim\limits_{n \to \infty} \left| \dfrac{(-1)^{n+1}(2x)^{n+1}}{(-1)^n (2x)^n} \right| = \lim\limits_{n \to \infty} \left| \dfrac{(-1)(2x)}{1} \right| = |2x| < 1 \;\Rightarrow\; |x| < \dfrac{1}{2}$

The radius of convergence $= \dfrac{1}{2}$

We know that the series is geometric with common ratio $-2x$.

If $|-2x| < 1 \rightarrow |2x| < 1$, then geometric series converges. $-1 < 2x < 1 \rightarrow -\dfrac{1}{2} < x < \dfrac{1}{2}$

AP Calculus BC Test 2 Answers — Section 1

Therefore, the interval of convergence is $-\dfrac{1}{2} < x < \dfrac{1}{2}$.

(c) $g(x) = \displaystyle\int_0^x 1 - (2t) + (2t)^2 - (2t)^3 + \cdots + (-1)^n (2t)^n + \cdots \, dt$

$= \left[t - t^2 + \dfrac{4}{3}t^3 - 2t^4 + \cdots \right]_0^x = x - x^2 + \dfrac{4}{3}x^3 - 2x^4 + \cdots$

Therefore, $P_3(x) = x - x^2 + \dfrac{4}{3}x^3$ and $P_3\left(\dfrac{1}{4}\right) = \dfrac{1}{4} - \left(\dfrac{1}{4}\right)^2 + \dfrac{4}{3}\left(\dfrac{1}{4}\right)^3 = \dfrac{1}{4} - \dfrac{1}{16} + \dfrac{1}{48} = \dfrac{12 - 3 + 1}{48} = \dfrac{5}{24}$.

(d) $g(x) = x - x^2 + \dfrac{4}{3}x^3 - 2x^4 + \cdots = g(0) + g'(0)x + \dfrac{g''(0)}{2!}x^2 + \dfrac{g'''(0)}{3!}x^3 + \dfrac{g^{(4)}(0)}{4!}x^4 + \cdots$

From the equation above, compare both sides.

Therefore, $\dfrac{g^{(4)}(0)}{4!}x^4 = -2x^4$ or $g^{(4)}(0) = -48$.

(e) $\left| P_3\left(\dfrac{1}{4}\right) - g\left(\dfrac{1}{4}\right) \right| \le \left| -2\left(\dfrac{1}{4}\right)^4 \right| = \dfrac{1}{128} < \dfrac{1}{100}$, because the series of g has terms that alternate,

decrease in absolute, and have a limit 0. Therefore, roughly we can say the error is bounded by the absolute value of the next term.

Or

$\left| P_3\left(\dfrac{1}{4}\right) - g\left(\dfrac{1}{4}\right) \right| \le \left| \dfrac{g^{(4)}(z)}{4!} \left(\dfrac{1}{4}\right)^4 \right| < \left| \dfrac{g^{(4)}(0)}{4!} \left(\dfrac{1}{4^4}\right) \right| = \dfrac{1}{128} < \dfrac{1}{100}$

For $0 \le z \le \dfrac{1}{4}$, $\left| g^{(4)}(0) \right|$ is the absolute maximum.

We can prove that $g'(x)$ has the absolute maximum at $x = 0$ on the interval $\left[0, \dfrac{1}{4}\right]$ as follows.

$\begin{cases} g'(x) = f(x) = \dfrac{1}{1+2x} \\ g''(x) = \dfrac{-2}{(1+2x)^2} \\ g'''(x) = \dfrac{8}{(1+2x)^3} \\ g^{(4)}(x) = \dfrac{-48}{(1+2x)^4} \end{cases}$

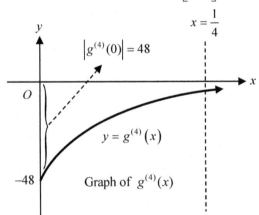

Graph of $g^{(4)}(x)$

Observe.

Actually $g(x) = \displaystyle\int_0^x \dfrac{1}{1+2t}\, dt = \dfrac{1}{2}\ln(1+2t)\Big|_0^x = \dfrac{1}{2}\ln(1+2x)$ and $g\left(\dfrac{1}{4}\right) = \dfrac{1}{2}\ln\left(1 + 2\cdot\dfrac{1}{4}\right) = 0.20273255$.

The exact value of the error is $\left| P_3\left(\dfrac{1}{4}\right) - g\left(\dfrac{1}{4}\right) \right| = \left| \dfrac{5}{24} - 0.20273255 \right| = 0.00560078$,

which is less than $\dfrac{1}{100}$.

AP CALCULUS BC

TEST 3

AP CALCULUS BC

A CALCULATOR **CANNOT BE USED ON PART A OF SECTION 1**. A GRAPHIC CALCULATOR FROM THE APPROVED LIST **IS REQUIRED FOR PART B OF SECTION 1 AND FOR PART A OF SECTION II** OF THE EXAMINATION.

SECTION 1
Time: 1 hour and 45 minutes
All questions are given equal weight.
Percent of total grade – 50

Part A: 55 minutes, 28 multiple-choice questions
A calculator is NOT allowed.
Part B: 50 minutes, 17 multiple-choice questions
A graphing calculator is required.

Part A

CALCULUS BC
SECTION 1, PART A
Time — 55 minutes
Number of questions — 28

A CALCULATOR MAY NOT BE USED ON THIS PART OF THE EXAMINATION.

Directions: Solve each of the following problems, using the available space for scratchwork. After examining the form of the choices, decide which is the best of the choices given and fill in the corresponding oval on the answer sheet. No credit will be given for anything written in the test book. Do not spend too much time on any one problem.

In this test:

(1) Unless otherwise specified, the domain of a function f is assumed to be the set of all real numbers x for which $f(x)$ is a real number.

(2) The inverse of a trigonometric function f may be indicated using the inverse function notation f^{-1} or with the prefix "arc" (e.g, $\sin^{-1} x = \arcsin x$)

AP Calculus BC Test 3   Section 1

1. Let f be the function given by $f(x) = xe^{-x}$. Which of the following is the x-coordinate of the point of inflection on the graph of f?

   (A) −2   (B) −1   (C) 0   (D) 2   (E) e

2. $\int_0^3 \dfrac{x}{\sqrt{1+x}}\,dx =$

   (A) $\dfrac{4}{3}$ (B) 2   (C) $\dfrac{7}{3}$ (D) $\dfrac{8}{3}$ (E) 3

3. If $x = \sqrt{y} + 1$, then what is the value of $\dfrac{dy}{dx}$ at $x = 3$?

   (A) −2   (B) −1   (C) 1   (D) 2   (E) 4

4. $\lim\limits_{h \to 0} \dfrac{\arctan\left(\dfrac{1}{4}+h\right) - \arctan\left(\dfrac{1}{4}\right)}{h} =$

   (A) $\dfrac{4}{5}$ (B) $\dfrac{16}{17}$ (C) 1   (D) $\dfrac{5}{4}$ (E) ∞

5. If the function f defined by $f(x) = \begin{cases} 1 + \ln x, & x \geq e \\ ax - b, & x < e \end{cases}$ is differentiable at $x = e$, where a and b are constants, what is the value of b?

   (A) −2   (B) −1   (C) $\dfrac{1}{e}$ (D) e (E) $2e$

6. If the function $f(x) = \dfrac{4x - k}{x^2 + 1}$ has a relative maximum at $x = 2$, which of the following is the value of k?

   (A) −3   (B) −2   (C) 1   (D) 2   (E) 3

AP Calculus BC Test 3　　　　Section 1

7. If the parametric equations are $x(t) = e^t$ and $y(t) = t$, which of the following could be the graph of $y = f(x)$?

(A)

(B)

(C)

(D)

(E)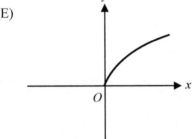

---

8. $\lim\limits_{x \to 1} \left(\sum\limits_{n=0}^{\infty} \dfrac{x^n}{n!} \right) =$

(A) $-\infty$　　(B) 0　　(C) 1　　(D) e　　(E) ∞

AP Calculus BC Test 3

Section 1

9. What is the average value of $\dfrac{2}{1+x^2}$ over the closed interval $[0, 1]$?

   (A) $\dfrac{\pi}{8}$ (B) 1 (C) $\dfrac{\pi}{4}$ (D) $\dfrac{\pi}{2}$ (E) π

---

10. Which of the following is the interval of convergence of the series $\displaystyle\sum_{n=0}^{\infty} \dfrac{(x-2)^n}{(n+1)^2}$?

    (A) $0 < x < 3$
    (B) $1 < x < 3$
    (C) $1 \le x < 3$
    (D) $1 < x \le 3$
    (E) $1 \le x \le 3$

---

11. If $f(x) = \displaystyle\int_0^{\ln x} e^{-t}\, dt$, then $f'(x) =$

    (A) $\dfrac{e}{x}$ (B) $e^{-\ln x}$ (C) $\dfrac{1}{x^2}$ (D) $\ln x$ (E) $\dfrac{1}{x}$

---

12. If $x = \sin y$ for $0 \le y \le \dfrac{\pi}{2}$, which of the following is the value of $\dfrac{dy}{dx}$ at $x = \dfrac{1}{2}$?

    (A) $-\dfrac{\sqrt{3}}{2}$ (B) $-\dfrac{1}{2}$ (C) $\dfrac{1}{2}$ (D) $\dfrac{\sqrt{3}}{2}$ (E) $\dfrac{2\sqrt{3}}{3}$

Dr. John Chung's AP Calculus AB/BC

AP Calculus BC Test 3 — Section 1

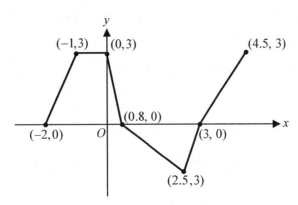

13. The graph of the piecewise linear function f is shown in the figure above. If $g(x) = \int_3^x f(t)\,dt$, what is value of $g(-1)$?

(A) −1.5
(B) −1.2
(C) −0.9
(D) 0.9
(E) 1.2

---

14. Which of the following series converges?

(A) $\sum_{n=1}^{\infty} \dfrac{n}{2n+3}$

(B) $\sum_{n=1}^{\infty} \dfrac{5(3)^n}{2^n}$

(C) $\sum_{n=1}^{\infty} \dfrac{2^n+1}{2^{n+1}}$

(D) $\sum_{n=1}^{\infty} \dfrac{1}{n+2}$

(E) $\sum_{n=1}^{\infty} \dfrac{(-1)^n}{n}$

---

15. If $f(x) = \ln\left(\dfrac{1-\cos x}{1+\cos x}\right)$, then $f'\left(\dfrac{\pi}{6}\right) =$

(A) −4 (B) $-\dfrac{1}{4}$ (C) $\dfrac{1}{4}$ (D) 2 (E) 4

AP Calculus BC Test 3

x	0	2	3	5
$f'(x)$	5	0	-2	3

16. Let f be the function differentiable over all real numbers x. If the function f has selected values of its first derivative f' given in the table above, which of the following statements must be true?

    (A) f is increasing on the interval $(0, 2)$.
    (B) f is decreasing on the interval $(0, 2)$.
    (C) f has a local maximum at $x = 2$.
    (D) f has a local minimum at $x = 2$.
    (E) f has at least one local minimum on the interval $(2, 5)$.

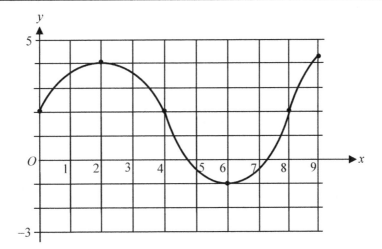

17. The graph of a differentiable function f on the closed interval $[0, 9]$ is shown in the figure above. If $g(x) = \int_0^x f(t)\,dt$, on what interval(s) is the graph of g concave downward?

    (A) $0 < x < 4$ only
    (B) $0 < x < 4$ and $8 < x < 9$
    (C) $2 < x < 6$ only
    (D) $0 \leq x < 2$ and $6 < x \leq 9$
    (E) $2 < x \leq 4$ only

AP Calculus BC Test 3 Section 1

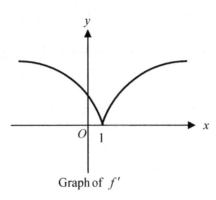

Graph of f'

18. The graph of f', the derivative of f is shown in the figure above. Which of the following is true about the graph of f at $x=1$?

   (A) f is not differentiable at $x=1$.
   (B) There exists a point of inflection at $x=1$.
   (C) f is discontinuous at $x=1$.
   (D) f has a local minimum at $x=1$.
   (E) f has a local maximum at $x=1$.

---

19. What is the sum of the series $\left(\dfrac{\pi}{3}\right) - \dfrac{(\pi/3)^3}{3!} + \dfrac{(\pi/3)^5}{5!} - \dfrac{(\pi/3)^7}{7!} + \cdots + \dfrac{(-1)^n (\pi/3)^{2n+1}}{(2n+1)!}$?

   (A) $\dfrac{1}{2}$ (B) $\dfrac{\sqrt{3}}{2}$ (C) $\dfrac{\pi}{2}$ (D) π (E) e

---

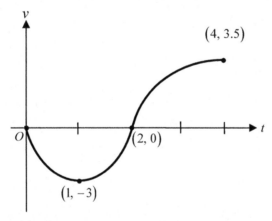

20. A particle moves on the x-axis so that the graph of its velocity at any time $0 < t < 5$ is shown above. For what values of t is the acceleration of the particle increasing?

   (A) $0 < t < 2$
   (B) $0 < t < 4$
   (C) $1 < t < 4$
   (D) $2 < t < 4$
   (E) $1 < t < 2$

450

AP Calculus BC Test 3   Section 1

21. Let $P(x) = 5 - 2(x-1) + 3(x-1)^2 - 2(x-1)^3$ be the fourth–degree Taylor polynomial for the function f about $x = 1$. If f has derivatives of all orders for all real numbers, which of the following is the third degree Taylor polynomial for $g(x) = \int_1^x f(t)\,dt$ about $x = 1$?

(A) $1 - (x-1) + (x-1)^2 - (x-1)^3$

(B) $5x - (x-1)^2 + (x-1)^3$

(C) $5(x-1) - \dfrac{2}{3}(x-1)^2 + \dfrac{3}{2}(x-1)^3$

(D) $5x - x^2 + x^3$

(E) $5(x-1) - (x-1)^2 + (x-1)^3$

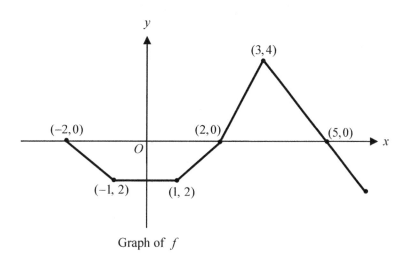

Graph of f

22. The graph of the piecewise linear function f is shown in the figure above. If $g(x) = \int_0^x f(t)\,dt$, which of the following is the expression for the line tangent to the graph of g at $x = 3$?

(A) $y = 4x - 13$

(B) $y = 4x - 9$

(C) $y = 2x - 6$

(D) $y = 2x - 8$

(E) $y = 4x - 8$

23. If a population modeled by a function P increases according to the logistic differential equation $\dfrac{dP}{dt} = 0.4P - 0.001P^2$, then $\lim\limits_{t \to \infty} P(t) =$

(A) 0
(B) 100
(C) 250
(D) 400
(E) 500

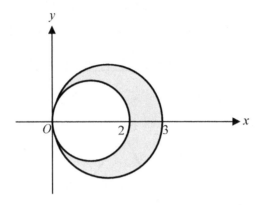

24. The graphs of the polar curves $r = 2\cos\theta$ and $r = 3\cos\theta$ are shown in the figure above. Which of the following expressions gives the area of the shaded region?

(A) $\dfrac{1}{2}\displaystyle\int_0^{\pi/2} \cos^2\theta\, d\theta$

(B) $5\displaystyle\int_0^{\pi/2} \cos^2\theta\, d\theta$

(C) $5\displaystyle\int_0^{\pi/2} \cos^4\theta\, d\theta$

(D) $10\displaystyle\int_0^{\pi/2} \cos^2\theta\, d\theta$

(E) $10\displaystyle\int_0^{\pi/2} \cos^4\theta\, d\theta$

25. $\lim\limits_{x \to 0} \dfrac{\displaystyle\int_e^{e+x} \ln(t^3)\, dt}{x} =$

(A) 0
(B) 1
(C) 2
(D) 3
(E) 4

AP Calculus BC Test 3

Section 1

26. A slope field is shown above. Which of the following could be the differential equation for the slope field?

(A) $\dfrac{dy}{dx} = xy - y$

(B) $\dfrac{dy}{dx} = \dfrac{x-1}{y-1}$

(C) $\dfrac{dy}{dx} = xy + y$

(D) $\dfrac{dy}{dx} = \dfrac{x+1}{y}$

(E) $\dfrac{dy}{dx} = \dfrac{x+1}{x}$

---

27. If n is a positive integer, then which of the following could be an expression for $\displaystyle\lim_{n \to \infty} \dfrac{1^2 + 2^2 + 3^2 + \cdots + n^2}{n^3}$?

(A) $\displaystyle\int_0^1 x^2\, dx$ (B) $\displaystyle\int_0^1 \dfrac{1}{x}\, dx$ (C) $\displaystyle\int_0^1 x^3\, dx$ (D) $\displaystyle\int_0^1 \dfrac{1}{x^2}\, dx$ (E) $\displaystyle\int_0^1 \dfrac{1}{x^2}\, dx$

Dr. John Chung's AP Calculus AB/BC

28. A particle moves along the curve defined by the equation $y = \frac{1}{2}(x^2 - 2x)$. If $\frac{dx}{dt} = 2t - 1$ for $t \geq 0$ with initial condition $x(0) = 1$, which of the following is the speed of the particle at time $t = 2$?

(A) $\frac{1}{2}$

(B) $2\sqrt{3}$

(C) $3\sqrt{5}$

(D) $\frac{9}{2}$

(E) $5\sqrt{3}$

END OF PART A OF SECTION 1

AP Calculus BC Test 3 Section 1

CALCULUS BC
SECTION 1, PART B
Time — 50minutes
Number of questions — 17

A GRAPHING CALCULATOR IS REQUIRED FOR SOME QUESTIONS ON THIS PART OF THE EXAMINATION.

Directions: Solve each of the following problems, using the available space for scratchwork. After examining the form of the choices, decide which is the best of the choices given and fill in the corresponding oval on the answer sheet. No credit will be given for anything written in the test book. Do not spend too much time on any one problem.

In this test:

(1) The exact numerical value of the correct answer does not always appear among the choices given. When this happens, select from among the choices the number that best approximates the exact numerical value.

(2) Unless otherwise specified, the domain of a function f is assumed to be the set of all real numbers x for which $f(x)$ is a real number.

(3) The inverse of a trigonometric function f may be indicated using the inverse function notation f^{-1} or with the prefix "arc" (e.g, $\sin^{-1} x = \arcsin x$)

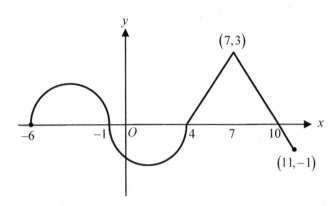

76. The graph of the function f on the closed interval $-6 \le x \le 11$ consists of two semicircles and two line segments as shown above. Which of the following is not true?

(A) $f'(-3.5) = 0$
(B) $f'(-1)$ exists
(C) $f'(4)$ does not exist.
(D) f has a point of inflection at $x = -1$.
(E) f has a relative maximum at $x = 7$.

---

77. Let P be the fourth-degree Taylor polynomial for the function f about $x = 2$ given by
$P(x) = (x-2) - \frac{1}{2}(x-2)^2 + \frac{1}{6}(x-2)^3 - \frac{1}{24}(x-2)^4$. What is the value of $f'''(2)$?

(A) 1
(B) 6
(C) 12
(D) 24
(E) 36

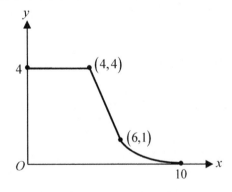

78. The graph of f is shown above. If $\int_6^{10} f(x)dx = 1.8$ and $g'(x) = f(x)$, then $g(10) - g(4) =$

(A) 5   (B) 5.3   (C) 6.8   (D) 14.8   (E) 22.8

79. Let f be the function whose graph goes through the point $(2, 3)$ and whose derivative is given by $\dfrac{dy}{dx} = \dfrac{e^x}{x^2}$. If the equation of the line tangent to the graph of f at $x = 2$ is used to approximate $f(2.1)$, then what is the value of the approximation?

(A) 1.852 (B) 2.654 (C) 3.185 (D) 3.561 (E) 5.125

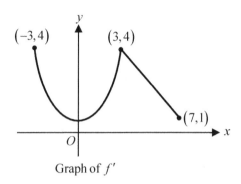

Graph of f'

80. The graph of f', the derivative of a function f on the closed interval $[-3, 7]$, is shown above. If $f''(0) = 0,$ which of the following statements is true?

(A) The graph of f has a point of inflection at $x = 0$ only.
(B) The graph of f has a relative minimum at $x = 0$.
(C) f is not differentiable at $x = 3$.
(D) The graph of f is continuous at $x = 3$.
(E) The graph of f is concave up on the interval $(-3, 0)$.

81. The function of f is given by $f(x) = x \sin\left(\dfrac{x^2}{2}\right)$. How many points of inflection does the graph of f have on the open interval $(0, 5)$?

(A) None (B) One (C) Two (D) Three (E) Four

AP Calculus BC Test 3

82. Which of the following series converge?

   I. $\sum_{n=1}^{\infty} \frac{(-1)^{n+1}}{n+2}$

   II. $\sum_{n=1}^{\infty} \frac{10n}{\ln n}$

   III. $\sum_{n=1}^{\infty} n \ln\left(1 + \frac{1}{n}\right)$

   (A) I only   (B) II only   (C) I and II only   (D) II and III only   (E) I, II, and III

---

83. A 20 feet-long ladder is leaning against the wall of a house and sliding down along the wall at the rate of 2 feet per second. How fast is the bottom of the ladder moving away from the wall when the top of the ladder is 10 feet from the bottom of the wall?

   (A) 1.115 feet/second
   (B) 1.155 feet/second
   (C) 1.254 feet/second
   (D) 1.356 feet/second
   (E) 2.315 feet/second

AP Calculus BC Test 3

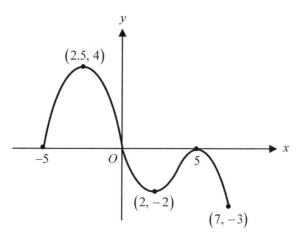

84. The graph of f', the derivative of a function f is shown above. If $f(-5) = 0$, $\int_{-5}^{0} f'(x)\,dx = 12.8$, $\int_{0}^{5} f'(x)\,dx = -6.8$, and $\int_{5}^{7} f'(x)\,dx = -2.3$, which of the following could be the graph of f?

(A)

(B)

(C)

(D)

(E)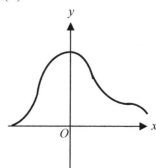

AP Calculus BC Test 3

Section 1

85. The base of a solid is the region R enclosed by the graphs $y = \ln(1+x^2)$ and $y = 2\cos x$. If each cross section of the solid perpendicular to the x-axis is a square, which of the following is an expression of the volume of the solid?

(A) $\pi \int_0^{0.834} \left(\ln(1+x^2)^2 - 4\cos^2 x \right) dx$

(B) $2\pi \int_0^{1.141} \left(\ln(1+x^2)^2 - 4\cos^2 x \right) dx$

(C) $\int_0^{1.141} \left(\ln(1+x^2)^2 - 4\cos^2 x \right) dx$

(D) $\int_0^{1.141} \left[\ln(1+x^2) - 2\cos x \right]^2 dx$

(E) $2 \int_0^{1.141} \left[\ln(1+x^2) - 2\cos x \right]^2 dx$

---

x	$f(x)$	$f'(x)$	$f''(x)$	$g(x)$	$g'(x)$	$g''(x)$
-1	1	3	2	3	4	3
0	3	2	1	-3	3	2
3	5	1	-2	2	2	-1
5	8	-1	-3	5	1	-2

86. The table of values of f, f', f'', g, g', and g'' at selected values of x is shown above. If $K(x) = f(g(x))$, then which of the following is the value of $K''(-1)$?

(A) -29
(B) -15
(C) 7
(D) 9
(E) 11

---

87. If the function f is given by $f(x) = \int_0^x (e^{-t} + t^2) dt$, what is the average rate of change of f over $[1, 4]$?

(A) 0.578   (B) 3.765   (C) 4.516   (D) 7.117   (E) 21.350

AP Calculus BC Test 3

Section 1

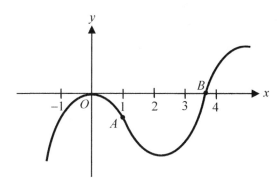

88. The graph of f is shown above. If the graph have points of inflection at points A and B, which of the following must be true?

   I. $f'(-1) > f'(1)$
   II. $f''(-1) > f''(2)$
   III. $f''(0) > f''(1)$

(A) I only   (B) II only   (C) I and II only   (D) I and III only   (E) I, II, and III

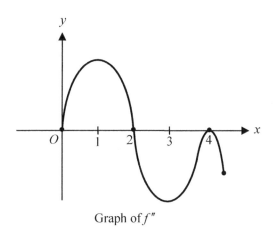

Graph of f''

89. The graph of f'', a second derivative of a function, is shown above for $0 \leq x \leq 5$. If $f'(0) = 0$ and $\int_0^2 f''(x)\,dx + \int_2^4 f''(x)\,dx = 0$, at which of the following values of x does f have a relative maximum?

(A) 1 only   (B) 2 only   (C) 3 only   (D) 4 only   (E) none

AP Calculus BC Test 3

Section 1

90. A particle moves along the x-axis so that its velocity at any time $t \geq 0$ is given by $v(t) = \sin(t^3 + 1)$. If the position of the particle is 5 at time $t = 0$, which of the following is the closest value to the position of the particle when the particle is first time at rest?

(A) 1.8
(B) 3.5
(C) 4.3
(D) 6.1
(E) 8.4

---

91. Let $y = P(t)$ be the particular solution to the logistic differential equation $\dfrac{dy}{dt} = \dfrac{y}{10}(100 - y)$. Which of the following is an expression for $P(t)$ with initial condition $P(0) = 20$?

(A) $P(t) = 20e^{-10t}$

(B) $P(t) = 100 - 80e^{-10t}$

(C) $P(t) = \dfrac{100}{1 + 4e^{-10t}}$

(D) $P(t) = \dfrac{200}{1 + 9e^{-10t}}$

(E) $P(t) = \dfrac{100}{2 + 3e^{-10t}}$

AP Calculus BC Test 3

Section 1

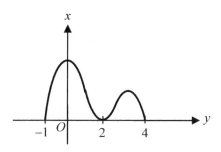

92. The graph of f', the derivative of a function f is shown above. If $f(-1) = 0$, Which of the following could be the graph of f?

(A)

(B)

(C)

(D)

(E)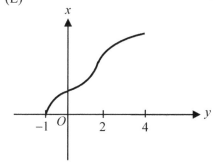

END OF SECTION 1

AP Calculus BC Test 3　　　　　　　　　　　　　　　　　　　　　Section II

CALCULUS BC

SECTION II
Time: 1 hour and 30 minutes
Percent of total grade – 50

Part A: 30 minutes, 2 problems
Part B: 60 minutes, 4 problems

Part A: A graphing calculator is required for some problems or parts of problems.
Part B: No calculator is allowed for these problems.

GENERAL INSTRUCTIONS FOR SECTION II PART A AND PART B

For each part of section II, you may wish to look over the problems before starting to work on them, since it is not expected that everyone will be able to complete all parts of all problems. All problems are given equal weight, but the parts of a particular problem are not necessary given equal weight.

- YOU SHOULD WRITE ALL WORK FOR EACH PART OF EACH PROBLEM IN THE SPACE PROVIDED FOR THAT PART IN THE TEST BOOKLET. Be sure to write clearly and legibly. If you make an error, you may save time by crossing it out rather than trying to erase it. Erased or cross-out work will not be graded.
- Show all work. Clearly label any functions, graphs, tables, or other object that you use. You will be graded on the correctness and completeness of your methods as well as your answers. Answers without supporting work may not receive credit.
- Justifications required that you give mathematical (noncalculator) reasons
- Your work must be expressed in standard mathematical notation rather than calculator syntax. For example, $\int_{1}^{5} x^2 \, dx$ may not be written as $\text{fnInt}(X^2, X, 1, 5)$
- Unless otherwise specified, answers (numeric or algebraic) need not be simplified.
- If you use decimal approximations in calculations, you will be graded on accuracy. Unless otherwise specified, your final answers should be accurate to three places after the decimal point.
- Unless otherwise specified, the domain of a function f is assumed to be the set of all real numbers x for which $f(x)$ is a real number.

CALCULUS BC
SECTION II, PART A
Time — 30 minutes
Number of questions — 2

A Graphing calculator is required for some problems or parts of problems.

Question 1

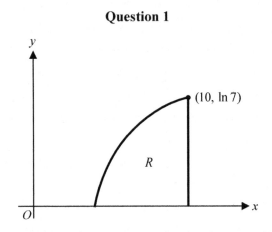

Let R be the region enclosed by the x-axis and the graphs of $y = \ln(x-3)$ and $x = 10$ as shown in the figure above.

(a) Find the area of R.

(b) The base of a solid is region R. Each cross section of the solid perpendicular to the y-axis is a semicircle whose diameter lies in the xy-plane. Write an expression involving one or more integrals that gives the volume of the solid.

(c) The vertical line $x = k$ divides region R into two equal areas. Write, but do not solve, an equation involving one or more integrals in terms of k.

(d) Find the volume of the solid generated when R is revolved about the vertical line $x = 2$.

AP Calculus BC Test 3

Section II

Question 2

t (hours)	0	3	5	9	15
$G(t)$ (gallons)	220	260	320	350	440

The amount of water flows into a swimming pool through a pipe at time t is strictly increasing. At time $t = 0$, the amount of water in the swimming pool is 220 gallons. The amount of water, $G(t)$, is twice-differentiable function is measured in gallons and t is measured in hours. The table above shows the amount of water at selected time t over 15 hour period.

(a) Use the data from the table to find an approximation for $G'(7)$. Show the computations that lead to your answer. Indicate units of measure.

(b) Approximate the average amount of the water over the time interval $0 \le t \le 15$ hours by using a right Riemann sum with the four subintervals in the table. Does this approximation overestimate or under estimate the average amount of the water over 15 hours? Explain your reasoning.

(c) Evaluate $\int_0^{15} G'(t)\,dt$. Using correct units, interpret the meaning of $\int_0^{15} G'(t)\,dt$, and the meaning of $G(0) + \int_0^{15} G'(t)\,dt$.

(d) For $0 \le t \le 15$, the function G is given by $G(t) = 220 + te^{\frac{t}{5}}$, as a model for the amount of the water in the swimming pool at time t. Find the average value of the amount of the water in the pool over the time interval $0 \le t \le 15$.

(e) For $t \ge 15$, the amount of the water in the pool is increasing with the rate $G'(t) = \dfrac{50 \ln x}{x}$. Based on the model, find $G(20)$?

END OF PART A OF SECTION II

**CALCULUS BC
SECTION II, PART B
Time — 60 minutes
Number of questions — 4**

No calculator is required for these problems.

Question 3

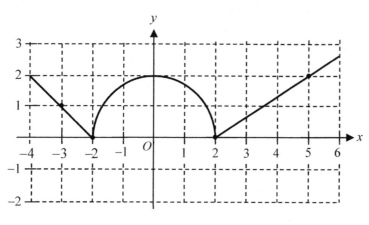

Graph of f

3. Let g be the function given by $g(x) = 3 + \int_{2}^{x} f(t)\,dt$. The figure above shows the graph of f, the derivative of a function g defined on the interval $(-4, 6)$. The graph consists of two line segments and a semicircle centered at origin.

   (a) Find the values of $g(0)$ and $g(5)$.

   (b) Find the values of $g'(-2)$, $g''(-2)$. If the value does not exist, explain your reasoning.

   (c) Write an equation for the line tangent to the graph of g at $x = 5$.

   (d) For $-4 < x < 6$, find all values of x for which the graph of g has a point of inflection.

   (e) Find the average value of g' over the closed interval $[0, 5]$.

Question 4

Let f be the function defined by $f(x) = \dfrac{x}{1+x^2}$ for all $x \geq 0$.

(a) Find the absolute maximum value of f on the open interval $(0, \infty)$. Justify your answer.

(b) What is the range of f?

(c) Let R be the region enclosed by the x-axis, the vertical lines $x = \dfrac{1}{2}$ and $x = 2$, and the graph of $y = f(x)$. Write an expression involving integral that gives the volume of the solid when R is revolved about the y-axis. Do not evaluate.

(d) Consider the function $g(x) = \dfrac{1}{x} - f(x)$. Find the value of $\displaystyle\lim_{b \to \infty} \int_1^b g(x)\,dx$.

Question 5

The power series for the function f is given by $f(x) = \displaystyle\sum_{n=1}^{\infty} \dfrac{(-1)^n x^n}{n}$ on its interval of convergence.

(a) Find the radius of convergence for the power series of f.

(b) Find the interval of convergence for the power series of f.

(c) Find the value of $f^{(5)}(0)$.

(d) Let $P_4(x)$ be the fourth-degree Taylor polynomial for f about $x = 0$. Show that $|P_4(0.5) - f(0.5)| < \dfrac{1}{100}$. Justify your answer.

Question 6

The differential equation is given by $\dfrac{dy}{dx} = \dfrac{x-2}{y}$.

(a) Sketch a slope field for the given differential equation $\dfrac{dy}{dx} = \dfrac{x-2}{y}$ on the axis at the given points.

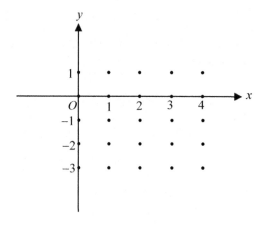

(b) Find the particular solution $y = f(x)$ to the differential equation $\dfrac{dy}{dx} = \dfrac{x-2}{y}$ with $f(0) = -2$. Sketch the graph of $y = f(x)$ in the xy-plane provided above.

(c) Find the values of $\lim\limits_{x \to 2} f(x)$ and $f'(2)$.

(d) Does the graph of $y = f(x)$ have a local maximum? Explain your reasoning.

END OF EXAM

No material on this page

AP Calculus BC Test 3 Answers

SECTION 1, PART A

1. **D**

   A point of inflection occurs at the point c of the graph of f $f''(c)=0$ or f'' is undefined at $x = c$.

   $f'(x) = e^{-x} - xe^{-x} \rightarrow f''(x) = -e^{-x} - \left(e^{-x} - xe^{-x}\right) = -2e^{-x} + xe^{-x}$

   Because $e^{-x}(x-2) = 0$, then $x = 2$.

2. **D**

   Change Variable: $\begin{cases} u = 1+x \rightarrow x = u-1 \\ du = dx \end{cases}$

   $\int_0^3 \frac{x}{\sqrt{1+x}}\,dx = \int_1^4 \frac{u-1}{\sqrt{u}}\,du = \int_1^4 \left(u^{\frac{1}{2}} - u^{-\frac{1}{2}}\right)du = \left[\frac{2}{3}u^{\frac{3}{2}} - 2u^{\frac{1}{2}}\right]_1^4$

   $= \left(\frac{2}{3}(4)^{\frac{3}{2}} - 2(4)^{\frac{1}{2}}\right) - \left(\frac{2}{3} - 2\right) = \left(\frac{16}{3} - 4\right) - \frac{2}{3} + 2 = \frac{8}{3}$

3. **E**

   Take a derivative of the equation.

   $1 = \frac{1}{2}y^{-\frac{1}{2}} \cdot \frac{dy}{dx}$ or $\frac{dy}{dx} = 2\sqrt{y}$ and when $x = 3$, $y = 4$.

   Therefore, $\left.\frac{dy}{dx}\right|_{y=4} = 2\sqrt{4} = 4$.

   Or $\frac{dy}{dx} = 2\sqrt{y} = 2(x-1)$ because $\sqrt{y} = x-1$. Then $\left.\frac{dy}{dx}\right|_{x=3} = 2(3-1) = 4$.

4. **B**

   $\lim\limits_{h \to 0} \dfrac{\arctan\left(\frac{1}{4}+h\right) - \arctan\left(\frac{1}{4}\right)}{h} = f'\left(\frac{1}{4}\right)$, where $f(x) = \arctan x$ and $f'(x) = \dfrac{1}{1+x^2}$.

   Therefore, $\lim\limits_{h \to 0} \dfrac{\arctan\left(\frac{1}{4}+h\right) - \arctan\left(\frac{1}{4}\right)}{h} = f'\left(\frac{1}{4}\right) = \left.\dfrac{1}{1+x^2}\right|_{x=\frac{1}{4}} = \dfrac{16}{17}$.

5. **B**

   Continuity: $1 + \ln e = \lim\limits_{x \to e^-}(ax-b) \rightarrow 2 = ae - b$

   Differentiable: $\left.(1+\ln x)'\right|_{x=e} = \left.(ax-b)'\right|_{x=e} \rightarrow \frac{1}{e} = a$ or $ae = 1$.

   Therefore, $2 = 1 - b$ or $b = -1$.

6. **E**

   In order to have a relative maximum, it follows that $f'(2) = 0$.

   $f'(x) = \dfrac{4(x^2+1) - (4x-k)(2x)}{(x^2+1)^2} = \dfrac{-4x^2 + 2kx + 4}{(x^2+1)^2}$

   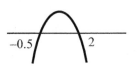

   Graph of f'

   Therefore, $f'(2) = -16 + 4k + 4 = 0$ or $k = 3$.

7. **D**

   Because $x = e^t$ and $t = y$, $x = e^y$.

AP Calculus BC Test 3 Answers

Therefore, $y = \ln x$.

E is the graph of $y = \ln x$.

8. **D**

$$\lim_{x \to 1}\left(\sum_{n=0}^{\infty}\frac{x^n}{n!}\right) = \lim_{x \to 1}\left(1 + x + \frac{x^2}{2!} + \frac{x^3}{3!} + \cdots\right) = \lim_{x \to 1} e^x = e$$

Because Taylor series of e^x is $e^x = 1 + x + \frac{x^2}{2!} + \frac{x^3}{3!} + \frac{x^4}{4!} + \cdots$.

9. **D**

The average value of $\frac{2}{1+x^2}$ is

$$\frac{1}{1-0}\int_0^1 \frac{2}{1+x^2}dx = \left[2\tan^{-1}x\right]_0^1 = 2\tan^{-1}1 - 2\tan^{-1}0 = 2\left(\frac{\pi}{4}\right) - 2(0) = \frac{\pi}{2}.$$

10. **E**

$$\text{Ratio} = \lim_{n \to \infty}\left|\frac{(x-2)^{n+1}}{(n+2)^2} \cdot \frac{(n+1)^2}{(x-2)^n}\right| = \lim_{n \to \infty}\left|\frac{(x-2)}{1} \cdot \frac{(n+1)^n}{(n+2)^n}\right| = |x-2| < 1$$

Since $-1 < x - 2 < 1$, the interval of convergence is $1 < x < 3$.

Now we apply end point test at $x = 1$ and $x = 3$.

$\begin{cases} x = 1 \to \sum_{n=0}^{\infty}\frac{(-1)^n}{(n+1)^2} = 1 - \frac{1}{2^2} + \frac{1}{3^2} - \frac{1}{4^2} + \cdots \to \text{alternating } p\text{-series; The series converges.} \end{cases}$

$\begin{cases} x = 1 \to \sum_{n=0}^{\infty}\frac{(1)^n}{(n+1)^2} = 1 + \frac{1}{2^2} + \frac{1}{3^2} + \frac{1}{4^2} + \cdots \to p\text{-series; The series converges.} \end{cases}$

Therefore, the interval of convergence is $1 \le x \le 3$.

11. **C**

We know that $\frac{d}{dx}\int_a^{u(x)} f(t)dt = f(u)u'$.

Therefore, $f'(x) = e^{-\ln x} \cdot (\ln x)' = e^{\ln\frac{1}{x}} \cdot \frac{1}{x} = \frac{1}{x} \cdot \frac{1}{x} = \frac{1}{x^2}$.

12. **E**

$1 = \cos(y) \cdot \frac{dy}{dx}$ or $\frac{dy}{dx} = \frac{1}{\cos y} \to$ We need to find the value of y when $x = \frac{1}{2}$.

If $x = \frac{1}{2}$, then $\frac{1}{2} = \sin y \to y = \frac{\pi}{6}$.

Therefore, $\left.\frac{dy}{dx}\right|_{y=\pi/6} = \frac{1}{\cos(\pi/6)} = \frac{2}{\sqrt{3}} = \frac{2\sqrt{3}}{3}$

Dr. John Chung's AP Calculus AB/BC

AP Calculus BC Test 3 Answers

13. **C**

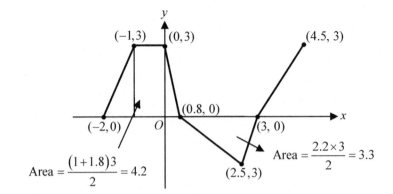

From the graph in figure above, $g(-1) = \int_3^{-1} f(t)dt = -\int_{-1}^{3} f(t)dt = -(4.2 - 3.3) = -0.9$.

14. **E**

(A) $\sum_{n=1}^{\infty} \dfrac{n}{2n+3} \rightarrow \lim_{n \to \infty} \dfrac{n}{2n+3} = \dfrac{1}{2}$ → The series diverges by nth term test.

(B) $\sum_{n=1}^{\infty} \dfrac{5(3)^n}{2^n} \rightarrow \sum_{i=1}^{n} \dfrac{5(3)^n}{2^n} = 5\sum_{n=1}^{\infty}\left(\dfrac{3}{2}\right)^n$ → The series diverges by geometric series test $\left(r = \dfrac{3}{2} > 1\right)$.

(C) $\sum_{n=1}^{\infty} \dfrac{2^n+1}{2^{n+1}} \rightarrow \lim_{n \to \infty} \dfrac{2^n+1}{2^{n+1}} = \dfrac{1}{2}$ → The series diverges by nth term test.

(D) $\sum_{n=1}^{\infty} \dfrac{1}{n+2} = \dfrac{1}{3}+\dfrac{1}{4}+\dfrac{1}{5}+\cdots$ → The series diverges because the series is harmonic series.

(E) $\sum_{n=1}^{\infty} \dfrac{(-1)^n}{n} = \dfrac{-1}{1}+\dfrac{1}{2}-\dfrac{1}{3}+\cdots$ → The series converges because it is alternating harmonic series.

15. **E**

$f(x) = \ln(1-\cos x) - \ln(1+\cos x) \Rightarrow f'(x) = \dfrac{\sin x}{1-\cos x} - \dfrac{-\sin x}{1+\cos x} = \dfrac{2\sin x}{1-\cos^2 x} = \dfrac{2}{\sin x}$

Therefore, $f'\left(\dfrac{\pi}{6}\right) = \dfrac{2}{\sin(\pi/6)} = \dfrac{2}{1/2} = 4$.

16. **E**

(A) f is increasing on the interval $(0, 2)$. False

(B) f is decreasing on the interval $(0, 2)$. False

(C) f has a local maximum at $x = 2$. False

(D) f has a local minimum at $x = 2$. False

(E) f has at least one local minimum on the interval $(0, 3)$. True (by Intermediate theorem)

AP Calculus BC Test 3 Answers

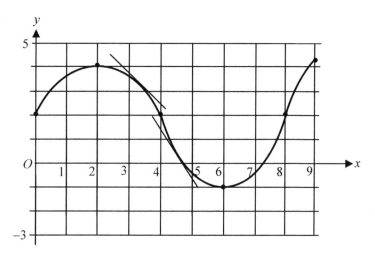

17. C

$g''(x)$ is the slope of the line tangent to the graph of $f(x)$ because $g''(x) = f'(x)$.
Since the graph of the function f on the open interval $(2, 6)$ has a negative slope,
C is correct.

18. B

(A) f is not differentiable at $x = 1$. False, because $f'(1) = 0$
(B) There exists a point of inflection at $x = 1$. True
(C) f is discontinuous at $x = 1$. False
(D) f has a local minimum at $x = 1$. False
(E) f has a local maximum at $x = 1$. False

Consider the graphs of $f'(x) = (x-1)^{\frac{2}{3}}$ and $f(x) = \frac{3}{5}(x-1)^{\frac{5}{3}}$ with $f(1) = 0$.

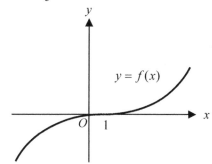

19. B

We know that $\sin x = x - \dfrac{x^3}{3!} + \dfrac{x^5}{5!} - \dfrac{x^7}{7!} + \cdots + \dfrac{(-1)^n x^{2n+1}}{(2n+1)!} + $

Because $\sin\left(\dfrac{\pi}{3}\right) = \left(\dfrac{\pi}{3}\right) - \dfrac{(\pi/3)^3}{3!} + \dfrac{(\pi/3)^5}{5!} - \dfrac{(\pi/3)^7}{7!} + \cdots + \dfrac{(-1)^n (\pi/3)^{2n+1}}{(2n+1)!} + $,

then $\sin\left(\dfrac{\pi}{3}\right) = \dfrac{\sqrt{3}}{2}$.

AP Calculus BC Test 3 Answers

20. A

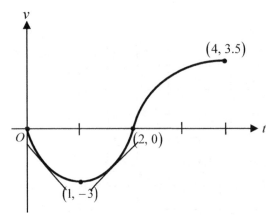

Acceleration is the slope of the line tangent to the graph of v.
We can see changes in slopes as follows.
For $0 < t < 2$, $-4, -3, -2, -1, 0, 1, 2, 3, 4 \to$ Acceleration is increasing.
For $2 < t < 4$, $5, 4, 3, 2, 1 \to$ Acceleration is decreasing.

21. E

Because $P(x) = 5 - 2(x-1) + 3(x-1)^2 - 2(x-1)^3$,

$$g_3(x) = \int_1^x 5 - 2(t-1) + 3(t-1)^2 \, dt = \left[5t - (t-1)^2 + (t-1)^3 \right]_1^x = 5(x-1) - (x-1)^2 + (x-1)^3.$$

22. A

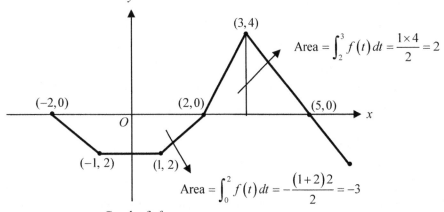

Graph of f

We know that $g'(3) = f(3) = 4$ and $g(3) = \int_0^3 f(t) \, dt = -3 + 2 = -1$.

Therefore, the equation of the tangent is $y - (-1) = 4(x-3)$ or $y = 4x - 13$.

23. D

Express the logistic differential equation into a standard form $\dfrac{dP}{dt} = kP\left(1 - \dfrac{P}{m}\right)$.

$$\frac{dP}{dt} = 0.4P - 0.001P^2 = 0.4P\left(1 - \frac{0.001P}{0.4}\right) = 0.4P\left(1 - \frac{P}{400}\right)$$

We can see the carrying capacity is 400.
Therefore, $\lim\limits_{t \to \infty} P(t) = 400$.

AP Calculus BC Test 3 Answers

24. B

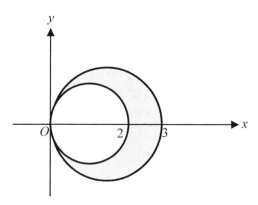

Therefore, $\text{Area} = \frac{1}{2}\int_{-\pi/2}^{\pi/2}\left(9\cos^2\theta - 4\cos^2\theta\right)d\theta = \int_{0}^{\pi/2} 5\cos^2\theta\, d\theta$ or $5\int_{0}^{\pi/2}\cos^2\theta\, d\theta$ because the both figures are symmetric to the respect to the *x*-axis.

25. D

We apply L'Hopital's rule, because $\lim\limits_{x \to 0} \dfrac{\int_{e}^{e+x} \ln(t^3)\, dt}{x} = \dfrac{0}{0}$.

Therefore, $\lim\limits_{x \to 0} \dfrac{\int_{e}^{e+x} \ln(t^3)\, dt}{x} = \lim\limits_{x \to 0} \dfrac{\ln(e+x)^3}{1} = \lim\limits_{x \to 0} 3\ln(e+x) = 3$,

because $\dfrac{d}{dx}\int_{e}^{e+x} \ln(t^3)\, dt = \ln(e+x)^3 (e+x)' = \ln(e+x)^3 = 3\ln(e+x)$.

26. D

$\dfrac{dy}{dx} = \dfrac{x+1}{y}$ because $\dfrac{dy}{dx} = 0$ at $x = -1$ and $\dfrac{dy}{dx}$ is undefined at $y = 0$.

27. A

$\lim\limits_{n \to \infty} \dfrac{1^2 + 2^2 + 3^2 + \cdots + n^2}{n^3} = \lim\limits_{n \to \infty} \sum\limits_{k=1}^{n} \dfrac{1}{n}\left(\dfrac{k}{n}\right)^2 = \int_{0}^{1} x^2\, dx$?

The length of each subinterval is $\dfrac{1}{n}$ and the graph of the curve is the graph of $y = x^2$ as follow.

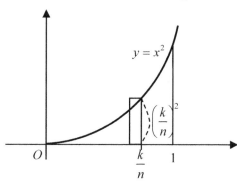

AP Calculus BC Test 3 Answers

28. C

$x(t) = t^2 - t + 1$, $x = 3$ at time $t = 2$. $\left.\dfrac{dx}{dt}\right|_{t=2} = 3$

$\dfrac{dy}{dt} = \dfrac{1}{2}\left(2x\dfrac{dx}{dt} - 2\dfrac{dx}{dt}\right) = \dfrac{1}{2}(6\times 3 - 2\times 3) = 6$

Speed: $\sqrt{(x'(t))^2 + (y'(t))^2} = \sqrt{9+36} = \sqrt{45} = 3\sqrt{5}$

SECTION 1, PART B

76. B

(A) $f'(-3.5) = 0$: True

(B) $f'(-1)$ exists: Not differentiable at $x = -1$ because f has a vertical tangent: False

(C) $f'(4)$ does not exist.: True

(D) f has a point of inflection at $x = -1$. It is true because the graph of f is continuous and $f''(x)$ is undefined at $x = -1$.

(E) f has a relative maximum at $x = 7$. True

77. A

We know that $f'''(2) = P'''(2)$ and $P'''(x) = 1 - (x - 2)$.

Therefore, $P'''(2) = 1$.

Or, use Taylor series for $P(x)$.

$P(x) = P(2) + P'(2)(x-2) + \dfrac{P''(2)}{2!}(x-2)^2 + \dfrac{P'''(2)}{3!}(x-2)^3 + \cdots$

From the series, $\dfrac{P'''(2)}{3!} = \dfrac{1}{6}$ or $P'''(2) = 1$.

78. C

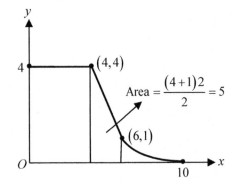

$g(10) - g(4) = \displaystyle\int_4^{10} f(x)dx = \int_4^6 f(x)dx + \int_6^{10} f(x)dx = 5 + 1.8 = 6.8$

79. C

The slope of the tangent is $\left.\dfrac{dy}{dx}\right|_{x=2} = \dfrac{e^2}{4}$.

Equation of tangent is $y - 3 = \dfrac{e^2}{4}(x-2)$ and $f(2.1) = \dfrac{e^2}{4}(2.1-2) + 3 \approx 3.185$.

Or you can you linear approximation. $f(2.1) = f(1) + 0.1 f'(0)$.

AP Calculus BC Test 3 Answers

80. **D**

(A) The graph of f has a point of inflection at $x = 0$ only. False: Another point of inflection at $x = 3$

(B) The graph of f has a relative minimum at $x = 0$. False

(C) f is not differentiable at $x = 3$. False

(D) The graph of f is continuous at $x = 3$. True because $f'(3) = 4$.

(E) The graph of f is concave up on the interval $(-3, 0)$. False: Concave down because $f''(x) < 0$.

81. **E**

$f'(x) = \sin\left(\dfrac{x^2}{2}\right) + x^2 \cos\left(\dfrac{x^2}{2}\right)$

Graph using a graphing calculator.
There are four points of inflection where $f''(x) = 0$.

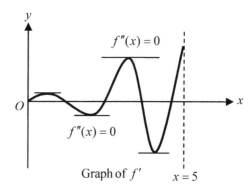

Graph of f'

82. **A**

I. $\displaystyle\sum_{n=1}^{\infty} \dfrac{(-1)^{n+1}}{n+2} = \dfrac{1}{3} - \dfrac{1}{4} + \dfrac{1}{5} - \dfrac{1}{6} \cdots$: The series converges because it is alternate harmonic series.

II. $\displaystyle\sum_{n=1}^{\infty} \dfrac{10n}{\ln n}$: The series diverges because of nth term test: $\displaystyle\lim_{n\to\infty} \dfrac{10n}{\ln n} = \lim_{n\to\infty} \dfrac{10}{1/n} = \lim_{n\to\infty} 10n = \infty$.

We applied L'Hopital's rule.

III. $\displaystyle\sum_{n=1}^{\infty} n \ln\left(1 + \dfrac{1}{n}\right)$: The series diverges because of nth term test:

$\displaystyle\lim_{n\to\infty} n\ln\left(1 + \dfrac{1}{n}\right) = \lim_{n\to\infty} \ln\left(1 + \dfrac{1}{n}\right)^n = \lim_{n\to\infty} \ln e = 1$

83. **B**

The primary equation is $x^2 + y^2 = 400$ and its derivative is

$2x\dfrac{dx}{dt} + 2y\dfrac{dy}{dt} = 0$, where $\dfrac{dy}{dt} = -2$.

When $y = 10$, the value of y is $x^2 + 100 = 400$ or $x = 10\sqrt{3}$.
By substituting all numbers into the derivative,

$2(10\sqrt{3})\dfrac{dx}{dt} + 2(10)(-2) = 0$ or $\dfrac{dx}{dt} = \dfrac{40}{20\sqrt{3}} \approx 1.155$.

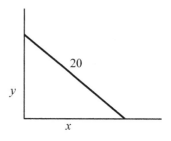

AP Calculus BC Test 3 Answers

84. E

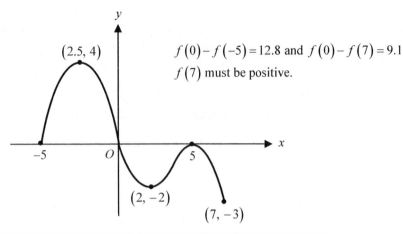

$f(0) - f(-5) = 12.8$ and $f(0) - f(7) = 9.1$
$f(7)$ must be positive.

$-5 \leq x < -2.5$	$f' > 0 \Rightarrow f$ is increasing $f'' > 0 \Rightarrow$ Concave up
$x = -2.5$	$f''(-2.5) = 0 \Rightarrow$ Point of inflection
$-2.5 < x < 0$	$f' > 0 \Rightarrow f$ is increasing $f'' < 0 \Rightarrow$ Concave down
$x = 0$	$f' = 0 \Rightarrow$ Local maximum
$0 < x < 2$	$f' < 0 \Rightarrow$ Decreasing $f'' < 0 \Rightarrow$ Concave down
$x = 2$	$f''(2) = 0 \Rightarrow$ Point of inflection
$2 < x < 5$	$f' < 0 \Rightarrow$ Decreasing $f'' > 0 \Rightarrow$ Concave up
$x = 5$	$f''(5) = 0 \Rightarrow$ Point of inflection
$5 < x \leq 7$	$f' < 0 \Rightarrow$ Decreasing $f'' < 0 \Rightarrow$ Concave down

85. E

Draw the graphs of $y = \ln(1 + x^2)$ and $y = 2\cos x$.

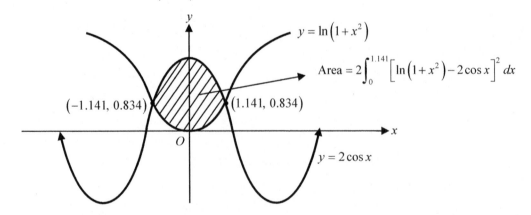

$$\text{Area} = 2\int_0^{1.141} \left[\ln(1+x^2) - 2\cos x\right]^2 dx$$

AP Calculus BC Test 3 Answers

86. **A**

$K'(x) = f'(g(x))g'(x)$

$K''(x) = f''(g(x))g'(x)g'(x) + f'(g(x))g''(x)$

$K''(-1) = f''(g(-1))g'(-1)g'(-1) + f'(g(-1))g''(-1) = f''(3)\cdot 4 \cdot 4 + f'(3)\cdot 3$

$= -2\cdot 4\cdot 4 + 1\cdot 3 = -29$

87. **D**

The average rate of change of f over $[1, 4]$ is $\dfrac{f(4)-f(1)}{4-1} = \dfrac{1}{3}\int_1^4 \left(e^{-t}+t^2\right)dt = 7.11652 \approx 7.117$.

88. **A**

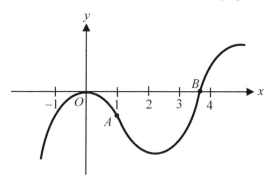

I. $f'(-1) > f(1)$: True because $f'(-1) > 0$ and $f'(1) < 0$.

II. $f''(-1) > f''(2)$: False because $f''(-1) < 0$ and $f''(2) > 0$.

III. $f''(0) > f''(1)$: False because $f''(0) < 0$ and $f''(1) = 0$.

89. **D**

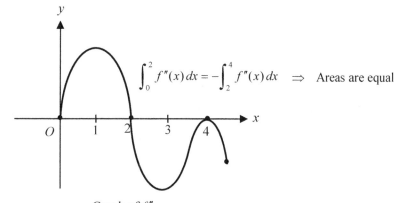

$\int_0^2 f''(x)\,dx = -\int_2^4 f''(x)\,dx \quad \Rightarrow \quad$ Areas are equal

Graph of f''

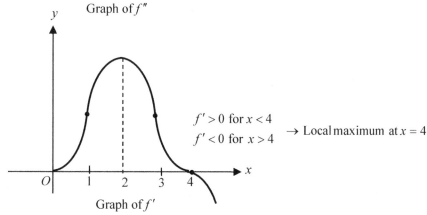

$f' > 0$ for $x < 4$

$f' < 0$ for $x > 4$

$\to$ Local maximum at $x = 4$

Graph of f'

Dr. John Chung's AP Calculus AB/BC

AP Calculus BC Test 3 Answers

From the graph of f', we can see the graph of f has a relative maximum at $x = 4$.

90. D

Because the velocity of the particle at rest is 0, we can see $v(t) = 0$ at $t = 1.289$ on the graph of v.

Draw the graph of v and find the zero.

Therefore, $x(1.289) = 5 + \int_0^{1.289} \sin(t^3 + 1)\, dt \approx 6.069$

91. C

Logistic differential equation: $\dfrac{dy}{dt} = ky\left(1 - \dfrac{y}{m}\right) = 10y\left(1 - \dfrac{y}{100}\right)$, where $m = 100$.

We got $m = 100$, $k = 10$.

Therefore, the particular solution is $y = \dfrac{100}{1 + Ae^{-10t}}$

With initial condition: $20 = \dfrac{100}{1 + A} \rightarrow A = 4$

Now the particular solution is $y = \dfrac{100}{1 + 4e^{-10t}}$.

92. D

The graph of f is increasing on the interval $[-1, 4]$.

For $-1 < x < 0$, the graph of f is increasing and concave up because the slope of the line tangent to the graph is positive.

For $0 < x < 2$, the graph of f is increasing and concave down because the slope of the line tangent to the graph is negative and so on.

At $x = 0, 2, 3$, points of inflection.

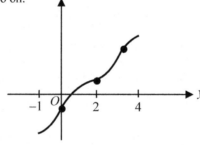

AP Calculus BC Test 3 Answers

SECTION II, PART A

Question 1

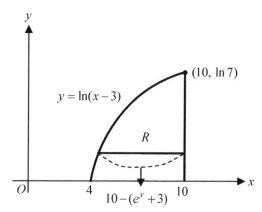

(a) Area $= \int_4^{10} \ln(x-3)\,dx \approx 7.621$

(b) $y = \ln(x-3)$ can be expressed in the form $x - 3 = e^y$ or $x = e^y + 3$.

The length of the diagonal of the semicircle is $10 - (e^y + 3)$ or $(7 - e^y)$.

The radius is $\dfrac{(7 - e^y)}{2}$ and the area of a semicircle is $A = \dfrac{\pi r^2}{2}$.

Therefore, the volume of the solid is $\dfrac{\pi}{2}\int_0^{\ln 7}\left(\dfrac{7-e^y}{2}\right)^2 dy \approx 13.882$

(c) $\int_4^k \ln(x-3)\,dx = \dfrac{13.882}{2}$

(d) Volume $= \pi \int_0^{\ln 7}\left[(10-2)^2 - (e^y + 3 - 2)^2\right]dy \approx 272.038$

Cylindrical Shell: Volume $= 2\pi \int_4^{10} (x-2)\ln(x-3)\,dx \approx 272.038$

AP Calculus BC Test 3 Answers

Question 2

(a) $G'(7) \approx \dfrac{G(9)-G(5)}{9-5} = \dfrac{350-320}{4} = 7.5$ gallons per hour

(b) Average value of G is $\dfrac{1}{15-0}\int_0^{15} G(t)\,dt$.

$\dfrac{1}{15}\int_0^{15} G(t)\,dt \approx \dfrac{1}{15}(3\times 260 + 320\times 2 + 350\times 4 + 440\times 6) = 364$ gallons

Because $G(t)$ is strictly increasing on the interval $[0, 15]$, it is an overestimate.

(c) $\int_0^{15} G'(t)\,dt = G(15) - G(0) = 440 - 220 = 220$ gallons

The amount of the water increased over the interval from $t = 0$ to $t = 15$ hours.

$G(0) + \int_0^{15} G'(t)\,dt$ is the amount of the water in the swimming pool at time $t = 15$.

(d) $\dfrac{1}{15}\int_0^{15}\left(220 + te^{t/5}\right) dt \approx 288.618$ gallons

(e) $G(20) = G(15) + \int_{15}^{20} G'(t)\,dt$

$G(20) = 440 + \int_{15}^{20}\left(\dfrac{50\ln t}{t}\right)dt \approx 481.022$ gallons

Question 3

(a) The area of a quarter of the circle is $\dfrac{\pi r^2}{4}$ or $\dfrac{\pi(2)^2}{4} = \pi$.

$g(0) = 3 + \int_2^0 f(t)\,dt = 3 - \int_0^2 f(t)\,dt = 3 - \pi$

$g(5) = 3 + \int_2^5 f(t)\,dt = 3 + \dfrac{3\times 2}{2} = 6$

(b) We know that $g'(x) = f(x)$.

$g'(2) = f(2) = 0$ but $g''(2) = f'(2)$ does not exist because there is a sharp corner at $x = 2$.

(c) The slope of the tangent is $g'(5) = f(5) = 2$ and the tangent passes through point $(5, g(5))$ or $(5, 6)$.

Therefore, the equation of the tangent is $y - 6 = 2(x - 5)$.

(d) For $-4 < x < 6$, find all values of x for which the graph of g has a point of inflection.

The graph of g has a point of inflection at $x = -2, x = 0,$ and $x = 2$ because $g''(x)$ changes sign at each of these values.

(e) Find the average value of g' over the closed interval $[0, 5]$.

$\dfrac{1}{5-0}\int_0^5 g'(x)\,dx = \dfrac{1}{5}[g(5) - g(0)] = \dfrac{1}{5}[6 - (3 - \pi)] = \dfrac{3 + \pi}{5}$

AP Calculus BC Test 3 Answers

Question 4

(a) Because $f'(x) = \dfrac{(1+x^2)-2x^2}{(1+x^2)^2} = \dfrac{1-x^2}{(1+x^2)^2} = \dfrac{(1+x)(1-x)}{(1+x^2)^2} = 0$, there is a critical point at $x = 1$.

At $x = 1$, the graph of f has a local maximum because $f' > 0$ for $x < 1$ and $f' < 0$ for $x > 1$.

Since the graph of f is strictly increasing on the interval $[0,1]$ and decreasing on the interval $[1, \infty)$.

Therefore, the local maximum of f at $x = 1$ becomes absolute maximum..

Maximum value $= f(1) = \dfrac{1}{1+1^2} = \dfrac{1}{2}$

(b) The range of f is $0 \leq y \leq \dfrac{1}{2}$ because $f(0) = 0$ and $\lim\limits_{x \to \infty} \dfrac{x}{1+x^2} = 0$.

(c) We apply cylindrical shell method.

Volume $= 2\pi \int_a^b xy\,dx = 2\pi \int_{1/2}^2 x\left(\dfrac{x}{1+x^2}\right)dx = 2\pi \int_{1/2}^2 \left(\dfrac{x^2}{1+x^2}\right)dx$

(d) Consider the function $g(x) = \dfrac{1}{x} - f(x)$. Find the value of $\lim\limits_{b \to \infty} \int_1^b g(x)\,dx$.

$\lim\limits_{b \to \infty} \int_1^b \left(\dfrac{1}{x} - \dfrac{x}{1+x^2}\right) = \lim\limits_{b \to \infty} \left[\ln(x) - \dfrac{1}{2}\ln(1+x^2)\right]_1^b = \lim\limits_{b \to \infty} \left[\ln\left(\dfrac{x}{\sqrt{1+x^2}}\right)\right]_1^b = \lim\limits_{b \to \infty} \left[\ln\left(\dfrac{b}{\sqrt{1+b^2}}\right) - \ln\dfrac{1}{\sqrt{2}}\right]$

$= \ln 1 + \ln\sqrt{2} = \ln\sqrt{2}$

Question 5

(a) We need the derivative function of f in order to find critical points on the open interval $(0, \infty)$.

Because $f'(x) = \dfrac{(1+x^2)-2x^2}{(1+x^2)^2} = \dfrac{1-x^2}{(1+x^2)^2} = \dfrac{(1+x)(1-x)}{(1+x^2)^2} = 0$, there is a critical point at $x = 1$.

At $x = 1$, the graph of f has a local maximum because $f' > 0$ for $x < 1$ and $f' < 0$ for $x > 1$.

Since the graph of f is strictly increasing on the interval $[0,1]$ and decreasing on the interval $[1, \infty)$.

Therefore, the local maximum of f at $x = 1$ becomes absolute maximum..

Maximum value $= f(1) = \dfrac{1}{1+1^2} = \dfrac{1}{2}$

(b) The range of f is $0 \leq y \leq \dfrac{1}{2}$ because $f(0) = 0$ and $\lim\limits_{x \to \infty} \dfrac{x}{1+x^2} = 0$.

(c) We apply cylindrical shell method.

Volume $= 2\pi \int_a^b xy\,dx = 2\pi \int_{1/2}^2 x\left(\dfrac{x}{1+x^2}\right)dx = 2\pi \int_{1/2}^2 \left(\dfrac{x^2}{1+x^2}\right)dx$

(d) Consider the function $g(x) = \dfrac{1}{x} - f(x)$. Find the value of $\lim\limits_{b \to \infty} \int_1^b g(x)\,dx$.

$\lim\limits_{b \to \infty} \int_1^b \left(\dfrac{1}{x} - \dfrac{x}{1+x^2}\right) = \lim\limits_{b \to \infty} \left[\ln(x) - \dfrac{1}{2}\ln(1+x^2)\right]_1^b = \lim\limits_{b \to \infty} \left[\ln\left(\dfrac{x}{\sqrt{1+x^2}}\right)\right]_1^b = \lim\limits_{b \to \infty} \left[\ln\left(\dfrac{b}{\sqrt{1+b^2}}\right) - \ln\dfrac{1}{\sqrt{2}}\right]$

$= \ln 1 + \ln\sqrt{2} = \ln\sqrt{2}$

AP Calculus BC Test 3 Answers

Question 6

(a) Sketch a slope field for the given differential equation $\dfrac{dy}{dx} = \dfrac{x-2}{y}$ on the axis at the given points.

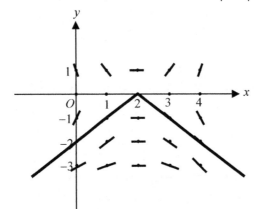

		0	1	2		3	4
	1	-2	-1	0		1	2
	0	undefined	undefined	undefined		undefined	undefined
y	-1	2	1	0		-1	-2
	-2	1	1/2	0		$-1/2$	-1
	-3	2/3	1/3	0		$-1/3$	$-2/3$

Values of $\dfrac{dy}{dx}$

(b) We separate the variables and solve as follows.

$y\,dy = (x-2)\,dx \;\to\; \int y\,dy = \int (x-2)\,dx \;\to\; \dfrac{y^2}{2} = \dfrac{x^2}{2} - 2x + C$

Substitute the initial condition $f(0) = -2$ into the equation.

$\dfrac{(-2)^2}{2} = 0 - 0 + C$ or $C = 2$

$\dfrac{y^2}{2} = \dfrac{x^2}{2} - 2x + 2 \;\to\; y^2 = x^2 - 4x + 4 \;\to\; y = -\sqrt{x^2 - 4x + 4}$ because $y(0) = -2$.

Therefore, particular solution is $y = -\sqrt{x^2 - 4x + 4} = -|x - 2|$

(c) $\lim\limits_{x \to 2} f(x) = 0$

$f'(2)$ is undefined, because the graph of f has a sharp corner at $x = 2$.

(d) 1) f is continuous.

2) The graph of f has a local maximum at $x = 2$ because there is change in sign of f' from positive to the negative at $x = 2$.

AP CALCULUS BC

TEST 4

AP CALCULUS BC

A CALCULATOR **CANNOT BE USED ON PART A OF SECTION 1**. A GRAPHIC CALCULATOR FROM THE APPROVED LIST **IS REQUIRED FOR PART B OF SECTION 1 AND FOR PART A OF SECTION II** OF THE EXAMINATION.

SECTION 1
Time: I hour and 45 minutes
All questions are given equal weight.
Percent of total grade – 50

Part A: 55 minutes, 28 multiple-choice questions
A calculator is NOT allowed.
Part B: 50 minutes, 17 multiple-choice questions
A graphing calculator is required.

CALCULUS BC
SECTION 1, PART A
Time— 55minutes
Number of questions— 28

A CALCULATOR MAY NOT BE USED ON THIS PART OF THE EXAMINATION.

Directions: Solve each of the following problems, using the available space for scratchwork. After examining the form of the choices, decide which is the best of the choices given and fill in the corresponding oval on the answer sheet. No credit will be given for anything written in the test book. Do not spend too much time on any one problem.

In this test:

(1) Unless otherwise specified, the domain of a function f is assumed to be the set of all real numbers x for which $f(x)$ is a real number.

(2) The inverse of a trigonometric function f may be indicated using the inverse function notation f^{-1} or with the prefix "arc" (e.g, $\sin^{-1} x = \arcsin x$)

AP Calculus BC Test 4   Section 1

1. At what values of x does $f(x) = 2x^5 - 5x^4 + 2x - 5$ have a point of inflection?

   (A) 0 only (B) $\frac{3}{2}$ only (C) $-\frac{3}{2}$ only (D) 0 and $\frac{3}{2}$ only (E) $-\frac{3}{2}$, 0, and $\frac{3}{2}$

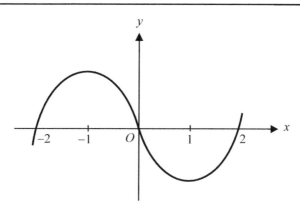

2. The graph of f' is shown above. Which of the following statements is true about f?

   (A) f is decreasing for $-1 \leq x \leq 0$.
   (B) f is increasing for $1 \leq x \leq 2$.
   (C) f has a local maximum at $x = -1$.
   (D) f has a local minimum at $x = -1$.
   (E) f has a local maximum at $x = 0$.

3. $\displaystyle\int_0^{\sqrt{5}} \frac{2x}{\sqrt{x^2+4}} dx =$

   (A) $\frac{1}{2}$ (B) 1 (C) $\frac{3}{2}$ (D) 2 (E) 3

4. $\displaystyle\lim_{x \to 1} \frac{1}{x-1} \int_1^x \left(3t^2 - 4t + 2e^{t-1}\right) dt =$

   (A) -2 (B) -1 (C) 0 (D) 1 (E) nonexistent

Dr. John Chung's AP Calculus AB/BC

AP Calculus BC Test 4   Section 1

5. Let f be defined by
$$f(x) = \begin{cases} \dfrac{x^2 - 8x + 16}{x - 4} & \text{for } x \neq 4 \\ k & \text{for } x = 4. \end{cases}$$
What is the value of k for which f is continuous for all real x?

(A) 0
(B) 1
(C) 2
(D) 3
(E) 4

6. What are all values of p for which $\displaystyle\int_1^\infty \frac{x^2}{x^p}\,dx$ converges?

(A) $p < -1$ (B) $-1 < p < 1$ (C) $0 < p < 3$ (D) $p > 3$ (E) none

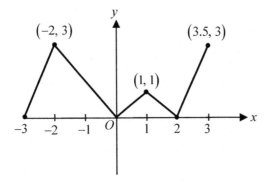

7. Let f be the continuous function given by the graph on the closed interval $[-3, 3]$, consisting of line segments. If $g(x) = \displaystyle\int_{-2}^x f(t)\,dt$, then $g''(-1) =$

(A) $-\dfrac{3}{2}$ (B) $-\dfrac{1}{2}$ (C) $\dfrac{1}{2}$ (D) $\dfrac{3}{2}$ (E) nonexistent

8. If $f(x) = \displaystyle\int_0^x (t \ln t)\,dt$, then $\displaystyle\lim_{h \to 0} \frac{f(e - h) - f(e)}{h} =$

(A) $-\infty$ (B) $-e$ (C) e (D) $2e$ (E) ∞

490

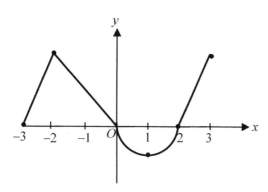

9. Let f be the continuous function defined on $[-3, 3]$ whose graph, consisting thee line segments and a semicircle, is given above. If $g(x) = \int_0^x f(t)\,dt$, which of the following statements must be true?

(A) The graph of g has a relative maximum at $x = -2$.
(B) The graph of g is not differentiable at $t = 0$.
(C) The graph of g has a point of inflection at $x = 1$ only.
(D) The graph of g is concave upward on the interval $(0, 2)$.
(E) The graph of g is concave upward on the interval $(-3, -2)$.

10. Which of the following is the radius of convergence of $\sum_{n=0}^{\infty} n!x^n$?

(A) 0   (B) 1   (C) 2   (D) 5   (E) ∞

11. The Taylor series for $\dfrac{1}{1+x}$ about $x = 0$ is $\sum_{n=0}^{\infty}(-1)^n x^n$. Witch of the following is a power series expansion for $\dfrac{x}{1-x}$?

(A) $1 + x + x^2 + x^3 + \cdots$
(B) $x + x^2 + x^3 + x^4 + \cdots$
(C) $x - x^2 + x^3 - x^4 + \cdots$
(D) $x^2 + x^4 + x^6 + x^8 + \cdots$
(E) $x - x^3 + x^4 - x^5 + \cdots$

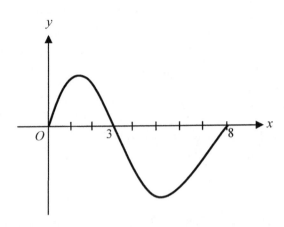

12. The graph of the function $y = f(x)$ is shown in the figure above. If the area of the region enclosed by the graphs of f and the x-axis for $0 \le x \le 3$ is 5, and the area of the region enclosed by the graphs of f and the x-axis for $3 \le x \le 8$ is 12, then $\int_0^8 (2f(x)+5)\,dx =$

(A) –36 (B) –9 (C) 26 (D) 39 (E) 74

13. If $f(x) = \sum_{n=0}^{\infty} \dfrac{(-1)^n x^{2n+1}}{2n+1}$ for $|x| < 1$, then $f'(x) =$

(A) $\dfrac{-1}{1-x^2}$ (B) $\dfrac{1}{1+x^2}$ (C) $\dfrac{-1}{1+x^2}$ (D) $\dfrac{-x}{1-x^2}$ (E) $\dfrac{x}{1+x^2}$

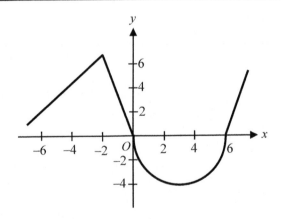

Graph of f

14. The graph of a function f is shown above. If $g(x) = \int_0^x f(t)\,dt$, at which of the following value of x does g have the greatest?

(A) $g(-4)$ (B) $g(-2)$ (C) $g(0)$ (D) $g(2)$ (E) $g(4)$

AP Calculus BC Test 4

Section 1

15. Let f be the function given by $f(x) = \cos x - 1$. What is the value of the estimation for $f(1)$ using the first two nonzero terms of the Taylor series for f about $x = 0$?

   (A) -1 (B) $-\dfrac{1}{2}$ (C) $-\dfrac{11}{24}$ (D) 0 (E) 1

---

16. For $x > 0$, if $y = x^x$, then $\dfrac{dy}{dx} =$

   (A) x^{x-1}
   (B) $x^x(x+1)$
   (C) $x^x(1 + \ln x)$
   (D) $x^{x-1}(1 + x + \ln x)$
   (E) $x^x(1 + x + \ln x)$

---

17. If the line $y = 4x$ is tangent to the graph of $y = x^2 + k$, then what is the value of k?

   (A) 0 (B) 1 (C) 2 (D) 3 (E) 4

---

18. What are all values of x for which the infinite series $\displaystyle\sum_{n=0}^{\infty} \dfrac{(x-2)^{n+1}}{n+1}$ converges?

   (A) $-1 < x < 1$ (B) $1 \le x \le 3$ (C) $1 < x < 3$ (D) $1 < x \le 3$ (E) $1 \le x < 3$

---

19. A curve is defined by the parametric equations $x = t^2 - 2t - 4$ and $y = \dfrac{t^3 + 1}{3}$. What is an equation of the line tangent to the curve at $t = 2$?

   (A) $y = x - 10$
   (B) $y = 2x + 11$
   (C) $y = 2x - 10$
   (D) $y = 3x - 7$
   (E) $y = 3x + 13$

AP Calculus BC Test 4   Section 1

20. If $g'(x) = 3g(x)$ and $g(-1) = e$, then $g(x)$ is

(A) e^{x+2} (B) e^{x^2-1} (C) e^{2x+3} (D) e^{3x+4} (D) e^{4x+5}

---

21. Which of the following is an expression for infinite power series

$$\lim_{n \to \infty} \frac{1}{n}\left\{\left(2 + \frac{1}{n}\right)^2 + \left(2 + \frac{2}{n}\right)^2 + \left(2 + \frac{3}{n}\right)^2 + \cdots + \left(2 + \frac{n}{n}\right)^2\right\}?$$

(A) $\int_0^1 x^2\,dx$ (B) $\int_1^2 (2+x)^2\,dx$ (C) $\int_2^3 x^2\,dx$ (D) $\int_2^4 (2+x)^2\,dx$ (E) $\int_2^3 (2+x)^2\,dx$

---

22. $\int_0^2 |x^2 - x|\,dx =$

(A) 1   (B) $\frac{3}{2}$ (C) 2   (D) $\frac{5}{2}$ (E) 3

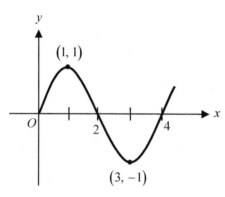

Graph of f

23. The graph of f is shown above. If $g(x) = \int_0^x f(t)\,dt$ and $\int_0^4 f(x)\,dx = 0$, then which of the following must be true?

   I. $g(3) > 0$ II. $g'(3) > 0$ III. $g''(3) > 0$

(A) I only   (B) II only   (C) III only   (D) I and II only   (E) I, II, and III

AP Calculus BC Test 4

Section 1

24. $\int x^3 e^{-x} dx =$

(A) $\dfrac{x^4}{4} e^{-x} + C$

(B) $e^{-x}(x^2 + 3x + 6) + C$

(C) $e^{-x}(x^3 - 3x^2 + 6x - 6) + C$

(D) $-e^{-x}(x^3 + 3x^2 + 6x + 6) + C$

(E) $-e^{-x}(x^3 - 3x^2 + 6x - 6) + C$

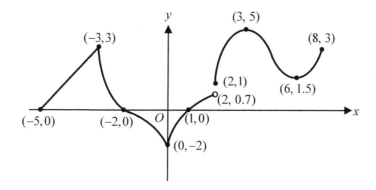

25. The graph of f is shown above. What are all critical points of the graph on the open interval $-5 < x < 8$?

(A) $(3, 8)$ and $(8, 3)$ only
(B) $(-5, 0), (-2, 0)$, and $(1, 0)$
(C) $(-5, 0), (-2, 0), (1, 0), (3, 5)$, and $(6, 1.5)$
(D) $(-3, 3), (0, -2), (3, 5)$, and $(6, 1.5)$
(E) $(-3, 3), (0, -2), (3, 5), (2, 1)$ and $(6, 1.5)$

26. Let f and g be differential functions such that $f(2) = 4$, $g(2) = 3$, $f'(3) = -5$, $f'(2) = -4$, $g(2) = 3$ and $g'(2) = -3$. If $h(x) = f(g(x))$, then what is the value of $h'(2)$?

(A) -15
(B) -12
(C) 12
(D) 15
(E) 25

AP Calculus BC Test 4

Section 1

27. If $g(x) = \dfrac{d}{dx}\left(\displaystyle\int_0^{x^2} 2\ln(t^3)\,dt\right)$, then $g(e) =$

   (A) 6
   (B) 12
   (C) $12e$
   (D) $12e^2$
   (E) $24e$

28. What is the coefficient of x^5 in the Taylor series for e^{x+1} about $x = 0$?

   (A) $\dfrac{1}{24}$ (B) $\dfrac{e}{120}$ (C) $\dfrac{1}{2}$ (D) $120e$ (E) 240

END OF PART A OF SECTION 1

AP Calculus BC Test 4

Section 1

CALCULUS BC
SECTION 1, PART B
Time— 50minutes
Number of questions— 17

A GRAPHING CALCULATOR IS REQUIRED FOR SOME QUESTIONS ON THIS PART OF THE EXAMINATION.

Directions: Solve each of the following problems, using the available space for scratchwork. After examining the form of the choices, decide which is the best of the choices given and fill in the corresponding oval on the answer sheet. No credit will be given for anything written in the test book. Do not spend too much time on any one problem.

In this test:

(1) The exact numerical value of the correct answer does not always appear among the choices given. When this happens, select from among the choices the number that best approximates the exact numerical value.

(2) Unless otherwise specified, the domain of a function f is assumed to be the set of all real numbers x for which $f(x)$ is a real number.

(3) The inverse of a trigonometric function f may be indicated using the inverse function notation f^{-1} or with the prefix "arc" (e.g, $\sin^{-1} x = \arcsin x$)

AP Calculus BC Test 4 — Section 1

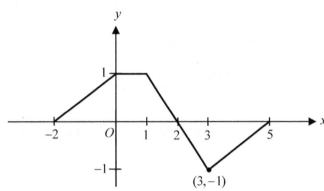

76. The graph of f is shown above for $-2 \le x \le 5$. What is the value of $\int_{-2}^{5}(f(x)+2)\,dx$?

(A) 1   (B) 4   (C) 6   (D) 15   (E) 19

---

77. Let $P(t)$ represent the number of elephants in a population at time t years, when time $t \ge 0$. The population $P(t)$ is increasing and $P(t)$ is given by $\dfrac{dP}{dt} = \dfrac{1}{10}(100 - P)$. What is the particular solution to the differential equation with initial condition $P(0) = 20$?

(A) $P(t) = 20 + 80e^{-10t}$

(B) $P(t) = 100 - 80e^{\frac{1}{10}t}$

(C) $P(t) = 100 - 80e^{-\frac{1}{10}t}$

(D) $P(t) = 100 - 80t$

(E) $P(t) = 100 - 80e^{-t}$

---

78. The radius of a sphere is increasing at a constant rate of 0.5 meters per second. What is the rate of increase in the surface area of the sphere at the instant when the volume of the sphere is 36π cubic meters?

(A) $18.850 \text{ m}^2/\sec$

(B) $20.125 \text{ m}^2/\sec$

(C) $24.897 \text{ m}^2/\sec$

(D) $37.700 \text{ m}^2/\sec$

(E) $47.567 \text{ m}^2/\sec$

AP Calculus BC Test 4 Section 1

x	f(x)	f'(x)	f''(x)	g(x)	g'(x)	g''(x)
-1	-3	3	-6	2	4	3
0	-4	4	-4	-2	3	2
1	-5	6	-2	-2.5	2	1
2	-6	8	-1	-3	1	0

79. The table above gives values of f, f', f'', g, g', and g'' at selected values of x. If $h(x) = f'\left(\dfrac{f(x)}{g(x)}\right)$, then $h'(0) =$

(A) -3 (B) -1 (C) $\dfrac{3}{4}$ (D) $\dfrac{8}{9}$ (E) 3

---

80. The velocity, in meters per second, of a runner is given by $v(t) = \dfrac{180e^{-0.1t}}{10 - e^{-5t}} - 10$, where time t is measured in seconds. Which of the following is the total distance run by the runner during the time interval $0 \le t \le 10$?

(A) 14.154 meters
(B) 29.033 meters
(C) 33.578 meters
(D) 46.012 meters
(E) 65.363 meters

---

81. A particle moves along the x-axis so that its position at any time is given by $x(t) = 3\cos(2t) - \ln(4t - 3) + \sin(2t)$ on the closed interval $[1, 5]$. How many times does the particle change direction?

(A) 1   (B) 2   (C) 3   (D) 4   (E) 5

---

82. The function $f(x)$ is continuous and differentiable on the closed interval $[0, 5]$. If $f(0) = -6$ and $f(5) = 4$, which of the following statements must be true?

(A) The maximum value of f on $[0, 5]$ is 4.
(B) $f'(x) < 0$ for $0 < x < 5$.
(C) $f(x)$ has a point of inflection on $[0, 5]$.
(D) There exists c, with $0 < c < 5$, for which $f(c) = 2$.
(E) There exists c, with $0 < c < 5$, for which $f'(c) = 0$.

AP Calculus BC Test 4 Section 1

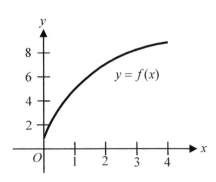

83. The graph of $y = f(x)$ is shown above. Which of the following must be true?

   I. A right Riemann sum over approximates $\int_0^4 f(x)\,dx$.

   II. A left Riemann sum under approximates $\int_0^4 f(x)\,dx$.

   III. A trapezoidal sum under approximates $\int_0^4 f(x)\,dx$.

   (A) I only   (B) II only   (C) I and II only   (D) I and III only   (E) I, II, and III

84. Which of the following is the value of $\lim\limits_{n\to\infty}\left\{\dfrac{(1+n)^3+(2+n)^3+(3+n)^3+\cdots+(n+n)^3}{n^4}\right\}$?

   (A) $\dfrac{2}{3}$ (B) $\dfrac{10}{7}$ (C) $\dfrac{15}{4}$ (D) $\dfrac{25}{4}$ (E) $\dfrac{51}{4}$

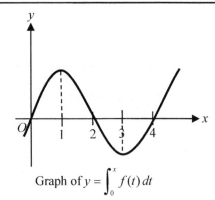

Graph of $y = \int_0^x f(t)\,dt$

85. Let f be the continuous function. If the graph of $y = \int_0^x f(t)\,dt$ is shown above, which of the following must be true?

   (A) $f(0) = 0$ (B) $f(1) > 0$ (C) $f(2) < 0$ (D) $f(3) < 0$ (E) $f(4) = 0$

AP Calculus BC Test 4   Section 1

86. Let f be the function given by $f(x) = \int_0^{2x} \cos(2-t^2)dt$ on the interval $0 \leq x \leq 2$. How many points of inflection does the graph of f have on this interval?

    (A) 1   (B) 2   (C) 3   (D) 5   (E) 7

---

87. Let f be the function defined by $f(x) = x - \ln x$. What is the value of c for which the instantaneous rate of change at $x = c$ is equal to the average rate of change of f over the closed interval $[1, e]$?

    (A) 1.016   (B) 1.718   (C) 2.315   (D) 2.562   (E) nonexistent

---

88. If the function is given by $f(x) = x^3 - x^2 + x - 1$, then $\lim\limits_{x \to 1} \dfrac{1}{(x-1)^2} \int_1^x (t^3 - t^2 + t - 1)dt =$

    (A) 1   (B) 2   (C) 5   (D) ∞ (E) undefined

---

89. Let f be the function defined by $f(x) = (x-k)(x+1)^2$, where k is a constant. What are all values of x for which the function f has a relative minimum at $x = -1$?

    (A) $k < -1$ (B) $-1 < k < 1$ (C) $k > 0$ (D) $1 < k < 3$ (E) $k > 1$

---

90. Let f be the function given by the infinite series $f(x) = \sum\limits_{n=1}^{\infty} \dfrac{x^n}{n}$. Which of the following is the intervals of convergence of $f'(x)$?

    (A) $-\infty < x < -1$
    (B) $-1 \leq x \leq 1$
    (C) $-1 \leq x < 1$
    (D) $-1 < x < 1$
    (E) $x < -1$ or $x > 1$

91. The function f is given by the Maclaurin series $f(x) = \sum_{n=0}^{\infty} \dfrac{(-1)^n x^{2n+1}}{(2n+1)!}$. The graph of the function f intersects the graph of $y = e^x - 2$ at $x =$

(A) 1.054
(B) 1.517
(C) 2.718
(D) 3.141
(E) 4.771

---

92. Let f be the function given by $f(x) = \sin x$. The third-degree Maclaurin polynomial of f is used to approximate values of $f(x)$. Which of the following is the greatest value of x for which the error resulting from the approximation is less than 0.3?

(A) 1.9 (B) 2.0 (C) 2.1 (D) 2.2 (E) 2.3

END OF SECTION 1

CALCULUS BC

SECTION II
Time: 1 hour and 30 minutes
Percent of total grade – 50

Part A: 30 minutes, 2 problems
Part B: 60 minutes, 4 problems

Part A: A graphing calculator is required for some problems or parts of problems.
Part B: No calculator is allowed for these problems.

GENERAL INSTRUCTIONS FOR SECTION II PART A AND PART B

For each part of section II, you may wish to look over the problems before starting to work on them, since it is not expected that everyone will be able to complete all parts of all problems. All problems are given equal weight, but the parts of a particular problem are not necessary given equal weight.

- YOU SHOULD WRITE ALL WORK FOR EACH PART OF EACH PROBLEM IN THE SPACE PROVIDED FOR THAT PART IN THE PINK TEST BOOKLET. Be sure to write clearly and legibly. If you make an error, you may save time by crossing it out rather than trying to erase it. Erased or cross-out work will not be graded.
- Show all work. Clearly label any functions, graphs, tables, or other object that you use. You will be graded on the correctness and completeness of your methods as well as your answers. Answers without supporting work may not receive credit.
- Justifications required that you give mathematical (noncalculator) reasons
- Your work must be expressed in standard mathematical notation rather than calculator syntax. For example, $\int_{1}^{5} x^2 \, dx$ may not be written as $\text{fnInt}(X^2, X, 1, 5)$
- Unless otherwise specified, answers (numeric or algebraic) need not be simplified.
- If you use decimal approximations in calculations, you will be graded on accuracy. Unless otherwise specified, your final answers should be accurate to three places after the decimal point.
- Unless otherwise specified, the domain of a function f is assumed to be the set of all real numbers x for which $f(x)$ is a real number.

CALCULUS BC
SECTION II, PART A
Time — 30 minutes
Number of questions — 2

A Graphing calculator is required for some problems or parts of problems.

Question 1

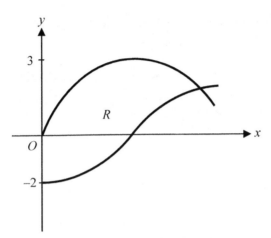

Let R be the region enclosed by the y-axis and the graphs of $y = 3\sin\left(\dfrac{\pi}{4}x\right)$ and $y = -2\cos\left(\dfrac{\pi}{3}x\right)$, as shown in the figure above.

(a) Find the area of R.

(b) Find the volume of the solid generated when R is revolved about the y-axis.

(c) The region R is the base of a solid. For this region, each cross section perpendicular to the x-axis is a square. Find the volume of the solid.

AP Calculus BC Test 4

Question 2

A particle moves in the xy-plane so that its position at any time t, $0 \le t \le 2$, is given by

$$x(t) = \sin(e^t) - \ln(1+t) \text{ and } y(t) = 2(1 - \cos(t))$$

(a) At what time of t, $0 \le t \le 2$, does $x(t)$ attain its minimum value? Find the position $(x(t), y(t))$ of the particle at this time.

(b) At what time of t, $0 \le t \le 2$, at which the line tangent to the path of the particle is vertical?

(c) Find the speed and the acceleration vector of the particle at time $t = 1$.

(d) Sketch the path of the particle in the xy-plane below. Indicate the direction of motion along the path and the coordinates of the point at which the line tangent to the path of the particle is vertical.

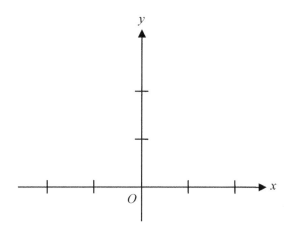

END OF PART A OF SECTION II

AP Calculus BC Test 4 — Section II

CALCULUS BC
SECTION II, PART B
Time — 60 minutes
Number of questions — 4

No calculator is required for these problems.

Question 3

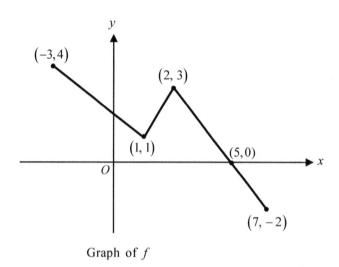

Graph of f

Let f be the continuous function defined on $[-3, 7]$ whose graph, consisting of three line segments is shown above. Let g be the function given by $g(x) = \int_1^x f(t)\,dt$.

(a) Find the values of $g(-3)$ and $g(7)$.

(b) Find the instantaneous rate of change of g at $x = 3$.

(c) Find the absolute maximum value of g on the closed interval $[-3, 7]$. Explain your reasoning.

(d) For $-3 < x < 7$, find all values of x-coordinates of points of inflection of the graph of g. Explain your reasoning.

(e) Find the equation of the line tangent to the graph of g at $x = 2$.

AP Calculus BC Test 4

Question 4

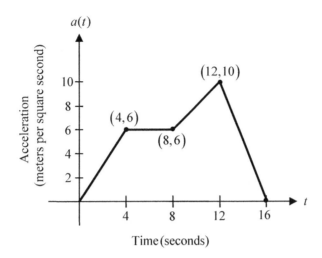

A particle moves along the x-axis. The acceleration $a(t)$ of the particle at time t, in meters per square second, is modeled by the piecewise-linear function defined by the graph above.

(a) Find $\int_0^{16} a(t)\,dt$. Explain the meaning of $\int_0^{16} a(t)\,dt$.

(b) Given that $v(t)$ is the velocity of the particle at time t, find the average rate of change of v over the interval $[0, 16]$. Indicate units of measure.

(c) For $0 \le t \le 16$, find the piecewise-defined function for $v(t)$.

(d) Given that $v(t)$ is the velocity of the particle at time t and that $v(0) = 4$ meters/second, find the value of $v(8)$. Indicate units of measure.

AP Calculus BC Test 4   Section II

Question 5

The function f is differentiable for all real numbers. The slope at each point (x,y) on the graph is given by $\dfrac{dy}{dx} = 4y^2 - 2xy^2$ and the point $(2,5)$ is on the graph of $y = f(x)$.

(a) Write an equation for the line tangent to the curve at $(2,5)$.

(b) Find $\dfrac{d^2y}{dx^2}$ and determine whether f has a local maximum, a local minimum, or neither at the point $(2,5)$. Using $\dfrac{d^2y}{dx^2}$, justify your answer.

(c) Find the particular solution $y = f(x)$ to the differential equation $\dfrac{dy}{dx} = 4y^2 - 2xy^2$.

Question 6

Let f be the function given by $f(x) = \dfrac{\ln(x+1)}{x}$. The Maclaurin series for $\ln x$ is

$$\ln x = \sum_{n=1}^{\infty} \dfrac{(-1)^{n-1}(x-1)^n}{n}.$$

(a) Write the first four nonzero terms and the general term of the Taylor series for f about $x = 0$.

(b) Find the interval of convergence of the Maclaurin series for f.

(c) Let $P_4(x)$ is fourth-degree Taylor polynomial for f about $x = 0$ to approximate $f\left(\dfrac{1}{2}\right)$. Explain why the estimate $P_4\left(\dfrac{1}{2}\right)$ differs from the actual value of $f\left(\dfrac{1}{2}\right)$ by less than $\dfrac{1}{150}$.

END OF EXAM

No material on this page

AP Calculus BC Test 4 Answers

Section 1

SECTION 1, PART A

1. **B**

   We need $f''(x) = 0$ to have a point of inflection as follow.

   $f'(x) = 10x^4 - 20x^3 + 2 \;\rightarrow\; f''(x) = 40x^3 - 60x^2 = 20x^2(2x-3) = 0$

   $f''(x) = 0$ at $x = 0$ or $x = 3/2$.

   But f'' changes in sign at $x = \dfrac{3}{2}$ only.

   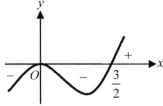
   Graph of f''

2. **(E)**

   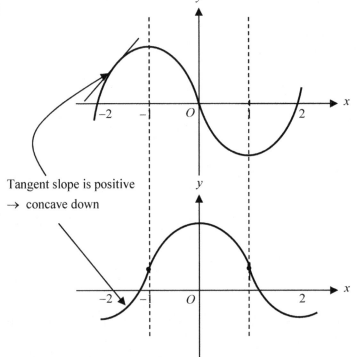

   Tangent slope is positive → concave down

   Possible graph of f

   (A) f is decreasing for $-1 \le x \le 0$. $\Rightarrow$ For $-1 \le x \le 0$, $f'(x) > 0$ $\Rightarrow$ $f(x)$ is increasing. False
   (B) f is increasing for $1 \le x \le 2$. $\Rightarrow$ For $1 \le x \le 2$, $f'(x) < 0$ $\Rightarrow$ $f(x)$ is decreasing. False
   (C) f has a local maximum at $x = -1$. $\Rightarrow$ At $x = -1$, $f(x)$ has a point of inflection. False
   (D) f has a local minimum at $x = -1$. False
   (E) f has a local maximum at $x = 0$. For $x < 0, f' > 0$, and for $x > 0, f' < 0$. True

3. **D**

   We apply u-substitution: $\begin{cases} u = x^2 + 4 \\ du = 2x\,dx \end{cases}$

   $\displaystyle\int_0^{\sqrt{5}} \dfrac{2x}{\sqrt{x^2+4}}\,dx = \int_4^9 \dfrac{du}{\sqrt{u}} = \left[2u^{1/2}\right]_4^9 = 2\sqrt{9} - 2\sqrt{4} = 2$

AP Calculus BC Test 4 Answers — Section 1

4. D

$$\lim_{x\to 1}\frac{1}{x-1}\int_1^x \left(3t^2-4t+2e^{t-1}\right)dt = \lim_{x\to 1}\frac{3x^2-4x+2e^{x-1}}{1} = \frac{3-4+2}{1} = 1 : \text{L'Hopital's Rule}$$

5. A

In order for f to be continuous, it is satisfied that $\lim_{x\to 4}\dfrac{x^2-8x+16}{x-4}=k$.

Therefore, $\lim_{x\to 4}\dfrac{x^2-8x+16}{x-4} = \lim_{x\to 4}\dfrac{(x-4)^2}{x-4} = \lim_{x\to 4}(x-4)=0$ and $k=0$.

6. D

$\int_1^\infty \dfrac{x^2}{x^p}dx = \int_1^\infty \dfrac{1}{x^{p-2}}dx$. If $p-2>1$ or $p>3$, the integral converges.

7. A

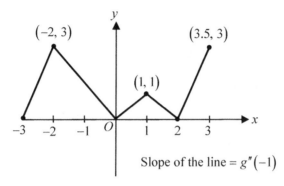

Slope of the line $= g''(-1)$

Because $g'(x)=f(x)$ and $g''(x)=f'(x)$, $g''(-1)=f'(-1)$ is the slope of the line segment at $x=-1$.

Therefore, $g''(-1)=f'(-1)=\dfrac{3-0}{-2-0}=-\dfrac{3}{2}$.

8. B

$$f(x)=\int_0^x (t\ln t)\,dt \;\to\; f'(x)=x\ln x$$

$$\lim_{h\to 0}\frac{f(e-h)-f(e)}{h} = \lim_{h\to 0}-\frac{f(e+(-h))-f(e)}{(-h)} = -f'(e) = -e\ln e = -e$$

9. E

(A) The graph of g has a relative maximum at $x=-2$. False
(B) The graph of g is not differentiable at $t=0$. False
(C) The graph of g has a point of inflection at $x=1$ only. False (Point of inflection at $x=-2$)
(D) The graph of g is concave upward on the interval $(0,2)$. False $\to g''<0$ and $g''>0$
(E) The graph of g is concave upward on the interval $(-3,-2)$. True $\to g''>0$
   Since g'' is the slope of the line that is positive.

10. A

Using ratio test, $\sum_{n=0}^\infty n!x^n \;\to\; \lim_{n\to\infty}\left|\dfrac{(n+1)!x^{n+1}}{n!x^n}\right| = \lim_{n\to\infty}|(n+1)x|<1 \to |\infty\times 0|<1$

For $x=0$ only, the power series converge. Therefore, the radius of convergence R is zero.

11. B

If $\dfrac{1}{1+x}=1-x+x^2-x^3+x^4-\cdots$, then $\dfrac{1}{1-x}=1+x+x^2+x^3+x^4+\cdots$

AP Calculus BC Test 4 Answers — Section 1

Multiplying both sides by x, we got $x\left(\dfrac{1}{1-x}\right) = x + x^2 + x^3 + x^4 + \cdots$.

12. C

From the graph, the area is

$$\int_0^8 (2f(x)+5)\,dx = 2\int_0^3 f(x)\,dx + 2\int_3^8 f(x)\,dx + \int_0^8 5\,dx = 2(5) + 2(-12) + 40 = 26.$$

13. B

We have to recognize that it is the power series of arctan x.

$$f(x) = \sum_{n=0}^{\infty} \dfrac{(-1)^n x^{2n+1}}{2n+1} = x - \dfrac{x^3}{3} + \dfrac{x^5}{5} - \dfrac{x^7}{7} + \cdots = \arctan x$$

Therefore, $f'(x) = \dfrac{1}{1+x^2}$.

14. C

Because $g(0) = 0$, the graph of g has the greatest value at $x = 0$.
The values of $g(-4), g(-2), g(2),$ and $g(4)$ are negative.

15. C

The Taylor series of $\cos x$ is $\cos x = 1 - \dfrac{x^2}{2!} + \dfrac{x^4}{4!} - \dfrac{x^6}{6!} + \cdots$.

Hence, $\cos x - 1 = -\dfrac{x^2}{2!} + \dfrac{x^4}{4!} - \dfrac{x^6}{6!} + \cdots$.

Therefore, $P_2(x) = -\dfrac{x^2}{2!} + \dfrac{x^4}{4!}$ or $P_2(1) = -\dfrac{1}{2} + \dfrac{1}{24} = -\dfrac{11}{24}$.

16. C

$\ln y = x \ln x \;\to\; \dfrac{y'}{y} = \ln x + 1 \;\to\; y' = y(\ln x + 1) \;\to\; y' = x^x(\ln x + 1)$

17. E

The slope of the tangent line at point (x_1, y_1) equals 4. $f'(x) = 2x \;\to\; 2x_1 = 4$ or $x_1 = 2$ and $y_1 = 4(2) = 8$.

Hence, point of the tangency is $(2, 8)$. Substituting $(2, 8)$ into the equation,

$8 = 2^2 + k$ or $k = 4$.

18. E

$$\lim_{n \to \infty} \left| \dfrac{(x-2)^{n+2}}{n+2} \cdot \dfrac{n+1}{(x-2)^{n+1}} \right| = \lim_{n \to \infty} \left| \dfrac{(n+1)}{(n+2)} \cdot (x-2) \right| = |x-2| < 1 \text{ or } 1 < x < 3 : \text{(Ratio Test)}$$

For $x = 1$, $\sum_{n=0}^{\infty} \dfrac{(-1)^{n+1}}{n+1} = -1 + \dfrac{1}{2} - \dfrac{1}{3} + \dfrac{1}{4} - \cdots$ Alternating Harmonic Series: The series converges.

For $x = 3$, $\sum_{n=0}^{\infty} \dfrac{(1)^{n+1}}{n+1} = 1 + \dfrac{1}{2} + \dfrac{1}{3} + \cdots$ Harmonic Series: The series diverges.

Therefore, the interval of convergence is $1 \le x < 3$.

19. B

The slope of the tangent line is $\dfrac{dy}{dx} = \dfrac{dy/dt}{dx/dt} = \dfrac{t^2}{2t-2}$, $\left.\dfrac{dy}{dx}\right|_{t=2} = \dfrac{4}{2} = 2$.

When $t = 2$, $x(2) = -4$ and $y(2) = 3$ or $(x, y) = (-4, 3)$.

The equation of the line tangent to the curve at $t = 2$ is $y - 3 = 2(x+4)$ or $y = 2x + 11$.

AP Calculus BC Test 4 Answers Section 1

20. D

Solve the differential equation. $\dfrac{g'}{g} = 3 \to \ln|g| = 3x + C_1 \to g = \pm e^{C_1} e^{3x}$ or $g = Ce^{3x}$.

With initial condition: $Ce^{-3} = e$ or $C = e^4$

Therefore, $g(x) = e^4 e^{3x} = e^{3x+4}$.

21. C

$$\lim_{n \to \infty} \dfrac{1}{n}\left\{\left(2+\dfrac{1}{n}\right)^2 + \left(2+\dfrac{2}{n}\right)^2 + \left(2+\dfrac{3}{n}\right)^2 + \cdots + \left(2+\dfrac{n}{n}\right)^2\right\} = \sum_{k=1}^{\infty} \dfrac{1}{n}\left(2+\dfrac{k}{n}\right)^2$$

$$= \int_0^1 (2+x)^2 \, dx \text{ or } \int_2^3 x^2 \, dx$$

22. A

The solutions of $x^2 - x = x(x-1) = 0$ are $x = 0$ or $x = 1$.

Hence, $|x^2 - x| = \begin{cases} x^2 - x, & x \geq 1 \\ -x^2 + x, & 0 \leq x < 1 \end{cases}$.

Therefore, $\int_0^2 |x^2 - x| \, dx = \int_0^1 (-x^2 + x) \, dx + \int_1^2 (x^2 - x) \, dx$

$= \left[\dfrac{-x^3}{3} + \dfrac{x^2}{2}\right]_0^1 + \left[\dfrac{x^3}{3} - \dfrac{x^2}{2}\right]_1^2 = \left[-\dfrac{1}{3} + \dfrac{1}{2}\right] + \left[\left(\dfrac{8}{3} - \dfrac{4}{2}\right) - \left(\dfrac{1}{3} - \dfrac{1}{2}\right)\right] = 1$.

23. A

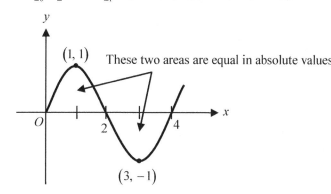

Graph of f

I. $g(3) = \int_0^3 f(t) \, dt$ must be positive. True

II. $g'(3) > 0$: False because $g'(3) = f(3) = -1$.

III. $g''(3) > 0$: False because $g''(3) = f'(3) = 0$ which is a point of inflection.

24. D

Use Tabular integration by parts

(D) $-e^{-x}(x^3 + 3x^2 + 6x + 6) + C$

25. D

Definition of critical point:

Let f be defined (continuous) at c. If $f'(c) = 0$ or if f' is undefined at c, then $(c, f(c))$ is a critical point of f.

(D) $(-3, 3), (0, -2), (3, 5)$, and $(6, 1.5)$

26. D

AP Calculus BC Test 4 Answers — Section 1

Since $h'(x) = f'(g(x)) \cdot g'(x)$, we have that $h'(2) = f'(g(2)) \cdot g'(2) = f'(3)g(2) = (-5)(-3) = 15$.

27. E

$$g(x) = \frac{d}{dx}\left(\int_0^{x^2} 2\ln(t^3)\,dt\right) = 2(2x)\ln(x^6) = 4x\ln(x^6)$$

$$g(e) = 4e\ln e^6 = 24e$$

28. B

Since $e^x = 1 + x + \frac{x^2}{2!} + \frac{x^3}{3!} + \frac{x^4}{4!} + \frac{x^5}{5!} + \cdots$, we got that $e^{x+1} = e(e^x) = e + ex + \cdots + e\frac{x^5}{5!} + \cdots$.

Therefore, the coefficient of x^5 in the Taylor series for e^{x+1} about $x = 0$ is $\frac{e}{5!}$ or $\frac{e}{120}$.

Or,

The term in Taylor series is $\frac{f^{(5)}(0)}{5!}x^5$ and $f^{(5)}(0) = e$ because $f^{(5)}(x) = e^{x+1}$.

Therefore, $\frac{e}{5!}$ is the coefficient of x^5.

SECTION 1, PART B

76. D

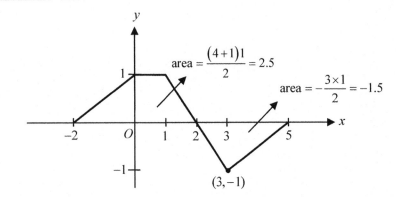

area = $\frac{(4+1)1}{2} = 2.5$

area = $-\frac{3 \times 1}{2} = -1.5$

$$\text{Area} = \int_{-2}^{5}(f(x)+2)\,dx = \int_{-2}^{5} f(x)\,dx + \int_{-2}^{5} 2\,dx = (2.5 - 1.5) + [2x]_{-2}^{5} = 1 + (10+4) = 15$$

77. C

Solve the differential equation. $\frac{dP}{dt} = \frac{1}{10}(100-P) \rightarrow \int \frac{dP}{(100-P)} = \int \frac{1}{10}\,dt \rightarrow -\ln|100-P| = \frac{1}{10}t + C_1$

$\rightarrow \ln|100-P| = -\frac{1}{10}t + C_2 \rightarrow 100 - P = Ce^{-\frac{1}{10}t} \rightarrow P = 100 - Ce^{-\frac{1}{10}t}$

With initial condition: $20 = 100 - C \rightarrow C = 80$

Therefore, the particular solution is $P(t) = 100 - 80e^{-\frac{1}{10}t}$.

78. D

We know that surface area of a sphere is $A = 4\pi r^2$ and the volume of the sphere is $V = \frac{4\pi r^3}{3}$.

AP Calculus BC Test 4 Answers   Section 1

Because $\dfrac{dA}{dt} = 8\pi r \dfrac{dr}{dt}$, we need to find the value of r at the instant when $V = \dfrac{4\pi r^3}{3} = 36\pi$.

Now we got that $r = 3$ and $\dfrac{dr}{dt} = 0.5$.

Therefore, $\dfrac{dA}{dt} = 8\pi(3)(0.5) = 12\pi \approx 37.700$

79. B

$h(x) = f'\left(\dfrac{f(x)}{g(x)}\right) \rightarrow h'(x) = f''\left(\dfrac{f(x)}{g(x)}\right)\left(\dfrac{f(x)}{g(x)}\right)' = f''\left(\dfrac{f(x)}{g(x)}\right)\left(\dfrac{f'(x)g(x)-f(x)g'(x)}{g^2(x)}\right)$

Therefore, $h''(0) = f''\left(\dfrac{f(0)}{g(0)}\right)\left(\dfrac{f'(0)g(0)-f(0)g'(0)}{g^2(0)}\right) = f''\left(\dfrac{-4}{-2}\right)\left(\dfrac{(4)(-2)-(-4)(3)}{(-2)^2}\right)$

$= f''(2)(1) = (-1)(1) = -1.$

80. B

Using a graphing calculator, draw the graph of v as follows.

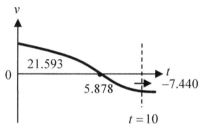

Find the value of t at which the velocity is 0. $t = 5.878$
Therefore, the travelled distance d is

$d = \int_0^{10}\left|\dfrac{180e^{-0.1t}}{10-e^{-5t}} - 10\right|dt = \int_0^{5.878}\left(\dfrac{180e^{-0.1t}}{10-e^{-5t}} - 10\right)dt - \int_{5.878}^{10}\left(\dfrac{180e^{-0.1t}}{10-e^{-5t}} - 10\right)dt$

$\approx 21.593 - (-7.440) = 29.033.$

81. C

At $v(t) = 0$, the particle changes direction. Find $v(t) = \dfrac{dx(t)}{dt}$.

$v(t) = -6\sin(2t) - \dfrac{4}{4t-3} + 2\cos(2t)$

The graph of v shows three zeros on the closed interval $[3, 5]$.

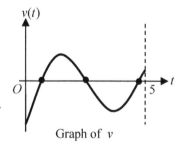

Graph of v

82. D

The Intermediate Value Theorem
There exists c for $0 < c < 5$, for which $f(c) = 2$.

83. E

I. A right Riemann sum over approximates $\int_0^4 f(x)\,dx$. True

II. A left Riemann sum under approximates $\int_0^4 f(x)\,dx$. True

III. A trapezoidal sum under approximates $\int_0^4 f(x)\,dx$. True

AP Calculus BC Test 4 Answers — Section 1

They are all true because the graph is strictly increasing and concave down on the closed interval $[0, 4]$.

84. C

Use the process of converting a Riemann sum to a definite integral.

$$\lim_{n\to\infty}\left\{\frac{(1+n)^3+(2+n)^3+(3+n)^3+\cdots+(n+n)^3}{n^4}\right\}=\sum_{k=1}^{\infty}\frac{1}{n}\left(1+\frac{k}{n}\right)^3=\int_1^2 x^3\,dx=\left[\frac{x^4}{4}\right]_1^2=\frac{15}{4}$$

Or $\sum_{k=1}^{\infty}\frac{1}{n}\left(1+\frac{k}{n}\right)^3=\int_0^1(1+x)^3 dx=\left[\frac{(1+x)^4}{4}\right]_0^1=\frac{2^4}{4}-\frac{1}{4}=\frac{16}{4}-\frac{1}{4}=\frac{15}{4}.$

85. C

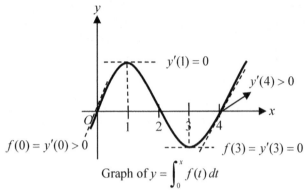

Graph of $y=\int_0^x f(t)\,dt$

Because $y'=f(x)$ is the slope of the line tangent to the graph,

(A) $f(0)=y'(0)>0$ (B) $f(1)=y'(1)=0$ (C) $f(2)=y'(2)<0$ True (D) $f(3)=y'(3)=0$
(E) $f(4)=y'(4)>0$

(C) is correct because the slope of the line tangent to the graph at $x=2$ is negative.

86. D

Draw the graph of f' using your graphing calculator.

$f'(x)=2\cos(2-4x^2)$

There are five points of inflection at which $f''(x)=0$.

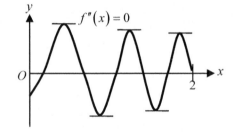

87. B

The instantaneous rate of change at $x=c$ is $f'(c)=1-\dfrac{1}{c}$ because $f'(x)=1-\dfrac{1}{x}$.

The average rate of change of f over the closed interval $[1, e]$ is $\dfrac{f(e)-f(1)}{e-1}$.

They are equal. $1-\dfrac{1}{c}=\dfrac{f(e)-f(1)}{e-1}$

Solve it. $1-\dfrac{1}{c}=\dfrac{e-\ln e-1}{e-1}\ \to\ 1-\dfrac{1}{c}=\dfrac{e-2}{e-1}\ \to\ \dfrac{1}{c}=1-\dfrac{e-2}{e-1}=\dfrac{1}{e-1}$

Therefore, $c=e-1\approx 1.718$.

88. A

$$\lim_{x\to 1}\frac{1}{(x-1)^2}\int_1^x(t^3-t^2+t+1)\,dt=\lim_{x\to 1}\frac{x^3-x^2+x-1}{2(x-1)}=\lim_{x\to 1}\frac{3x^2-2x+1}{2}=\frac{3-2+1}{2}=1$$

Apply L'Hôpital's rule two times because it is in the form $\dfrac{0}{0}$.

AP Calculus BC Test 4 Answers

89. A

$f'(x) = (x+1)^2 + 2(x-k)(x+1) = (x+1)(3x+1-2k)$

The graph of f' should be as follows.
In order to have a local minimum at $x = -1$,
The value of $\dfrac{2k-1}{3}$ should be less than -1.

Therefore, $\dfrac{2k-1}{3} < -1 \rightarrow 2k-1 < -3 \rightarrow 2k < -2$ or $k < -1$.

Or, we know that the graph of f will be as follows in order to have a local minimum at $x = -1$.

  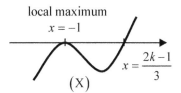

90. D

$f'(x) = \sum_{n=1}^{\infty} \left(\dfrac{x^n}{n}\right)' = \sum_{n=1}^{\infty} \dfrac{nx^{n-1}}{n} = \sum_{n=1}^{\infty} x^{n-1} = 1 + x + x^2 + x^3 + \cdots$

Because the series is a geometric series, the series converges if the common ratio $|x| < 1$.
Therefore, the interval of convergence is $-1 < x < 1$.

91. A

The power series equals $\sin x$ because $\sin x = x - \dfrac{x^3}{3!} + \dfrac{x^5}{5!} - \dfrac{x^7}{7!} + \cdots = \sum_{n=0}^{\infty} \dfrac{(-1)^n x^{2n+1}}{(2n+1)!}$

Using your graphing calculator, find the point of intersection of the graphs of $y = \sin x$ and $y = e^x - 2$.
The point of intersection is at $x = 1.054$.

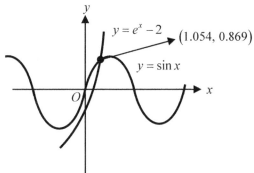

92. B

Because $P_3(x) = x - \dfrac{x^3}{3!}$ and $f(x) = \sin x$, the error will be $|\sin x - P_3| = \left|\sin x - x + \dfrac{x^3}{3!}\right|$.

Error $y = \sin x - \left(x - \dfrac{x^3}{6}\right) < 0.3$

From the graphs $y_1 = \sin x - x + \dfrac{x^3}{6}$ and $y_2 = 0.3$, we got an intersection at $x = 2.090$.

Therefore, the greatest value of x is 2.0 because the error resulting from the approximation is less than 0.3.

AP Calculus BC Test 4 Answers — Section 1

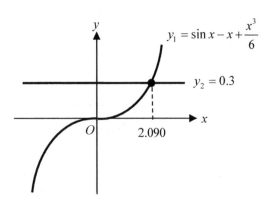

SECTION II, PART A

Question 1

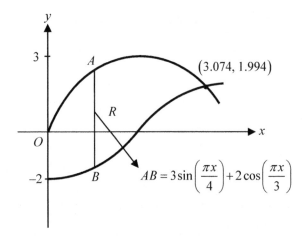

(a) Using your graphing calculator, draw the graphs of the two functions and find the intersection. The point of intersection is at $(3.074, 1.994)$. Therefore, the area of R is

$$\text{Area} = \int_0^{3.074} \left(3\sin\left(\frac{\pi x}{4}\right) - {}^-2\cos\left(\frac{\pi x}{3}\right)\right) dx = \int_0^{3.074} \left(3\sin\left(\frac{\pi x}{4}\right) + 2\cos\left(\frac{\pi x}{3}\right)\right) dx \approx 6.525.$$

(b) Cylindrical Shell Method is strictly recommended.

$$\text{Volume} = 2\pi \int_0^{3.074} xy\,dx = 2\pi \int_0^{3.074} x\left(3\sin\left(\frac{\pi x}{4}\right) + 2\cos\left(\frac{\pi x}{3}\right)\right) dx \approx 49.686$$

(c) Since the length of an edge of the square is $3\sin\left(\frac{\pi x}{4}\right) + 2\cos\left(\frac{\pi x}{3}\right)$, the volume of the solid is

$$\text{Volume} = \int_0^{3.074} \left(3\sin\left(\frac{\pi x}{4}\right) + 2\cos\left(\frac{\pi x}{3}\right)\right)^2 dx \approx 16.612.$$

Question 2

(a) Find the position $(x(t), y(t))$ of the particle at this time.

Using your calculator, graph $y = x'(t)$ such that $x'(t) = e^t \cos(e^t) - \dfrac{1}{t+1}$.

We can see that $x'(t)$ has a minimum at time $t = 1.567$.
Therefore, when $t = 1.567$,

$$x(1.567) = \sin(e^{1.567}) - \ln(1+1.567) \approx -1.940$$

$$y(1.567) = 2(1 - \cos(1.567)) \approx 1.992.$$

Therefore, the position of the particle is $(-1.940, 1.992)$ at time $t = 1.567$.

(b) The tangent line to the graph is vertical at the point where $\dfrac{dy}{dx}$ is undefined.

Therefore, $\dfrac{dx}{dt} = 0$ and $\dfrac{dy}{dt} \neq 0$ at time $t = 1.567$.

$\dfrac{dx}{dt} = e^t \cos(e^t) - \dfrac{1}{t+1} = 0$ at $t = 1.567$

$\dfrac{dy}{dt} = 2\sin t = 2\sin(1.567) \neq 0$

At time $t = 1.567$, the tangent line is vertical.

(c) Since $x'(1) = e\cos(e) - \dfrac{1}{2}$ and $y'(1) = 2\sin(1)$, the speed is

$$\text{Speed} = \sqrt{(e\cos(e) - 0.5)^2 + (2\sin(1))^2} \approx 2.996$$

Acceleration vector $\langle a_x, a_y \rangle$ at $t = 1$ as follows.

$$a_x(t) = x''(t) = e^t \cos(e^t) - e^{2t}\sin(e^t) + \dfrac{1}{(t+1)^2} \;\rightarrow\; a_x(1) \approx -5.264$$

$$a_y(t) = 2\cos t \;\Rightarrow\; a_y(1) = 2\cos(1) \approx 1.081$$

Therefore, the acceleration vector is $a = \langle -5.264, 1.081 \rangle$.

(d) Using your graphing calculator, draw the graph of the parametric functions.

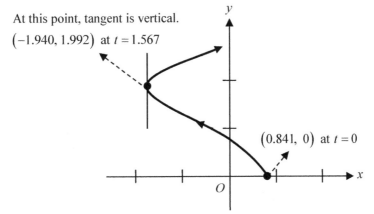

At this point, tangent is vertical.
$(-1.940, 1.992)$ at $t = 1.567$

$(0.841, 0)$ at $t = 0$

AP Calculus BC Test 4 Answers — Section 1

Question 3

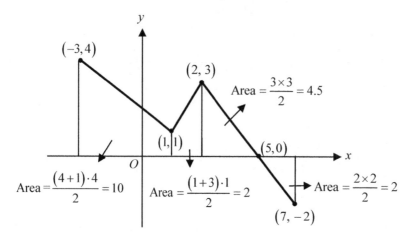

Graph of f

(a) $g(-3) = \int_1^{-3} f(t)dt = -\int_{-3}^{1} f(t)dt = -\left(\frac{(1+4)4}{2}\right) = -10$

$g(7) = \int_1^{7} f(t)dt = 2 + 4.5 - 2 = 4.5$

(b) Slope is $\frac{3-0}{2-5} = -1$ and $y = -x + b$. Using point $(2,3)$, $3 = -2 + b \rightarrow b = 5$.

The equation is $y = -x + 5$. If $x = 3$, then $y = -3 + 5 = 2$. Therefore, $g'(3) = f(3) = 2$.

(c) The absolute maximum value of g is $g(5)$ because $g(x)$ has the greatest area at $x = 5$ as follows.

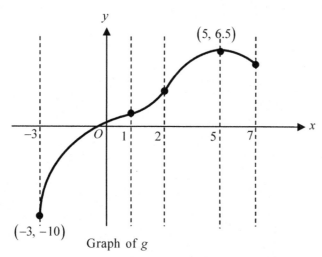

Graph of g

Therefore the absolute maximum is $g(5) = \int_1^{5} f(t)dt = 2 + 4 = 6.5$.

(d) The graph of g has a point of inflection at each of $x = 1$ and $x = 2$ because $g''(x)$ changes in sign at each of these values.

(e) Because $g(2) = \int_1^{2} f(t)dt = 2$, the tangent line passes through point $(2, 2)$.

The slope of the tangent line is $g'(2) = f(2) = 3$.

Therefore, the equation of the line tangent to the graph of g at $x = 2$ is $y - 2 = 3(x - 2)$.

AP Calculus BC Test 4 Answers

Question 4

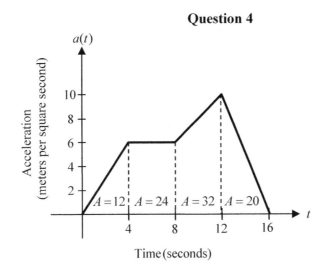

(a) $\int_0^{16} a(t)\,dt = 12 + 24 + 32 + 20 = 88$ meters per second

$\int_0^{16} a(t)\,dt$ is the particle's change in velocity in meters per second from $t = 0$ to $t = 16$ seconds.

(b) Average rate of change of v over the interval $[0, 16]$ is $\dfrac{v(16) - v(0)}{16}$ or $\dfrac{1}{16}\int_0^{16} a(t)\,dt$.

Therefore, average rate of change is $\dfrac{1}{16}\int_0^{16} a(t)\,dt = \dfrac{88}{16} = 5.5$ meters/second²

(c) For $0 \le t \le 8$, the piecewise-defined function for $a(t)$ is as follows.

$$a(t) = \begin{cases} y = \dfrac{3}{2}t & \text{for } 0 \le t < 4 \\ y = 6 & \text{for } 4 \le t < 8 \\ y = t - 2 & \text{for } 8 \le t < 12 \\ y = -\dfrac{5}{2}t + 40 & \text{for } 12 \le t \le 16 \end{cases}$$

(d) $v(8) = v(0) + \int_0^8 a(t)\,dt = 12 + (12 + 24) = 48$ meters per second

AP Calculus BC Test 4 Answers — Section 1

Question 5

(a) Since $\dfrac{dy}{dx} = 4y^2 - 2xy^2$, the slope of the tangent line is

$$\left.\dfrac{dy}{dx}\right|_{(2,5)} = 4(5)^2 - 2(2)(5)^2 = 0.$$

Therefore, the equation of the tangent line is $y = 5$.

(b) $\dfrac{d^2y}{dx^2} = 8yy' - (2y^2 + 4xyy') = -2y^2$ and $y' = 0$

$\left.\dfrac{d^2y}{dx^2}\right|_{(2,5)} = \left.-2y^2\right|_{(2,5)} = -50$ because $\left.y'\right|_{(2,5)} = 0$.

We know that $\dfrac{d^2y}{dx^2} < 0$ at point $(2, 5)$.

Hence the graph of f has a local maximum, because $f'(2) = 0$ and $f''(2) < 0$.

(c) $\dfrac{dy}{dx} = y^2(4 - 2x) \;\to\; \dfrac{dy}{y^2} = (4 - 2x)dx \;\to\; \int \dfrac{dy}{y^2} = \int (4 - 2x)dx$

$-\dfrac{1}{y} = 4x - x^2 + C_1 \;\to\; \dfrac{1}{y} = x^2 - 4x + C$

With initial condition: $\dfrac{1}{5} = 4 - 8 + C$ or $C = \dfrac{21}{5}$

Therefore, particular solution is $\dfrac{1}{y} = x^2 - 4x + \dfrac{21}{5}$ or $y = \dfrac{1}{x^2 - 4x + 4.2}$.

The graph of f is shown in the figure below.

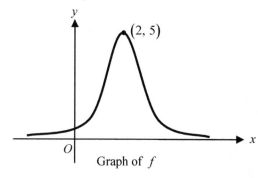

Graph of f

AP Calculus BC Test 4 Answers

Section 1

Question 6

(a) $\ln x = (x-1) - \dfrac{(x-1)^2}{2} + \dfrac{(x-1)^3}{3} - \dfrac{(x-1)^4}{4} + \cdots + \dfrac{(-1)^{n-1}(x-1)^n}{n} + \cdots$

$\ln(x+1) = (x) - \dfrac{(x)^2}{2} + \dfrac{(x)^3}{3} - \dfrac{(x)^4}{4} + \cdots + \dfrac{(-1)^{n-1}(x)^n}{n} + \cdots$

$\dfrac{\ln(x+1)}{x} = 1 - \dfrac{x}{2} + \dfrac{x^2}{3} - \dfrac{x^3}{4} + \cdots + \dfrac{(-1)^{n-1}x^{n-1}}{n} + \cdots$

(b) We know that $f(x) = \sum\limits_{n=1}^{\infty} \dfrac{(-1)^{n-1}x^{n-1}}{n}$ or $\sum\limits_{n=1}^{\infty} \dfrac{(-x)^{n-1}}{n}$ and $a_n = \dfrac{(-x)^{n-1}}{n}$.

We apply ratio and end point tests.

Ratio $= \lim\limits_{n \to \infty} \left| \dfrac{a_{n+1}}{a_n} \right| = \lim\limits_{n \to \infty} \left| \dfrac{(-x)^n}{n+1} \cdot \dfrac{n}{(-x)^{n-1}} \right| = \lim\limits_{n \to \infty} \left| \dfrac{n}{n+1} \cdot (-x) \right| = |x| < 1$ or $-1 < x < 1$

End point test:

For $x = -1$, $\sum\limits_{n=1}^{\infty} \dfrac{(-(-1))^{n-1}}{n} = \sum\limits_{n=1}^{\infty} \dfrac{1}{n} = 1 + \dfrac{1}{2} + \dfrac{1}{3} + \dfrac{1}{4} + \cdots$

The series diverges because it is harmonic series at $x = -1$.

For $x = 1$, $\sum\limits_{n=1}^{\infty} \dfrac{(-x)^{n-1}}{n} = \sum\limits_{n=1}^{\infty} \dfrac{(-x)^{n-1}}{n} = 1 - \dfrac{1}{2} + \dfrac{1}{3} - \dfrac{1}{4} + \cdots$

The series converges because it is alternating harmonic series.
Therefore, interval of convergence is $-1 < x \leq 1$.

(c) We need to show that $\left| P_4\left(\dfrac{1}{2}\right) - f\left(\dfrac{1}{2}\right) \right| < \dfrac{1}{150}$.

Error $= \left| P_4\left(\dfrac{1}{2}\right) - f\left(\dfrac{1}{2}\right) \right| < \left| \dfrac{(1/2)^5}{6} \right| = \dfrac{1}{192} < \dfrac{1}{150}$.

The series has term that alternate, decrease in absolute value and have a limit 0.
Hence the error is bounded by the absolute value of the fifth-degree term.

Observe.

Actual error $= \left| P_4\left(\dfrac{1}{2}\right) - f\left(\dfrac{1}{2}\right) \right| \approx 0.003653$

Made in the USA
Columbia, SC
29 April 2018